CAMBRIDGE STUDIES IN
ADVANCED MATHEMATICS 77

HODGE THEORY AND COMPLEX ALGEBRAIC GEOMETRY II

Already published; for full details see http://publishing.cambridge.org/stm/mathematics/csam/

11 J.L. Alperin *Local representation theory*
12 P. Koosis *The logarithmic integral I*
13 A. Pietsch *Eigenvalues and s-numbers*
14 S.J. Patterson *An introduction to the theory of the Riemann zeta-function*
15 H.J. Baues *Algebraic homotopy*
16 V.S. Varadarajan *Introduction to harmonic analysis on semisimple Lie groups*
17 W. Dicks & M. Dunwoody *Groups acting on graphs*
18 L.J. Corwin & F.P. Greenleaf *Representations of nilpotent Lie groups and their applications*
19 R. Fritsch & R. Piccinini *Cellular structures in topology*
20 H. Klingen *Introductory lectures on Siegel modular forms*
21 P. Koosis *The logarithmic integral II*
22 M.J. Collins *Representations and characters of finite groups*
24 H. Kunita *Stochastic flows and stochastic differential equations*
25 P. Wojtaszczyk *Banach spaces for analysts*
26 J.E. Gilbert & M.A.M. Murray *Clifford algebras and Dirac operators in harmonic analysis*
27 A. Frohlich & M.J. Taylor *Algebraic number theory*
28 K. Goebel & W.A. Kirk *Topics in metric fixed point theory*
29 J.E. Humphreys *Reflection groups and Coxeter groups*
30 D.J. Benson *Representations and cohomology I*
31 D.J. Benson *Representations and cohomology II*
32 C. Allday & V. Puppe *Cohomological methods in transformation groups*
33 C. Soulé et al *Lectures on Arakelov geometry*
34 A. Ambrosetti & G. Prodi *A primer of nonlinear analysis*
35 J. Palis & F. Takens *Hyperbolicity, stability and chaos at homoclinic bifurcations*
37 Y. Meyer *Wavelets and operators I*
38 C. Weibel *An introduction to homological algebra*
39 W. Bruns & J. Herzog *Cohen-Macaulay rings*
40 V. Snaith *Explicit Brauer induction*
41 G. Laumon *Cohomology of Drinfield modular varieties I*
42 E.B. Davies *Spectral theory and differential operators*
43 J. Diestel, H. Jarchow & A. Tonge *Absolutely summing operators*
44 P. Mattila *Geometry of sets and measures in euclidean spaces*
45 R. Pinsky *Positive harmonic functions and diffusion*
46 G. Tenenbaum *Introduction to analytic and probabilistic number theory*
47 C. Peskine *An algebraic introduction to complex projective geometry I*
48 Y. Meyer & R. Coifman *Wavelets and operators II*
49 R. Stanley *Enumerative combinatorics I*
50 I. Porteous *Clifford algebras and the classical groups*
51 M. Audin *Spinning tops*
52 V. Jurdjevic *Geometric control theory*
53 H. Voelklein *Groups as Galois groups*
54 J. Le Potier *Lectures on vector bundles*
55 D. Bump *Automorphic forms*
56 G. Laumon *Cohomology of Drinfeld modular varieties II*
57 D.M. Clarke & B.A. Davey *Natural dualities for the working algebraist*
59 P. Taylor *Practical foundations of mathematics*
60 M. Brodmann & R. Sharp *Local cohomology*
61 J.D. Dixon, M.P.F. Du Sautoy, A. Mann & D. Segal *Analytic pro-p groups, 2nd edition*
62 R. Stanley *Enumerative combinatorics II*
64 J. Jost & X. Li-Jost *Calculus of variations*
68 Ken-iti Sato *Lévy processes and infinitely divisible distributions*
71 R. Blei *Analysis in integer and fractional dimensions*
72 F. Borceux & G. Janelidze *Galois theories*
73 B. Bollobás *Random graphs*
74 R.M. Dudley *Real analysis and probability*
75 T. Sheil-Small *Complex polynomials*
76 C. Voisin *Hodge theory and complex algebraic geometry I*

HODGE THEORY AND COMPLEX ALGEBRAIC GEOMETRY II

CLAIRE VOISIN

Institut de Mathématiques de Jussieu

Translated by Leila Schneps

CAMBRIDGE UNIVERSITY PRESS
Cambridge, New York, Melbourne, Madrid, Cape Town,
Singapore, São Paulo, Delhi, Tokyo, Mexico City

Cambridge University Press
The Edinburgh Building, Cambridge CB2 8RU, UK

Published in the United States of America by Cambridge University Press, New York

www.cambridge.org
Information on this title: www.cambridge.org/9780521718028

First published 2003
Reprinted 2005

A catalogue record for this publication is available from the British Library

ISBN 978-0-521-80283-3 Hardback
ISBN 978-0-521-71802-8 Paperback

Contents

0
Introduction

The first volume of this book was devoted to the study of the cohomology of compact Kähler manifolds. The main results there can be summarised as follows. (Throughout this volume, we write for example vI.6.1 to refer to volume I, section 6.1.)

The Hodge decomposition (vI.6.1). If X is a compact Kähler manifold, then for each integer k, we have a canonical decomposition

$$H^k(X, \mathbb{C}) = \bigoplus_{p+q=k} H^{p,q}(X),$$

known as the Hodge decomposition, depending only on the complex structure of X. Every space $H^{p,q}(X) \subset H^k(X, \mathbb{C})$ can be identified with the set of cohomology classes representable in de Rham cohomology by a closed form which is of type (p, q) at every point of X, relative to the complex structure on X. In particular, we have the Hodge symmetry

$$H^{p,q}(X) = \overline{H^{q,p}(X)},$$

where $\alpha \mapsto \overline{\alpha}$ denotes the natural action of complex conjugation on $H^k(X, \mathbb{C}) = H^k(X, \mathbb{R}) \otimes \mathbb{C}$. The Hodge filtration F on $H^k(X, \mathbb{C})$ is the decreasing filtration defined by

$$F^i H^k(X, \mathbb{C}) = \bigoplus_{p \geq i} H^{p,k-p}(X).$$

The Lefschetz decomposition (vI.6.2). Let ω be a Kähler form on X. Then ω is a real closed 2-form of class $[\omega] \in H^2(X, \mathbb{R})$. We write

$$L : H^k(X, \mathbb{R}) \to H^{k+2}(X, \mathbb{R})$$

for the operator (known as the Lefschetz operator) obtained by taking the cup-product with the class $[\omega]$. For $n = \dim_{\mathbb{C}} X$, and for every $k \leq n$, we have isomorphisms

$$L^{n-k} : H^k(X, \mathbb{R}) \to H^{2n-k}(X, \mathbb{R})$$

(this result is known as the hard Lefschetz theorem), and thus we have the Lefschetz decomposition

$$H^k(X, \mathbb{R}) = \bigoplus_{2r \leq k} L^r H^{k-2r}(X, \mathbb{R})_{\text{prim}}, \quad k \leq n,$$

where the primitive cohomology $H^l(X, \mathbb{R})_{\text{prim}}$ for $l \leq n$ is defined by

$$H^l(X, \mathbb{R})_{\text{prim}} = \operatorname{Ker}(L^{n-l+1} : H^l(X, \mathbb{R}) \to H^{2n-l+2}(X, \mathbb{R})).$$

Mixed Hodge structures (vI.8.4). Let X be a compact Kähler manifold, and let $Z \subset X$ be a closed analytic subset. Let U be the open set $X - Z$. Then the cohomology groups $H^k(U, \mathbb{Q})$ are equipped with a mixed Hodge structure of weight n, i.e. with two filtrations W and F, an increasing filtration W defined over $\mathbb{Q}$, and a decreasing filtration F defined over $\mathbb{C}$, satisfying the condition:

The filtration F_i induced by F on each space $K_i := \operatorname{Gr}_i^W H^k(U, \mathbb{C})$ equips K_i with a pure Hodge structure of weight $n + i$.

This means that for every integer p, it satisfies the condition

$$F_i^p K_i \oplus \overline{F}_i^{n+i+1-p} K_i = K_i,$$

which implies the existence of a Hodge decomposition

$$K_i = \bigoplus_{p+q=n+i} K_i^{p,q}, \quad K_i^{p,q} = F_i^p K_i \cap \overline{F}_i^{n+i-p} K_i.$$

Variations of Hodge structure (vI.10.1). If $\phi : X \to Y$ is a proper holomorphic submersive map with Kähler fibres, the Hodge filtration on the cohomology of the fibres X_y of ϕ varies holomorphically in the following sense. By Ehresmann's theorem, locally over each point $0 \in Y$, the fibration ϕ admits differentiable trivialisations

$$F = (F_0, \phi) : X_U \cong X_0 \times U, \quad X_U := \phi^{-1}(U).$$

The map F_0 is a retraction of X onto the fibre X_0, and for each $y \in U$, it induces a diffeomorphism $X_y \cong X_0$. In particular, we have a canonical isomorphism when U is contractible, namely the isomorphism

$$H^k(X_y, \mathbb{Z}) \cong H^k(X_0, \mathbb{Z})$$

obtained by combining the two restriction isomorphisms

$$H^k(X_U, \mathbb{Z}) \cong H^k(X_0, \mathbb{Z}) \quad \text{and} \quad H^k(X_U, \mathbb{Z}) \cong H^k(X_y, \mathbb{Z}).$$

Letting r_i denote the integer $\dim F^i H^k(X_y, \mathbb{C})$ for all $y \in Y$, then for each integer i, we have the map

$$\mathcal{P} : U \to \mathrm{Grass}(r_i, H^k(X_0, \mathbb{C})),$$

which to $y \in U$ associates the subspace

$$F^i H^k(X_y, \mathbb{C}) \subset H^k(X_y, \mathbb{C}) = H^k(X_0, \mathbb{C}).$$

The fact that the Hodge filtration varies holomorphically with the complex structure on the fibres can be expressed by the fact that the so-called period map $\mathcal{P}$ is holomorphic for every k, i.

Transversality (vI.10.2). The period map defined above locally gives a holomorphic subbundle

$$F^i \mathcal{H}^k \subset \mathcal{H}^k,$$

where $\mathcal{H}^k = H^k(X_0, \mathbb{C}) \otimes \mathcal{O}_U$ is the sheaf of sections of the trivial holomorphic vector bundle with fibre $H^k(X_0, \mathbb{C})$. Let $\nabla : \mathcal{H}^k \to \mathcal{H}^k \otimes \Omega_U$ be the connection given by the usual differentiation of functions in the trivialisation above.

The Griffiths transversality condition is, without a doubt, the most important notion in the theory of variations of Hodge structure. It states that the Hodge bundles $F^i \mathcal{H}^k$ satisfy the property

$$\nabla F^i \mathcal{H}^k \subset F^{i-1} \mathcal{H}^k \otimes \Omega_Y.$$

Note that the data $(\mathcal{H}^k, F^i \mathcal{H}^k, \nabla)$ are in fact globally defined on Y, but they are only locally trivial; ∇ is known as the Gauss–Manin connection. In general, the Hodge bundle will be defined by

$$\mathcal{H}^k = H^k_{\mathbb{C}} \otimes \mathcal{O}_Y,$$

where $H^k_{\mathbb{C}} = R^k \phi_* \mathbb{C}$. The isomorphisms used above,

$$H^k_{\mathbb{C}}(U) \cong H^k(X_y, \mathbb{C}) \quad \text{for} \quad y \in U,$$

simply show that $H^k_{\mathbb{C}}$ is a local system, and give local trivialisations $H^k_{\mathbb{C}}$ of $\mathcal{H}^k$.

Cycle classes and the Abel–Jacobi map (vI.11.1, vI.12.1). Let $Z \subset X$ be a closed, reduced and irreducible analytic subset of codimension k of a compact Kähler manifold X. We have the cohomology class $[Z] \in H^{2k}(X, \mathbb{Z})$, which can be defined, for example, as the Poincaré dual class of $j_*[\widetilde{Z}]_{\mathrm{fund}}$, where $j : \widetilde{Z} \to Z \to X$ is a desingularisation of Z and $[\widetilde{Z}]_{\mathrm{fund}} \in H_{2\dim \widetilde{Z}}(\widetilde{Z}, \mathbb{Z})$ is the homology class of the smooth compact oriented manifold $\widetilde{Z}$. Then the image of the class $[Z]$ in $H^{2k}(X, \mathbb{C})$ lies in $H^{k,k}(X)$. Such a class is called a Hodge class.

Using Hodge theory, one can also define secondary invariants, called Abel–Jacobi invariants, for a cycle $Z = \sum_i n_i Z_i$ of codimension k which is homologous to 0, i.e. which is such that $\sum_i n_i [Z_i] = 0$ in $H^{2k}(X, \mathbb{Z})$. The Hodge decomposition gives a decomposition

$$H^{2k-1}(X, \mathbb{C}) = F^k H^{2k-1}(X) \oplus \overline{F^k H^{2k-1}(X)}.$$

We then define the kth intermediate Jacobian of X as the complex torus

$$J^{2k-1}(X) = H^{2k-1}(X, \mathbb{C})/(F^k H^{2k-1}(X) \oplus H^{2k-1}(X, \mathbb{Z})),$$

and we have the Abel–Jacobi invariant

$$\Phi_X^k(Z) \in J^{2k-1}(X)$$

defined by Griffiths. The Abel–Jacobi map generalises the Albanese map for 0-cycles given by

$$\mathrm{alb}_X : \mathcal{Z}_0(X)_{\mathrm{hom}} \to J^{2n-1}(X) = H^0(X, \Omega_X)^*/H_1(X, \mathbb{Z}), \quad n = \dim X$$

$$z \mapsto \int_\gamma \in H^0(X, \Omega_X)^*, \ \ \partial\gamma = z.$$

These results highlight the existence of relations between Hodge theory, topology, and the analytic cycles of a Kähler manifold. For example, the Hodge decomposition and the Hodge symmetry show that the Betti numbers $b_i(X) = \operatorname{rank} H^i(X, \mathbb{Z})$ are even whenever i is odd. The hard Lefschetz theorem shows that the Betti numbers b_{2i} are increasing for $2i \leq n = \dim X$, and that the Betti numbers b_{2i-1} are increasing for $2i - 1 \leq n = \dim X$. The cycle class map shows that the existence of interesting analytic cycles of codimension k is related to the existence of Hodge classes of degree $2k$, which can be seen on the Hodge structure on $H^{2k}(X)$. Finally, in the algebraic case, where we may assume that the class $[\omega]$ is integral and is even the cohomology class of a hypersurface $Y \subset X$, the hard Lefschetz theorem partly implies the Lefschetz theorem on hyperplane sections, which says that if $j : Y \hookrightarrow X$ is the inclusion of an ample hypersurface, then the restriction map

$$j^* : H^k(X, \mathbb{Z}) \to H^k(Y, \mathbb{Z})$$

is an isomorphism for $k < \dim Y$ and an injection for $k = \dim Y$. Indeed, by Kodaira's embedding theorem, the ampleness of Y is equivalent to the condition that the real cohomology class $[Y] \in H^2(X, \mathbb{R})$ is a Kähler class. As we have the equalities

$$j_* \circ j^* = L, \quad j^* \circ j_* = L_Y,$$

where L (resp. L_Y) is the Lefschetz operator associated to the Kähler class $[Y]$ (resp. $[Y]_{|Y}$), the hard Lefschetz theorem shows for example that the restriction map $j^* : H^k(X, \mathbb{Q}) \to H^k(Y, \mathbb{Q})$ is injective for $k \leq \dim Y$ and surjective for $k > \dim Y$.

The fact that the period map is holomorphic also gives relations between Hodge theory and algebraic geometry. For example, it enables us (at least partially) to study moduli spaces classifying the deformations of the complex structure on a polarised algebraic variety, and possibly, when the period map is injective, to realise these moduli spaces as subspaces of domains of global periods. Other subtler applications of the fact that the period map is holomorphic come from the study of the curvature of the Hodge bundles, which can make it possible to polarise the moduli space itself (see Viehweg 1995; Griffiths 1984). Finally, we also deduce that for a family of smooth projective or compact Kähler varieties $\phi : X \to Y$, the Hodge loci $Y_\lambda^k \subset Y$ for a section λ of the local system $R^{2k}\phi_*\mathbb{Z}$, which are defined by

$$Y_\lambda^k = \{y \in Y \mid \lambda_y \in F^k H^{2k}(X_y, \mathbb{C})\},$$

are analytic subsets of Y. This result agrees with the Hodge conjecture, which predicts that $y \in Y_\lambda^k$ if and only if a multiple of λ_y is the cohomology class of a cycle $Z_y \subset X_y$ of codimension k, so that Y_λ^k is the image in Y of a relative Hilbert scheme parametrising subvarieties in the fibres of ϕ.

The applications described above do not constitute particularly tight links between the topology of algebraic varieties, their algebraic cycles and their Hodge theory. The present volume is devoted to the description of much finer interactions between these three domains. We do not, however, propose an exhaustive description of these interactions here, and each of the three parts of this volume ends with a sketch of possible developments which lie beyond the scope of this course. The remainder of this introduction aims to give a synthetic picture of these interactions, which might otherwise be obscured by the separation of the volume into three seemingly independent parts.

Two themes which recur constantly throughout this volume are the Lefschetz theorems and Leray spectral sequences. In the first case, we compare the topology of an algebraic variety X with that of its hyperplane sections, and in the second case we study the topology of a variety X admitting a (usually proper and submersive) morphism $\phi : X \to Y$, using the topology of the fibres X_y, and more precisely in the submersive case, using the local systems $R^k\phi_*\mathbb{Z}$ on Y.

The Lefschetz theorem on hyperplane sections is proved using Morse theory on affine varieties, and does not require any arguments from Hodge theory. However, it does not yield the hard Lefschetz theorem, i.e. the Lefschetz decomposition, which is the only ingredient needed (in an entirely formal way) in the proof of Deligne's theorem:

Theorem 0.1 *The Leray spectral sequence of the rational cohomology of a projective fibration degenerates at E_2.*

Concretely, this result implies the following invariant cycles theorem for smooth projective fibrations:

Theorem 0.2 *If $\phi : X \to Y$ is a smooth projective fibration, then the restriction map*

$$H^k(X, \mathbb{Q}) \to H^k(X_y, \mathbb{Q})^\rho$$

is surjective.

Here, $H^k(X_y, \mathbb{Q})^\rho \subset H^k(X_y, \mathbb{Q})$ denotes the subspace of classes invariant under the monodromy action

$$\rho : \pi_1(Y, y) \to \mathrm{Aut}\, H^k(X_y, \mathbb{Q}).$$

This puts important constraints on the families of projective varieties. However, qualitatively speaking, it is not a very refined statement. Rather, it is Hodge theory which yields the true global invariant cycles theorem, which imposes qualitative constraints on the monodromy representation associated to a projective fibration. If $\phi : X \to Y$ is a dominant morphism between smooth projective varieties, and $U \subset Y$ is the Zariski open (dense) subset of regular values of ϕ, then we have a smooth and proper fibration $\phi : X_U := \phi^{-1}(U) \to U$, so we have a monodromy representation

$$\rho : \pi_1(U, y) \to \mathrm{Aut}\, H^k(X_y, \mathbb{Q}) \quad \text{for} \quad y \in U.$$

Then, we have the following result.

Theorem 0.3 *The restriction map*

$$H^k(X, \mathbb{Q}) \to H^k(X_y, \mathbb{Q})^\rho \quad \text{for} \quad y \in U$$

is surjective. In particular, $H^k(X_y, \mathbb{Q})^\rho$ is a Hodge substructure of $H^k(X_y, \mathbb{Q})$.

The main additional ingredient enabling us to deduce this theorem from the preceding one is the existence of mixed Hodge structures on the cohomology groups of a quasi-projective complex manifold, and the strictness of the morphisms of mixed Hodge structures.

These results, which illustrate the qualitative influence of Hodge theory on the topology of algebraic varieties, are the main object of the Part I of this volume, which is devoted to topology. It also contains an exposition of Picard–Lefschetz theory, which gives a precise description of the geometry of a Lefschetz degeneration. If $Y \overset{j}{\hookrightarrow} X$ is the inclusion of a smooth hyperplane section, the vanishing cohomology $H^*(Y, \mathbb{Z})_{\mathrm{van}}$ is defined as the kernel of the Gysin morphism

$$j_* : H^*(Y, \mathbb{Z}) \to H^{*+2}(X, \mathbb{Z}).$$

Picard–Lefschetz theory shows that the vanishing cohomology is generated by the vanishing cycles, which are classes of spheres contracting to a point when Y degenerates to a nodal hypersurface. Another important consequence of this study is the description of the local monodromy action (the Picard–Lefschetz formula). Combined with the preceding result, it gives the irreducibility theorem for the monodromy action on the vanishing cohomology for the universal family of smooth hyperplane sections of a smooth projective variety X.

This result has numerous consequences, in particular in the study of algebraic cycles; it is a key ingredient in Lefschetz' proof of the Noether–Lefschetz theorem, which says that the Picard group of a general surface Σ of degree ≥ 4 in $\mathbb{P}^3$ is generated by the class of the line bundle $\mathcal{O}_\Sigma(1)$. It also occurs in the proof of the Green–Voisin theorem on the triviality of the Abel–Jacobi map for general hypersurfaces of degree ≥ 6 in $\mathbb{P}^4$. Using the Picard–Lefschetz formula and the transitivity of the monodromy action on vanishing cycles, one can also show that the monodromy group is very large; indeed, it tends to be equal to the group of isomorphisms preserving the intersection form (see Beauville 1986b). This has important restrictive consequences on the Hodge structures of general hyperplane sections: apart from the applications mentioned above, Deligne (1972) uses the monodromy group (combined with the notion of the Mumford group of a Hodge structure) to show that the rational Hodge structure on the H^2 of a general surface of degree ≥ 5 in $\mathbb{P}^3$ is not a quotient of the Hodge structure on the H^2 of an abelian variety. All these results illustrate the influence of topology on Hodge theory.

The second part of this volume is devoted to the study of infinitesimal variations of Hodge structure for a family of smooth projective varieties $\phi : X \to Y$, and its applications, especially those concerning the case of complete families of hypersurfaces or complete intersections of a given variety X.

The Leray spectral sequence which comes into play here is the spectral sequence Ω^k_X of sheaf cohomology equipped with the Leray filtration

$$L_r\Omega^k_X := \phi^*\Omega^r_Y \wedge \Omega^{k-r}_X,$$

relative to the functor ϕ_*. Here, the sheaves we consider are sheaves of holomorphic differential forms; however, this Leray spectral sequence is related to the Leray spectral sequence of the morphism ϕ by the fact that the latter can be computed in de Rham cohomology as the spectral sequence of the complex of differential forms $(A^k_X, d) := \Gamma(X, \mathcal{A}^k_X)$, equipped with the filtration given by the global sections of the Leray filtration

$$L_r\mathcal{A}^k_X := \phi^*\mathcal{A}^r_Y \wedge \mathcal{A}^{k-r}_X.$$

One can show that the term $E_1^{p,q}$ of this filtration, which is a complex of coherent sheaves on Y equipped with the differential

$$d_1 : E_1^{p,q} \to E_1^{p+1,q},$$

is a complex which occurs naturally in the study of the variations of Hodge structure. Let us consider the Hodge bundles $\mathcal{H}^k := R^k\phi_*\mathbb{C} \otimes \mathcal{O}_Y$ introduced above, and their Hodge filtration $F^i\mathcal{H}^k$. The transversality property

$$\nabla F^i\mathcal{H}^k \subset F^{i-1}\mathcal{H}^k \otimes \Omega_Y$$

gives a filtration (also denoted by $F^{\cdot}$) on the de Rham complex

$$\mathrm{DR}(\mathcal{H}^k) : 0 \to \mathcal{H}^k \stackrel{\nabla}{\to} \mathcal{H}^k \otimes \Omega_Y \stackrel{\nabla}{\to} \cdots \to \mathcal{H}^k \otimes \Omega^N_Y \to 0,$$

where $N = \dim Y$ and ∇ denotes the Gauss–Manin connection. By Griffiths transversality, we can set

$$F^i\mathrm{DR}(\mathcal{H}^k) = 0 \to F^i\mathcal{H}^k \stackrel{\nabla}{\to} F^{i-1}\mathcal{H}^k \otimes \Omega_Y \stackrel{\nabla}{\to} \cdots \to F^{i-N}\mathcal{H}^k \otimes \Omega^N_Y \to 0.$$

The complex $\mathcal{K}_{p,q} := \mathrm{Gr}^p_F\mathrm{DR}(\mathcal{H}^k)$ for $p + q = k$ can be written

$$0 \to \mathcal{H}^{p,q} \stackrel{\overline{\nabla}}{\to} \mathcal{H}^{p-1,q+1} \otimes \Omega_Y \stackrel{\overline{\nabla}}{\to} \cdots \to \mathcal{H}^{p-N,q+N} \otimes \Omega^N_Y \to 0,$$

where $\mathcal{H}^{p,q} := F^p\mathcal{H}^k/F^{p+1}\mathcal{H}^k$ for $p + q = k$ and the differential $\overline{\nabla}$ of the complex $\mathcal{K}_{p,q}$ describes the infinitesimal variation of Hodge structure on the cohomology of the fibres X_y in a precise sense. An essential point is then the following result.

Proposition 0.4 *The complex $(E_1^{p,q}, d_1)$ relative to the bundle Ω^k_X equipped with its Leray filtration, graded by the degree p, can be identified with the complex $(\mathcal{K}_{k,q}, \overline{\nabla})$.*

The main result proved in this part is Nori's connectivity theorem, which is a strengthened Lefschetz theorem for complete families of hypersurfaces or complete intersections of sufficiently large degree of a smooth projective variety X. Let X be a smooth complex $(n+r)$-dimensional projective variety, and let $L_1, \ldots, L_r$ be sufficiently ample line bundles on X. Let $B \subset \prod_{i=1}^{i=r} H^0(X, L_i)$ be the open set parametrising the smooth complete intersections. For every morphism $\phi : T \to B$, let $j : \mathcal{Y}_T \subset T \times X$ denote the family of complete intersections parametrised by T,

$$\mathcal{Y}_T = \{(t, x) \in T \times X \mid x \in Y_{\phi(t)}\}.$$

Nori proved the following result.

Theorem 0.5 *For every quasi-projective smooth variety T and every submersive morphism $\phi : T \to B$, the restriction induced by the inclusion j,*

$$j^* : H^k(T \times X, \mathbb{Q}) \to H^k(\mathcal{Y}_T, \mathbb{Q}),$$

is an isomorphism for $k < 2n = 2 \dim Y_b$ and is injective for $k = 2n$.

Note that the usual Lefschetz theorem would prove this statement only up to ranks $k < n$ and $k = n$ respectively.

The proof of Nori's theorem splits naturally into two parts. The first involves a Hodge theory argument using the existence of a mixed Hodge structure on the relative cohomology $H^k(T \times X, \mathcal{Y}_T, \mathbb{Q})$. Using this, one can show that in order to obtain the connectivity statement above, it is enough to show that the restriction maps

$$j^* : H^p\left(T \times X, \Omega^q_{T \times X}\right) \to H^p\left(\mathcal{Y}_T, \Omega^q_{\mathcal{Y}_T}\right)$$

are isomorphisms for $p + q < 2n$, $p < n$, and are injective for $p + q \leq 2n$, $p \leq n$. (If the varieties $\mathcal{Y}_T$ and $T \times X$ were smooth projective varieties, this would follow immediately from the Hodge decomposition and the Hodge symmetry, but for quasi-projective varieties, the argument is subtler.)

The second step involves studying the restriction maps

$$j^* : H^p\left(T \times X, \Omega^q_{T \times X}\right) \to H^p\left(\mathcal{Y}_T, \Omega^q_{\mathcal{Y}_T}\right)$$

using proposition 0.4 above, and comparing the variations of Hodge structure of the two families $T \times X \to T$ and $\mathcal{Y}_T \to T$, or more precisely the cohomology of the associated complexes $\mathcal{K}_{r,s}$. We restrict ourselves here to the typical case of hypersurfaces of projective space, in which case the variations of Hodge structure, or more precisely the complexes $\mathcal{K}_{r,s}$ above, can be described via the theory of residues of meromorphic differential forms, using the

Koszul complexes (see Green 1984a) of the Jacobian rings of hypersurfaces. The exactness of these complexes in small degree, which concludes the proof of Nori's theorem in the case of hypersurfaces, then follows from a theorem of Mark Green (1984b) on the vanishing of the syzygies of projective space.

Part II of this volume contains one chapter, chapter 6, devoted to the infinitesimal variations of Hodge structure of hypersurfaces of projective space. Following Griffiths, we show that the primitive (or vanishing) cohomology of degree $n-1$ of a hypersurface Y of $\mathbb{P}^n$ defined by a homogeneous polynomial f of degree d is generated by the residues $\mathrm{Res}_Y \frac{P\Omega}{f^l}$, where Ω is a generator of $H^0(K_{\mathbb{P}^n}(n+1))$. Moreover, the minimal order l of the pole corresponds to the Hodge level via the relation $\mathrm{Res}_Y \frac{P\Omega}{f^l} \in F^{n-l}H^{n-1}(Y,\mathbb{C})_{\mathrm{prim}}$.

We obtain from this a simple description of the variations of Hodge structure of hypersurfaces in projective space in terms of multiplication in the Jacobian ring $R_f = S/J_f$, where S is the ring of polynomials in $n+1$ variables, and J_f is the ideal generated by the partial derivatives of f relative to a system of homogeneous coordinates on $\mathbb{P}^n$. These Jacobian rings possess remarkable algebraic properties (symmetriser lemma, Macaulay's theorem), some of whose geometric consequences we describe. In the language introduced above, Macaulay's theorem is essentially a statement on the exactness of the complexes $\mathcal{K}_{p,q}$ in degree 0 for $q < n-1$ and sufficiently large d, and the symmetriser lemma is a statement on the exactness of these complexes in degree 1 for $q < n-2$.

Chapter 6 contains results which are somewhat less general than those presented throughout the rest of the book. Its aim is to illustrate the fact that an important aspect of the Hodge theory of algebraic varieties is algebro-geometric in nature, and thus essentially computable using the methods of algebraic geometry. The transcendental nature of Hodge theory shows up clearly in the presence of integral or rational structures on the cohomology groups $H^k(X,\mathbb{C})$. These groups can be computed via the formula

$$H^k(X,\mathbb{C}) = \mathbb{H}^k(X,\Omega_X^\bullet),$$

with reference to the algebraic de Rham complex and the Zariski topology. Indeed, this formula follows from the analogous formula in analytic geometry, which itself follows from the fact that the holomorphic de Rham complex is a resolution of the constant sheaf (which is totally false in the Zariski topology; see Bloch & Ogus 1974), and from Serre's GAGA principle. In studying the variations of Hodge structure, the rational structure of the cohomology which allowed us to define the notion of Hodge structure is replaced by the notion of a locally constant class for the Gauss–Manin connection, which is defined algebraically on Hodge bundles, which are algebraic bundles as the formula above shows. Via the Riemann–Hilbert correspondence

Local systems, monodromy $\leftrightarrow$ *Vector bundles, flat connections,*

we may work solely with Hodge bundles and the Gauss–Manin connection, and transversality will be the essential tool from now on. Thus, it is important to realise that the very explicit description of the Hodge filtration of a hypersurface which we give here is perfectly adapted to infinitesimal computation, but does not enable us to describe the Hodge structure of a hypersurface; even the real structure of the cohomology of a hypersurface cannot be described simply in this way.

As mentioned above, the main result of part II is Nori's theorem, which is a strengthened Lefschetz theorem for complete families of complete intersections of large degree of a smooth variety. Whether or not this theorem is topological in nature is an interesting problem. As sketched above, the present proof makes abundant use of Hodge theory, both on the infinitesimal level and in its first step, which reduces the statement to one of pure algebraic geometry. The hypotheses are not in themselves topological, since one must consider sufficiently ample hypersurfaces, and locally complete families (corresponding to the fact that the morphism ϕ is submersive). However this may be, the common themes of Leray spectral sequences and Lefschetz theorems constitute a strict parallel between the first and second parts of this volume, and largely orient the second part towards the application of methods of variation of Hodge structure to the study of the topology of algebraic varieties.

Part II also contains numerous applications to algebraic cycles, and thus provides close links with part III, which is devoted to Chow groups. We first give a local description of the Hodge loci mentioned above, and show how Griffiths transversality occurs in the schematic structure of these loci. Furthermore, Macaulay's theorem mentioned above implies a generalised Noether–Lefschetz theorem on hypersurfaces of projective space. (The argument also generalises to hypersurfaces of large degree of any variety.) Let B be the open set parametrising the smooth hypersurfaces of degree d in $\mathbb{P}^n$, and let $\phi : \mathcal{Y}_B \to B$ be the universal hypersurface. For every open set $U \subset B$ and every section $0 \neq \lambda \in \Gamma(U, R^{n-1}\phi_*\mathbb{C}_{\text{prim}})$, we have the Hodge loci U_λ^p defined by

$$t \in U \Leftrightarrow \lambda_t \in F^p H^{n-1}(Y_y, \mathbb{C}).$$

Theorem 0.6 *If d is sufficiently large and $p > 0$, the Hodge loci are proper analytic subsets.*

Finally, the symmetriser lemma and the horizontality property satisfied by the Abel–Jacobi map for families of cycles in families of varieties imply the following result.

Theorem 0.7 *If X is a general hypersurface in $\mathbb{P}^{2n}$ of sufficiently large degree d, then the image of the Abel–Jacobi map*

$$\Phi_X^n : \mathrm{CH}^n(X)_{\mathrm{hom}} \to J^{2n-1}(X)$$

of X is contained in the torsion of $J^{2n-1}(X)$.

Nori's connectivity theorem also has considerable consequences in the theory of algebraic cycles. One of these consequences, which we describe explicitly here, is the existence of cycles homologous to 0 which are annihilated by the Abel–Jacobi map and none of whose multiples is algebraically equivalent to 0.

From a more general point of view, one of the ingredients introduced by Nori has had a great deal of influence recently on the study of algebraic cycles. Given a cycle $Z \subset X$ of codimension k in a fibred variety $\phi : X \to Y$, viewed as a family $(Z_y)_{y \in Y}$ of cycles in the fibres, Nori considers the class $[Z] \in H^k(X, \Omega_X^k)$. We can also consider the infinitesimal invariants of this class, obtained by studying the Leray spectral sequence introduced above, relative to the Leray filtration on Ω_X^k. This generalises the infinitesimal Griffiths invariant (see Griffiths 1983; Voisin 1988), and gives criteria for the generic non-triviality of the fibres $Z_y \in \mathrm{CH}^k(X_y)$. Examples of developments and applications of this idea can be found in Voisin (1994; 2002), and it also led to the notion of an 'arithmetic cycle class', developed independently by M. Green and the Japanese school (Asakura, Saito, and others).

The third and final part of this volume is devoted to the relations between Hodge theory and algebraic cycles. Using the results of the preceding part and essentially infinitesimal techniques, we will establish a certain number of results on the various 'cycle class' maps, and certain consequences concerning equivalence relations which are not as fine as rational equivalence (homological equivalence, Abel–Jacobi equivalence, algebraic equivalence). The theme of the influence of Chow groups – groups of cycles modulo rational equivalence – on the Hodge theory of smooth complex projective varieties pervades the whole of part III. The prototype of this relation is given in the following theorem, due to Mumford.

Theorem 0.8 *Let S be a smooth complex projective surface such that* $\mathrm{CH}_0(S)$ *is finite-dimensional. Then $H^{2,0}(S) = 0$.*

One of the characterisations of the notion of 'finite-dimensionality' appearing here is the fact that the difference map

$$\sigma_k : S^{(k)} \times S^{(k)} \to \mathrm{CH}_0(S)_{\mathrm{hom}}, \quad (Z_1, Z_2) \mapsto Z_1 - Z_2$$

is surjective for sufficiently large k. (We assume here that S is connected.) Perhaps the most striking result in the converse direction, that of determining the Chow groups via Hodge theory, is the following result due to Roitman.

Theorem 0.9 *Let X be a smooth projective variety such that $\mathrm{CH}_0(X)$ is finite-dimensional. Then the Albanese map*

$$\mathrm{alb}_X : \mathrm{CH}_0(X)_{\mathrm{hom}} \to \mathrm{Alb}\, X = H^0(X, \Omega_X)^*/H_1(X, \mathbb{Z}),$$

i.e. the Abel–Jacobi map for the 0-cycles, is an isomorphism.

Mumford's theorem has undergone various generalisations. Firstly, we have the following.

Theorem 0.10 *Let X be a smooth complex projective variety and $j : W \hookrightarrow X$ a subvariety of dimension k such that $j_* : \mathrm{CH}_0(W) \to \mathrm{CH}_0(X)$ is surjective. Then $H^0(X, \Omega_X^l) = 0$ for all $l > k$.*

Mumford's theorem is the case $k = 1$ of this statement. Indeed, we have the following, third characterisation of the notion of 'finite-dimensionality':

Proposition 0.11 *$\mathrm{CH}_0(X)$ is finite-dimensional if and only if there exists a curve $j : C \hookrightarrow X$ such that $j_* : \mathrm{CH}_0(C) \to \mathrm{CH}_0(X)$ is surjective.*

More generally, the Chow groups of all dimensions play a role in Hodge theory, as shown by the following theorem, which was proved independently by Lewis and Schoen.

Theorem 0.12 *Let X be a smooth projective complex variety such that the cycle class maps*

$$cl : \mathrm{CH}_l(X)_{\mathbb{Q}} \to H^{2n-2l}(X, \mathbb{Q})$$

are injective for $l \leq k$. Then $H^{p,q}(X) = 0$ for $p \neq q$, $q \leq k$.

We can obtain a similar result by replacing the above condition by the following condition:

The Abel–Jacobi map

$$\Phi_X^{n-l} : \mathrm{CH}_l(X)_{\mathrm{hom}} \to J^{2n-2l-1}(X)$$

is injective.

These results give precise information on the role played by Chow groups in Hodge theory. Their mutual relations are actually very mysterious, and form the main object of the Bloch–Beilinson conjectures, of which we present a simplified version here, not including the 'Beilinson formulae'. We restrict ourselves to the following geometric conjecture.

Conjecture 0.13 *For every X, there exists a filtration $F^i\mathrm{CH}^k(X)_{\mathbb{Q}}$ on the Chow groups with rational coefficients $\mathrm{CH}^k(X)_{\mathbb{Q}}$, having the following properties:*

(i) This filtration is stable under the action of correspondences.
(ii) We have $F^{k+1}\mathrm{CH}^k(X)_{\mathbb{Q}} = 0$.
(iii) Its associated graded group $\mathrm{Gr}^i_F\mathrm{CH}^k(X)_{\mathbb{Q}}$ is controlled by the group $H^{2k-i}(X)$ in the following sense: $\mathrm{Gr}^i_F\mathrm{CH}^k(X)_{\mathbb{Q}} = 0$ if $H^{2k-i}(X,\mathbb{Q}) = 0$, and more generally, for a correspondence $\Gamma \subset X \times Y$, the induced morphism

$$\Gamma_* : \mathrm{Gr}^i_F\mathrm{CH}^k(X)_{\mathbb{Q}} \to \mathrm{Gr}^i_F\mathrm{CH}^l(Y)_{\mathbb{Q}} \quad \text{with} \quad l = k + \operatorname{codim}\Gamma - \dim X$$

is zero if the morphism

$$[\Gamma]_* : H^{2k-i}(X,\mathbb{Q}) \to H^{2l-i}(X,\mathbb{Q})$$

of Hodge structures induced by the Hodge class $[\Gamma]$ is zero.

For example, $F^1\mathrm{CH}^k(X)_{\mathbb{Q}} = \mathrm{CH}^k(X)_{\mathrm{hom}} \otimes \mathbb{Q}$, so by definition the cycle class map

$$\mathrm{cl} : \mathrm{CH}^k(X)_{\mathbb{Q}} \to H^{2k}(X,\mathbb{Q})$$

induces an injection

$$\mathrm{Gr}^0_F\mathrm{CH}^k(X)_{\mathbb{Q}} \to H^{2k}(X,\mathbb{Q}).$$

The Beilinson conjectures concern the existence of a category of mixed motives MM, containing the Hodge structures, and such that we can express the groups

$\mathrm{Gr}_F^i \mathrm{CH}^k(X)_{\mathbb{Q}}$ by the formula

$$\mathrm{Gr}_F^i \mathrm{CH}^k(X)_{\mathbb{Q}} = \mathrm{Ext}^i_{\mathrm{MM}}(\mathbb{Q}(i), H^{2k-i}(X, \mathbb{Q})).$$

We refer to Janssen (1989; 1994) for a presentation of these developments, and restrict ourselves here to giving some results supporting conjecture 0.13. For example, this conjecture contains as a special case the converse of Mumford's theorem 0.8, known as the Bloch conjecture for surfaces, and we will describe the known results on this conjecture. We conclude this part with some computations in the Chow ring of an abelian variety, following Beauville and Bloch, which give structure results which partially confirm conjecture 0.13 for abelian varieties.

We refer the reader to the text itself for a fully detailed presentation of its entire contents, as each chapter begins with an introduction which explains its logical situation within the framework of the book, and summarises the main results it contains.

Acknowledgments. This book presents the contents of a course taught at Jussieu in 1998–1999 and 1999–2000. I would like to thank the University of Paris 6 for giving me the opportunity of teaching in its DEA programme, and particularly teaching an advanced course in the second year, whose content is that of the present volume. I would also like to thank the audience of those courses, for their interest and their assiduity which were extremely helpful and encouraging throughout the two years.

Although it is probably obvious to the reader, I would like to emphasise the immense influence of the work of Phillip Griffiths on the subject explored within this book. In a complementary way to Deligne's beautiful and powerful contribution based on the idea of mixed Hodge theory, it was Griffiths who established the founding theorems in the theory of variations of Hodge structures. His work modelled the domain, and suggested numerous new research directions; most of the results presented in the second part of this volume were inspired by his ideas.

Part I

The Topology of Algebraic Varieties

1

The Lefschetz Theorem on Hyperplane Sections

This chapter is devoted to a presentation of Morse theory on affine varieties, and its application to the proof of the famous Lefschetz theorem on hyperplane sections, which is the following statement.

Theorem 1.1 *Let X be a projective n-dimensional variety, and let $j : Y \hookrightarrow X$ be a hyperplane section such that $U = X - Y$ is smooth. Then the restriction map*

$$j^* : H^k(X, \mathbb{Z}) \to H^k(Y, \mathbb{Z})$$

is an isomorphism for $k < n - 1$ and is injective for $k = n - 1$.

This Lefschetz theorem follows from the vanishing of the cohomology with compact support in degree $> n$ of a smooth n-dimensional affine variety. Thus, most of this chapter concerns the study of the topology of affine varieties; the most general result we present is the following (see Andreotti & Frankel 1959; Milnor 1963).

Theorem 1.2 *A smooth affine variety X of (complex) dimension n has the homotopy type of a CW-complex of (real) dimension $\leq n$.*

This statement is obtained by applying the results of Morse theory to the square of the distance function $h_0(x) = d(x, 0)^2$ on X, where the metric is deduced from a Hermitian metric on the ambient space. The essential point is the following result.

Proposition 1.3 *The Morse index of the function h_0 at a non-degenerate critical point is at most equal to n.*

Theorem 1.2 then follows from this proposition and the following basic theorem of Morse theory.

Theorem 1.4 *If $f : X \to \mathbb{R}$ is a Morse function, and λ is a critical value corresponding to a unique critical point of index r, then the level set $X_{f\leq\lambda+\epsilon}$ has the homotopy type of the union of the level set $X_{f\leq\lambda-\epsilon}$ with a ball B^r.*

We give an introduction to Morse theory in the first section of this chapter, and this theorem is proved there. In the second section, we study the case of the square of the distance function on affine varieties, and deduce the Lefschetz theorem on hyperplane sections. Finally, in the last section, we give another proof of this result using Hodge theory and the vanishing theorems. This proof gives the result for rational cohomology under the hypothesis that Y and X are smooth.

1.1 Morse theory

1.1.1 Morse's lemma

Let X be a differentiable variety, and let f be a differentiable function on X. Assume that everything is C^∞, although in fact the result still holds under weaker hypotheses. We say that $0 \in X$ is a critical point of f if $df(0) = 0$. The value $f(0) \in \mathbb{R}$ is then called a critical value of f.

Let $x_1, \ldots, x_n$ be local coordinates on X centred at 0. The vector fields $\frac{\partial}{\partial x_i}$ for $i = 1, \ldots, n$ give a basis of $T_{X,x}$ for every x in a neighbourhood of 0.

Definition 1.5 *The Hessian of f at the point 0 is the bilinear form on $T_{X,0}$ defined by*

$$\mathrm{Hess}_0 f\left(\frac{\partial}{\partial x_i}, \frac{\partial}{\partial x_j}\right) = \left(\frac{\partial^2 f}{\partial x_i \partial x_j}\right)(0).$$

The formula for the derivatives of a composition of maps (the chain rule) shows immediately that $\mathrm{Hess}_0 f$ does not depend on the choice of coordinates, since 0 is a critical point of f. Moreover, the symmetry of partial derivatives shows that $\mathrm{Hess}_0 f$ is symmetric.

Definition 1.6 *We say that $0 \in X$ is a non-degenerate critical point of f if the quadratic (or symmetric bilinear) form $\mathrm{Hess}_0 f$ on $T_{X,0}$ is non-degenerate.*

Such a quadratic form Q can be diagonalised in a suitable basis $u_1, \ldots, u_n$ of $T_{X,0}$, i.e. there is a basis such that

$$Q(u_i, u_j) = \delta_{ij}\epsilon_i$$

with $\epsilon_i = \pm 1$.

Definition 1.7 *The Morse index of f at 0, written* $\mathrm{ind}_0 f$, *is the index of the quadratic form* $\mathrm{Hess}_0 f$, *i.e. the number of ϵ_i equal to -1 in any diagonalisation as above.*

The non-degenerate critical points are classified by their index up to diffeomorphism, as shown in the following proposition.

Proposition 1.8 (Morse's lemma) *If 0 is a non-degenerate critical point of a function f, then in a neighbourhood of 0, there exist coordinates $x_1, \ldots, x_n$ centred at 0 such that for $x = (x_1, \ldots, x_n)$, we have*

$$f(x) = f(0) - \sum_{i=1}^{r} x_i^2 + \sum_{i=r+1}^{n} x_i^2$$

with $r = \mathrm{ind}_0 f$.

The following result is a first corollary of proposition 1.8.

Corollary 1.9 *If 0 is a non-degenerate critical point of f, then 0 is an isolated critical point of f, and $f(0)$ is an isolated critical value of f restricted to a neighbourhood of 0.*

This corollary also follows immediately from the fact that $\mathrm{Hess}_0 f$ can be viewed as the differential of the map χ defined (using coordinates) by

$$x \mapsto \chi(x) = df_x \in (\mathbb{R}^n)^*.$$

If this differential is an isomorphism, i.e. when the Hessian is non-degenerate, the local inversion theorem shows that in a neighbourhood of 0, the set $\chi^{-1}(0)$ of critical points of f is reduced to $\{0\}$.

Proof of proposition 1.8 We proceed by induction on n. If $n > 0$, then clearly we can find a hypersurface $Y \subset X$ passing through 0, defined in the neighbourhood of 0, and smooth at 0, such that $f_{|Y}$ admits 0 as a non-degenerate critical point. Indeed, for this last condition to be satisfied, it suffices that the non-degenerate quadratic form $\mathrm{Hess}_0 f$ remain non-degenerate on the hyperplane $T_{Y,0} \subset T_{X,0}$. By hypothesis, there thus exist coordinates $x_1, \ldots, x_{n-1}$ on Y such that for $x = (x_1, \ldots, x_{n-1}) \in Y$, we have

$$f(x) = f(0) - \sum_{i=1}^{r'} x_i^2 + \sum_{i=r'+1}^{n-1} x_i^2$$

with $r' = \mathrm{ind}_0 f_{|Y}$. (Here r' can be equal to r or $r-1$.) The functions x_i can be extended to functions also called x_i on X. The function

$$f - f(0) - \left(-\sum_{i=1}^{r'} x_i^2 + \sum_{i=r'+1}^{n-1} x_i^2\right)$$

on X is thus $\mathcal{C}^\infty$, and vanishes along Y.

Hadamard's lemma then shows that if t is an equation defining Y, there exists a $\mathcal{C}^\infty$ function g such that

$$f - f(0) - \left(-\sum_{i=1}^{r'} x_i^2 + \sum_{i=r'+1}^{n-1} x_i^2\right) = tg.$$

The fact that 0 is a non-degenerate critical point of f is then expressed by the fact that the function g vanishes at 0 and has non-zero differential at 0.

Lemma 1.10 *If $g(0) = 0$, and if the functions $x_1, \ldots, x_{n-1}, t$ give a system of coordinates centred at 0, then there exist $\mathcal{C}^\infty$ functions α_i, $i = 1 \ldots, n-1$ and ϕ such that*

$$g(x_1, \ldots, x_{n-1}, t) = 2\sum_{i=1}^{n-1} x_i \alpha_i + t\phi.$$

Temporarily admitting this lemma, let us now set

$$x_i' = x_i + \epsilon_i \alpha_i t$$

for $i \leq n-1$, where ϵ_i is equal to -1 for $i \leq r'$ and to 1 otherwise. Then for $x = (x_i, t)$, we have

$$f(x) = f(0) - \sum_{i=1}^{r'} x_i'^2 + \sum_{i=r'+1}^{n-1} x_i'^2 + t^2\psi,$$

where ψ is a $\mathcal{C}^\infty$ function. The fact that 0 is a non-degenerate critical point of f then immediately implies that $\psi(0) \neq 0$, so in a neighbourhood of 0, we can write $t^2\psi = \pm(x'_n)^2$. Clearly the x'_i form a system of coordinates on X centred at 0 in which f has the desired expression (possibly after permuting the coordinates). □

Proof of lemma 1.10 We show it by induction, applying Hadamard's lemma. The lemma holds on Y by the induction hypothesis, so there exist $\mathcal{C}^\infty$ functions α_i on X such that $g - 2\sum_{i=1}^{n-1} x_i\alpha_i$ vanishes on Y. Then, by Hadamard's lemma, we have

$$g = 2\sum_{i=1}^{n-1} x_i\alpha_i + t\phi.$$

□

Now let X be a topological space, and $f : X \to \mathbb{R}$ a continuous function.

Definition 1.11 *We say that f is an exhaustion function if for every element $M \in \mathbb{R}$, the closed subset $f^{-1}(]-\infty, M]) \subset X$ is compact.*

Such a map is in particular proper, and the fibres $X_a := f^{-1}(a)$ are compact. In what follows, we will write $X_{\leq M}$ for the subset $f^{-1}(]-\infty, M])$ for every $M \in \mathbb{R}$, and $X_{[M_1,M_2]}$ for the subset $f^{-1}([M_1, M_2])$, where $M_1 \leq M_2 \in \mathbb{R}$. These sets are called level sets.

If f is a differentiable exhaustion function having only non-degenerate critical points, then every critical value corresponds to a finite number of critical points, and the set of critical values is discrete by corollary 1.9. In particular, there exists only a finite number of critical values in each interval $]-\infty, M]$, $M \in \mathbb{R}$.

Such a function is called a Morse function. We sometimes require that every fibre X_λ have at most one critical point, and that the indices $r(\lambda)$ for a critical value λ increase with λ.

1.1.2 Local study of the level sets

Let us consider the function

$$f(x) = -\sum_{i=1}^{r} x_i^2 + \sum_{i=r+1}^{n} x_i^2$$

defined on $\mathbb{R}^n$. Let B^η denote the ball of radius η centred at 0. Let $B^r_\epsilon \subset \mathbb{R}^n$

and $S_\epsilon^{r-1} \subset \mathbb{R}^n$ denote the ball and the sphere defined by

$$B_\epsilon^r := \left\{ (x_1, \ldots, x_n) \in B^\eta \;\middle|\; x_i = 0,\ i > r,\ \sum_{i \le r} x_i^2 \le \epsilon \right\} \subset \mathbb{R}^n_{\le \epsilon},$$
$$S_\epsilon^{r-1} := \left\{ (x_1, \ldots, x_n) \in B^\eta \;\middle|\; x_i = 0,\ i > r,\ \sum_{i \le r} x_i^2 = \epsilon \right\} \subset \mathbb{R}^n_{\le -\epsilon}. \tag{1.1}$$

We easily see that for $\epsilon \le \eta^2$, $B_\epsilon^r < B^\eta$, for $\epsilon = \eta^2$, we have

$$S_\epsilon^{r-1} = \partial B_\epsilon^r = B^\eta_{\le -\epsilon}.$$

Let B denote the ball of radius $\sqrt{2\epsilon}$ in $\mathbb{R}^n$, and S its boundary, the sphere of radius $\sqrt{2\epsilon}$ in $\mathbb{R}^n$. We propose to show the following result.

Proposition 1.12 *There exists a retraction by deformation of $B_{\le\epsilon}$ onto the union $B_{\le -\epsilon} \cup_{S_\epsilon^{r-1}} B_\epsilon^r$, which induces a retraction by deformation of $S_{\le \epsilon}$ onto $S_{\le -\epsilon}$.*

More precisely, we will exhibit a homotopy

$$(H_t)_{t\in[0,1]} : B_{\le\epsilon} \to B_{\le\epsilon}$$

such that

(i) $H_1 = \mathrm{Id}$, $H_t = \mathrm{Id}$ on $B_{\le -\epsilon} \bigcup_{S_\epsilon^{r-1}} B_\epsilon^r$.
(ii) H_0 has values in $B_{\le -\epsilon} \bigcup_{S_\epsilon^{r-1}} B_\epsilon^r$.
(iii) On $S_{[-\epsilon,\epsilon]}$, the homotopy H_t is given by a trivialisation of the fibration $f : S_{[-\epsilon,\epsilon]} \to [-\epsilon, \epsilon]$, i.e. a diffeomorphism

$$C = (C_0, f) : S_{[-\epsilon,\epsilon]} \cong S_{-\epsilon} \times [-\epsilon, \epsilon],$$

and by the retraction by deformation K_t' of the segment $[-\epsilon, \epsilon]$ onto the point $-\epsilon$ given by $K_t'(\alpha) = -(1-t)\epsilon + t\alpha$.

We have the following lemma.

Lemma 1.13 *For $\epsilon = \eta^2$, there exists a homotopy*

$$(R_t)_{t\in[0,1]} : B^\eta_{\le\epsilon} \to B^\eta_{\le\epsilon}$$

such that $R_1 = \mathrm{Id}$, and R_t is the identity on the ball B_ϵ^r defined in (1.1). Finally, $\mathrm{Im}\, R_0$ is contained in the ball B_ϵ^r, so R_0 is a retraction onto B_ϵ^r.

Note that as $B^\eta_{\le -\epsilon} = S_\epsilon^{r-1}$, the above homotopy is the identity on the level subset $B^\eta_{\le -\epsilon}$.

Proof Writing $x = (X_1, X_2)$ with $X_1 = (x_1, \ldots, x_r)$ and $X_2 = (x_{r+1}, \ldots, x_n)$, consider the map $R_t : B^\eta \to B^\eta$ given by

$$R_t(x) = (X_1, tX_2).$$

□

We will now show that up to increasing the radius of the ball, we can construct a retraction by deformation as above, which moreover preserves the boundary. Let us return to the ball B of radius $\sqrt{2\epsilon}$ in $\mathbb{R}^n$, and its boundary S, the sphere of radius $\sqrt{2\epsilon}$ in $\mathbb{R}^n$.

Lemma 1.14 *The restriction of the map f to S does not admit any critical points in the set*

$$S_{[-\epsilon,\epsilon]} := \{x \in S \mid -\epsilon \leq f(x) \leq \epsilon\}.$$

Proof A critical point of f in S is such that the functions $f = -f_1 + f_2$ and $\| \ \|^2 = f_1 + f_2$ have proportional differentials, where $f_1(x) = \|X_1\|^2$ and $f_2(x) = \|X_2\|^2$. This implies that df_1 and df_2 are proportional, and clearly this is not possible unless df_1 or $df_2 = 0$. But then f_1 or $f_2 = 0$, so $|f| = \| \ \|^2$. Now, this is impossible on $S_{[-\epsilon,\epsilon]}$, since on S, we have $\| \ \|^2 = 2\epsilon$. □

By Ehresmann's theorem (see vI, prop. 9.3), there exists a trivialisation

$$C = (C_0, f) : S_{[-\epsilon,\epsilon]} \cong S_{-\epsilon} \times [-\epsilon, \epsilon],$$

where $C_0 : S_{[-\epsilon,\epsilon]} \to S_{-\epsilon}$ is a differentiable map which induces a diffeomorphism $S_t \cong S_{-\epsilon}$ for every $t \in [-\epsilon, \epsilon]$.

In fact, it is easy to explicitly produce such a trivialisation. With notation as above, we take

$$C_0(x) = (\alpha(x)X_1, \beta(x)X_2),$$

where the positive functions α, β must satisfy the following conditions:

$$\begin{aligned} \alpha^2 f_1 + \beta^2 f_2 &= 2\epsilon, \\ -\alpha^2 f_1 + \beta^2 f_2 &= -\epsilon. \end{aligned} \tag{1.2}$$

The conditions (1.2) simply state that $C_0(x) \in S$ and $f(C_0(x)) = -\epsilon$. It is easy to see that the equations (1.2) have unique solutions in $S_{[-\epsilon,\epsilon]}$.

Using the above trivialisation, we construct a retraction by deformation $(K_t)_{t\in[0,1]}$ of $S_{[-\epsilon,\epsilon]}$ onto $S_{-\epsilon}$, compatible with f. This means that K_t induces the identity on $S_{-\epsilon}$ for every t, $K_1 = \mathrm{Id}$, and the image of K_0 is contained in $S_{-\epsilon}$. (The compatibility condition also says that $K_t(S_\alpha) \subset S_{K_t'(\alpha)}$ for a certain retraction by deformation K_t' of the segment $[-\epsilon, \epsilon]$ onto the point ϵ.) We simply set

$$K_t = C^{-1} \circ (C_0, (1-t)(-\epsilon) + tf).$$

In other words, in the trivialisation C, K_t is induced by the affine retraction K_t' of the segment $[-\epsilon, \epsilon]$ onto the point $-\epsilon : K_t'(u) = (1-t)(-\epsilon) + tu$. Note that K_t is of the form

$$K_t(x) = (\alpha_t(x)X_1, \beta_t(x)X_2), \tag{1.3}$$

where the positive functions α_t, β_t for $t \in [0, 1]$ are determined by the conditions

$$\alpha_t^2 f_1 + \beta_t^2 f_2 = 2\epsilon,$$
$$-\alpha_t^2 f_1 + \beta_t^2 f_2 = (1-t)(-\epsilon) + tf.$$

□

Proof of proposition 1.12 We construct the homotopy H_t by setting $H_t = R_t$ in the ball B^η of radius $\sqrt{\epsilon}$, and $H_t = K_t$ on $S_{[-\epsilon,\epsilon]}$. We then look for H_t of the form

$$H_t(x) = (\alpha_t'(x)X_1, \beta_t'(x)X_2)$$

in $B_{[-\epsilon,\epsilon]}$, where the functions α_t', β_t' now satisfy the conditions

$$\alpha_t'^2 f_1 + \beta_t'^2 f_2 \leq 2\epsilon, \quad -\epsilon \leq -\alpha_t'^2 f_1 + \beta_t'^2 f_2 \leq (-f_1 + f_2), \tag{1.4}$$

i.e. $H_t(x) \in B$, $f(H_t(x)) \leq f(x)$, and must coincide with α_t, β_t of (1.3) on $S_{[-\epsilon,\epsilon]}$ and with $1, t$ in $B^\eta_{\leq\epsilon}$. We set $H_t = \mathrm{Id}$ in $B_{\leq -\epsilon}$.

It is easy to check that we can construct such a pair (α_t', β_t'), which also satisfies the conditions

$$\mathrm{Im}\, H_0 \subset B_{\leq -\epsilon} \cup B_\epsilon^r$$

and

$$H_{t\,|B_{-\epsilon} \cup_{S_\epsilon^{r-1}} B_\epsilon^r} = \mathrm{Id},$$

already satisfied in B^η and $S_{[-\epsilon,\epsilon]}$. □

1.1.3 Globalisation

Now let X be a differentiable variety, and $f : X \to \mathbb{R}$ a differentiable exhaustion function. Let λ be a critical value, and let $\epsilon > 0$ be such that λ is the only critical value of f in $[\lambda - \epsilon, \lambda + \epsilon]$. Let 0_i, $i = 1, \ldots, r$ be the critical points of f on X_λ, and let r_i be their Morse indices. The local analysis above, together with proposition 1.8, allows us to prove the following theorem.

Theorem 1.15 *There exists a retraction by deformation of the level set $X_{\leq\lambda+\epsilon}$ onto the union of $X_{\leq\lambda-\epsilon}$ with r_i-dimensional balls B^{r_i} glued on $X_{\leq\lambda-\epsilon}$ along their boundaries, which are disjoint $(r_i - 1)$-dimensional spheres S^{r_i-1}.*

Proof As f is a fibration over $[\lambda - \epsilon, \lambda[$ and $]\lambda, \lambda + \epsilon]$, Ehresmann's theorem shows that it suffices to prove the theorem for very small ϵ. Morse's lemma then shows that in the neighbourhood of each 0_i, there exists a ball B_i in which f can be written as in the preceding subsection with $r = r_i$. We may of course assume that these balls are disjoint. By the preceding subsection, in each of these balls B^i, we now have a retraction by deformation H_t^i of $B^i_{\leq\lambda+\epsilon}$ onto $B^i_{\leq\lambda-\epsilon} \cup_{S_\epsilon^{r_i-1}} B_\epsilon^{r_i}$, which has the property of being induced by a trivialisation of the fibration f in the neighbourhood of $S^i_{[\lambda-\epsilon,\lambda+\epsilon]}$, where $S^i = \partial B^i$, and by the affine retraction of the segment $[\lambda - \epsilon, \lambda + \epsilon]$ onto the point $\lambda - \epsilon$. But since over the segment $[\lambda - \epsilon, \lambda + \epsilon]$, the restriction of f to $X - \bigcup_i B_0^i$ is a fibration of manifolds with boundary (where B_0^i denotes the interior of B^i), the trivialisations of the fibration f in the neighbourhood of $S^i_{[\lambda-\epsilon,\lambda+\epsilon]}$ extend to a trivialisation of the fibration f on $X - \bigcup_i B_0^i$ over $[\lambda - \epsilon, \lambda + \epsilon]$:

$$C = (C_0, f) : \left(X - \bigcup_i B_0^i\right)_{[\lambda-\epsilon,\lambda+\epsilon]} \cong \left(X - \bigcup_i B_0^i\right)_{\lambda-\epsilon} \times [\lambda - \epsilon, \lambda + \epsilon].$$

The $(H_t^i)_{|S^i} = K_t^i$ then extend to a retraction by deformation of $(X - \bigcup_i B_0^i)_{[\lambda-\epsilon,\lambda+\epsilon]}$ onto $(X - \bigcup_i B_0^i)_{\lambda-\epsilon}$, given in the trivialisation above by

$$H_t(x) = (C_0(x), K_t'(f(x))).$$

Clearly, H_t can be glued together with the H_t^i in B^i and with Id in $X_{\leq\lambda-\epsilon}$, which yields the desired retraction by deformation. □

1.2 Application to affine varieties

1.2.1 Index of the square of the distance function

Definition 1.16 *A (smooth) affine variety is a (smooth) closed analytic subvariety of $\mathbb{C}^N$ for some integer N.*

Let X be such a smooth, connected variety and let n denote its complex dimension. If h is a Hermitian metric on $\mathbb{C}^N$ and $0 \in \mathbb{C}^N$, we obtain a $\mathcal{C}^\infty$ function $f_0 : X \to \mathbb{R}$ by setting $f_0(x) = h(\overrightarrow{0x})$.

More generally, we can define such a 'square of the distance' function for any differentiable subvariety X of a Euclidean space $\mathbb{R}^N$. First consider the general situation. Obviously, f is always an exhaustion function.

Lemma 1.17 *Let $X \subset \mathbb{R}^N$ be a differentiable subvariety. Then for a general point 0 of $\mathbb{R}^N$, the corresponding function f_0 is a Morse function.*

Proof We have $df_{0,x}(u) = 2\langle\overrightarrow{0x}, u\rangle$ for $u \in T_{X,x}$. Thus, x is a critical point of f when $\overrightarrow{0x}$ is orthogonal to $T_{X,x}$. Let $Z \subset X \times \mathbb{R}^N$ be the set

$$Z = \{(x, 0) \in X \times \mathbb{R}^N \mid \overrightarrow{0x} \perp T_{X,x}\}.$$

Clearly, we have $\dim Z = N$. The second projection $\pi : Z \to \mathbb{R}^N$ is thus submersive at a point if and only if it is immersive at that point.

Lemma 1.18 *Let $(x, 0) \in Z$, and let $u \in T_{X,x}$. Then the tangent vector $(u, 0) \in T_{X,x} \times T_{\mathbb{R}^N,0}$ lies in $T_{Z,(x,0)}$ if and only if $u \in T_{X,x}$ lies in the kernel of the quadratic form* $\mathrm{Hess}_x f_0$.

Admitting this lemma, we see that the set of points 0 for which the function f_0 admits a degenerate critical point is the image under π of the set of points of Z where π is not an immersion, so not a submersion. By Sard's lemma (see Rudin 1966), this set has empty interior. □

Proof of lemma 1.18 The subset Z is defined in $X \times \mathbb{R}^N$ as the vanishing locus of the section σ of the vector bundle $\mathrm{pr}_1^*\Omega_X$ given by

$$\sigma_{(x,0)} = \langle\overrightarrow{0x}, \cdot\rangle = df_{0,x}.$$

Taking local coordinates x_i on X, σ can be written

$$\sum_i \frac{\partial f_0}{\partial x_i} dx_i.$$

The tangent space to Z at $(x, 0)$ is thus described by

$$T_{Z,(x,0)} = \left\{ (u, w) \in T_{X,x} \times T_{\mathbb{R}^N,0} \,\middle|\, d_u \frac{\partial f_0}{\partial x_i} + d_w \frac{\partial f_0}{\partial x_i} = 0, \ \forall i \right\}.$$

Writing $u = \sum_i u_i \frac{\partial}{\partial x_i}$, the first term is equal to $\sum_j u_j \frac{\partial^2 f_0}{\partial x_i \partial x_j}$. Thus, the vector $(u, 0)$ lies in $T_{Z,(x,0)}$ if and only if we have $\sum_j u_j \frac{\partial^2 f_0}{\partial x_i \partial x_j} = 0$ for every i, which means by definition that u lies in the kernel of $\mathrm{Hess}_x f_0$. □

Let us now return to the case where X is an n-dimensional complex analytic subvariety of $\mathbb{C}^N$, and the metric is Hermitian.

Proposition 1.19 *The index of the function $f = f_0$ is less than or equal to n at every critical point of f.*

Proof Let us introduce the second fundamental form

$$\Phi : S^2 T_{X,x} \to \mathbb{C}^N / T_{X,x},$$

which can be defined as the differential of the Gauss map

$$X \to \mathrm{Grass}(n, N), \ x \mapsto T_{X,x}.$$

By definition, $\Phi(u, v)$ can be computed as follows. Let V be a vector field on X defined in the neighbourhood of x, and whose value at x is v. Then, as $T_X \subset (T_{\mathbb{C}^N})_{|X}$, we can see V as a map with values in $\mathbb{C}^N$, and set

$$\Phi(u, v) = d_u V \bmod T_{X,x}.$$

One knows that this depends only on the vector v and not on the vector field V; see vI, lemma 10.7.

Let h be the Hermitian form, and $Q = \Re h = \langle , \rangle$ the corresponding Euclidean scalar product. The formula

$$df_x(u) = 2\langle \overrightarrow{0x}, u \rangle, \quad u \in T_{X,x}$$

shows immediately by differentiation that for a critical point x of f, we have

$$\mathrm{Hess}_x f(u, v) = 2(\langle \overrightarrow{0x}, \Phi(u, v) \rangle + \langle u, v \rangle).$$

To diagonalise $\mathrm{Hess}_x f$, it thus suffices to diagonalise the quadratic form

$$Q(u, v) = \langle \overrightarrow{0x}, \Phi(u, v) \rangle$$

in an orthonormal basis for $\langle\ ,\ \rangle$. But since X is a complex subvariety of $\mathbb{C}^N$, the second fundamental form Φ is $\mathbb{C}$-bilinear for the complex structures on $T_{X,x}$ and $\mathbb{C}^N / T_{X,x}$, so that Q is the real part of the $\mathbb{C}$-bilinear symmetric form $H(u, v) = h(\Phi(u, v), \overrightarrow{0x})$ on $T_{X,x} \cong \mathbb{C}^n \subset \mathbb{C}^N$.

Lemma 1.20 *Let H be a $\mathbb{C}$-bilinear symmetric form on $\mathbb{C}^n$, and let $\langle\ ,\ \rangle = \Re h$ be the Euclidean product associated to a Hermitian form on $\mathbb{C}^n$. Then the eigenvalues of the form $Q = \Re H$ with respect to the Euclidean form constitute a set which is stable under the involution $\lambda \mapsto -\lambda$ (including multiplicities).*

Admitting this lemma, we conclude that the eigenvalues of $\mathrm{Hess}_x f$ with respect to the Euclidean metric are of the form $2(1+\lambda_i)$, $2(1-\lambda_i)$, $i = 1, \ldots, n$. Thus, at most n of these eigenvalues are negative. □

Proof of lemma 1.20 The form H, and thus also the form $Q = \Re H$, are multiplied by -1 under the automorphism of $\mathbb{C}^n$ induced by multiplication by i (which is unitary). Thus, the set of eigenvalues is stable under multiplication by -1. The statement concerning the multiplicities follows easily, since every form $Q = \Re H$ is a specialisation of a form of the type $\Re H$ with distinct eigenvalues. □

Definition 1.21 *A compact CW-complex is a topological space which can be written as a finite union of closed 'cells' homeomorphic to closed balls of $\mathbb{R}^k$. The largest integer k appearing in this cellular decomposition is called the dimension of the CW-complex.*

We define a CW-complex as a topological space which is the countable union of increasing open sets U_i such that the closure $\overline{U}_i$ is a compact CW-complex. The following result is a consequence of theorem 1.15 and proposition 1.19 (see Andreotti & Frankel 1959; Milnor 1963).

Theorem 1.22 *An affine variety X of complex dimension n has the homotopy type of a CW-complex of real dimension $\leq n$.*

Proof Let $X \subset \mathbb{C}^N$, and take a Morse function on X of the form $h_0(x) = d(x, 0)^2$, for a Hermitian metric on $\mathbb{C}^N$. Then X can be written as the countable union of the interiors X_k^0 of the level sets $X_k = \{x \in X \mid h_0(x) \leq k\}$, and assuming that the integers k are not critical values, theorem 1.15 and proposition 1.19 show that each level set X_{k+1} has the homotopy type of the union of X_k with a finite number of balls of dimension $\leq k$. □

1.2.2 Lefschetz theorem on hyperplane sections

Let X be a smooth complex projective variety, and $Y \overset{j}{\hookrightarrow} X$ a smooth hyperplane section. The cohomology class $[Y] \in H^2(X, \mathbb{Z})$ is equal to $h := c_1(\mathcal{O}_X(1))$ (see vI, Thm 11.33). Let us first show that we have the identity

$$j^* \circ j_* = h_Y \cup : H^k(Y, \mathbb{Z}) \to H^{k+2}(Y, \mathbb{Z}), \tag{1.5}$$

where $h_Y := h_{|Y} = [Y]_{|Y}$.

This formula follows immediately from the description of the Gysin morphism j_* as the composition $k_* \circ (\cup \eta_Y) \circ \pi^*$, where

$$\begin{array}{ccc} T & \xrightarrow{\pi} & Y \\ {\scriptstyle k}\downarrow & & \\ X & & \end{array}$$

is an open tubular neighbourhood of Y in X, and $\eta_Y \in H_c^2(T, \mathbb{Z})$ is the cohomology class with compact support of Y in T. Indeed, forgetting the torsion we know that the Gysin morphism $j_* : H^k(Y, \mathbb{Z}) \to H^{k+2}(X, \mathbb{Z})$ is the Poincaré dual of the restriction morphism

$$j^* : H^{2n-k-2}(X, \mathbb{Z}) \to H^{2n-k-2}(Y, \mathbb{Z}).$$

Passing to real coefficients, we must check that for a closed k-form β on Y and a closed $(2n-k-2)$-form α on X, we have

$$\int_Y \beta \wedge j^*\alpha = \int_X \tilde{\eta}_Y \wedge \pi^*\beta \wedge \alpha.$$

Here, the form $\tilde{\eta}_Y$ is a de Rham representative of η_Y: it is a closed 2-form with support in T such that $\pi_*\tilde{\eta}_Y = 1_Y$ (see vI.11.1.2). The form $\tilde{\eta}_Y \wedge \pi^*\beta$ on T is extended to X by 0, and thus the right-hand term is equal to $\int_T \tilde{\eta}_Y \wedge \pi^*\beta \wedge \alpha$. Now, as T can be retracted by deformation onto Y by π, there exists a closed $(2n-k-2)$-form α' on Y and a $(2n-k-3)$-form α'' on T such that

$$\alpha_{|T} = \pi^*\alpha' + d\alpha''.$$

But then, applying Stokes and the fact that $\pi_* \tilde{\eta}_Y = 1_Y$, we find

$$\int_T \tilde{\eta}_Y \wedge \pi^*\beta \wedge \alpha = \int_T \tilde{\eta}_Y \wedge \pi^*\beta \wedge \pi^*\alpha'$$
$$= \int_Y \beta \wedge \alpha' = \int_Y \beta \wedge j^*\alpha,$$

which proves the equality $j_* = k_* \circ (\cup \eta_Y) \circ \pi^*$.

As h_Y is a Kähler class, we can now apply the hard Lefschetz theorem (vI, Thm 6.25) to conclude that:

The cup-product

$$\cup h_Y : H^k(Y, \mathbb{Q}) \to H^{k+2}(Y, \mathbb{Q})$$

is injective for $k < n - 1 := \dim Y$ *and surjective for* $k + 2 > n - 1 = \dim Y$.

As $\cup h_Y = j^* \circ j_*$, it follows that

$$j_* : H^k(Y, \mathbb{Q}) \to H^{k+2}(X, \mathbb{Q})$$

is injective for $k < n - 1$ and

$$j^* : H^k(X, \mathbb{Q}) \to H^k(Y, \mathbb{Q})$$

is surjective for $k > n - 1$.

Moreover, we also have the equality

$$j_* \circ j^* = [Y]\cup = h\cup : H^k(X, \mathbb{Z}) \to H^{k+2}(X, \mathbb{Z}), \tag{1.6}$$

which can be proved by the same argument as above.

Applying the Lefschetz theorem cited above to X, we conclude that $h\cup = j^* \circ j_*$ is injective on $H^k(X, \mathbb{Q})$ for $k < n = \dim X$, which implies that

$$j^* : H^k(X, \mathbb{Q}) \to H^k(Y, \mathbb{Q})$$

is injective for $k < n = \dim X$.

We will now use the topological analysis developed in the preceding section to obtain a much stronger statement, namely the Lefschetz theorem on hyperplane sections.

Theorem 1.23 *Let* $X \subset \mathbb{P}^N$ *be a (not necessarily smooth)* n*-dimensional algebraic subvariety, and let* $Y = \mathbb{P}^{N-1} \cap X$ *be a hyperplane section such that*

$U := X - Y$ is smooth and n-dimensional. Then the restriction morphism

$$j^* : H^k(X, \mathbb{Z}) \to H^k(Y, \mathbb{Z})$$

is an isomorphism for $k < n - 1$ and is injective for $k = n - 1$.

Proof Let us admit the fact, which is proved using triangulations, that Y admits a fundamental system of neighbourhoods Y_i in X which can be retracted by deformation onto Y. It follows that

$$H^k(X, Y, \mathbb{Z}) \cong \lim_{\to} H^k(X, Y_i, \mathbb{Z}). \tag{1.7}$$

By excision, we also obtain isomorphisms

$$H^k(X, Y_i, \mathbb{Z}) \cong H^k(U, Y_i \cap U, \mathbb{Z}).$$

But since U is an oriented differentiable $2n$-dimensional variety, Poincaré duality gives an isomorphism (which is canonical, depending only on the orientation)

$$H^k(U, U - K, \mathbb{Z}) \cong H_{2n-k}(K, \mathbb{Z}) \tag{1.8}$$

for every compact set $K \subset U$ having the property that K is the retraction by deformation of an open set of U (see Spanier 1996, 6.2).

If we now apply the Poincaré duality (1.8) to $K_i := U - Y_i \cap U$, we obtain an isomorphism

$$H^k(X, Y, \mathbb{Z}) \cong \lim_{\to} H^k(U, Y_i \cap U, \mathbb{Z}) \cong \lim_{\to} H_{2n-k}(K_i, \mathbb{Z}).$$

Since every singular chain is contained in one of the compact sets $K_i \subset U$, it is clear that we have

$$H_{2n-k}(U, \mathbb{Z}) = \lim_{\to} H_{2n-k}(K_i, \mathbb{Z}).$$

In conclusion, we have a natural isomorphism

$$H^k(X, Y, \mathbb{Z}) \cong H_{2n-k}(U, \mathbb{Z}). \tag{1.9}$$

Returning to our proof, we consider the long exact sequence of relative cohomology of the pair (X, Y):

$$\cdots \to H^k(X, Y, \mathbb{Z}) \to H^k(X, \mathbb{Z}) \to H^k(Y, \mathbb{Z}) \to H^{k+1}(X, Y, \mathbb{Z}) \to \cdots.$$

It follows that theorem 1.23 is equivalent to the vanishing of the groups $H^k(X, Y, \mathbb{Z})$ for $k \leq n - 1 = \dim Y$. Applying the isomorphism (1.9), this is equivalent to the vanishing of the groups $H_k(U, \mathbb{Z})$ for $k \geq n + 1$.

Now, U is an affine variety embedded in $\mathbb{C}^N$. Let us equip $\mathbb{C}^N$ with a Hermitian metric, and let $0 \in \mathbb{C}^N$ be such that the function f_0 is a Morse function on U. U can be written as the union of the increasing level sets $U_{\leq M}$, $M \in \mathbb{Z}$, with $U_{\leq -1} = \emptyset$. It follows immediately that

$$H_k(U, \mathbb{Z}) = \varinjlim_{M \to \infty} H_k(U_{\leq M}, \mathbb{Z}),$$

and it suffices to show that for every level set $U_{\leq M}$, we have

$$H_k(U_{\leq M}, \mathbb{Z}) = 0 \quad \text{for} \quad k > n.$$

But this follows immediately, by induction on M, from theorem 1.15 and proposition 1.19. Indeed, assuming for simplicity that M and $M+1$ are not critical values of f_0, there exists a finite number of critical values λ_i of the function f_0 contained between M and $M+1$, i.e.

$$M < \lambda_1 < \cdots < \lambda_i < \cdots \lambda_k < M+1.$$

For $1 \leq i \leq k-1$, let us choose $\lambda_i' \in]\lambda_i, \lambda_{i+1}[$, and set $M = \lambda_0'$, $M+1 = \lambda_k'$. Then by theorem 1.15, each level set $U_{\leq \lambda_i'}$ has the homotopy type of the union of $U_{\leq \lambda_{i-1}'}$ with balls $B_{i,j}^{r_j}$ of dimension r_j equal to the index of the critical point $x_{i,j}$ of critical value λ_i, glued along their boundary $S_{i,j}^{r_j - 1}$. By proposition 1.19, all of these indices are at most equal to n. Thus, for each of these balls, we have

$$H_k(B, S, \mathbb{Z}) = 0 \quad \text{for} \quad k > n,$$

which by excision implies that

$$H_k(U_{\leq M+1}, U_{\leq M}, \mathbb{Z}) = 0 \quad \text{for} \quad k > n.$$

Thus, by the long exact sequence of relative homology of the pair (U_{M+1}, U_M), if

$$H_k(U_{\leq M}, \mathbb{Z}) = 0 \quad \text{for} \quad k > n,$$

then also $H_k(U_{\leq M+1}, \mathbb{Z}) = 0$ for $k > n$. As $U_{\leq -1} = \emptyset$, we thus have

$$H_k(U_{\leq M}, \mathbb{Z}) = 0 \quad \text{for} \quad k > n$$

for every M. This proves theorem 1.23. □

1.2.3 Applications

Recall (see vI.7.2.1) that the cohomology of the projective space $\mathbb{P}^n$ is described by $H^i(\mathbb{P}^n, \mathbb{Z}) = 0$ for odd i and $H^{2i}(\mathbb{P}^n, \mathbb{Z}) = \mathbb{Z}h^i$, where $h = c_1(\mathcal{O}_{\mathbb{P}^n}(1)) \in H^2(\mathbb{P}^n, \mathbb{Z})$.

Moreover, using the dth Veronese embedding

$$\Phi_d : \mathbb{P}^n \to \mathbb{P}^N,$$

which is given by homogeneous polynomials of degree d, and which is such that the pullback by Φ_d of a linear form on $\mathbb{P}^N$ is a polynomial of degree d on $\mathbb{P}^n$, we can consider a hypersurface $X \subset \mathbb{P}^n$ of degree d as an intersection $\Phi_d(\mathbb{P}^n) \cap \mathbb{P}^{N-1}$. Thus, the following result for hypersurfaces of projective space follows from theorem 1.23.

Corollary 1.24 *Let $X \subset \mathbb{P}^n$ be a hypersurface. Then $H^k(X, \mathbb{Z}) = 0$ for k odd, $k < \dim X$, and $H^{2k}(X, \mathbb{Z}) = \mathbb{Z}h^k$ for $2k < \dim X$.*

If moreover X is smooth, then the next result follows from Poincaré duality.

Corollary 1.25 *Let $X \subset \mathbb{P}^n$ be a smooth hypersurface. Then $H^k(X, \mathbb{Z}) = 0$ for k odd, $k > \dim X$, and $H^{2k}(X, \mathbb{Z}) = \mathbb{Z}\alpha$ for $2k > \dim X$, where the class α has intersection with h^{n-1-k} equal to 1.*

Remark 1.26 *Let us take the case of a smooth hypersurface X in $\mathbb{P}^4$. The preceding corollary shows that $H_2(X, \mathbb{Z}) = H^4(X, \mathbb{Z})$ is generated by the unique class α such that $\langle \alpha, h \rangle = 1$. If X contains a line, the homology class of this line is thus equal to α. In general, if $d := \deg X > 5$, then X does not contain a line (at least if the equation of X is chosen generically). However, a curve $C = X \cap \mathbb{P}^2$ is of degree d and is thus of class $d\alpha$. Kollár (1990) showed that in general, for sufficiently large d, the class α is not the class of an algebraic cycle, although $d\alpha$ is. This is one of the counterexamples to the Hodge conjecture for integral cohomology.*

Corollaries 1.24 and 1.25 can be generalised immediately to complete intersections in projective space, by repeated applications of theorem 1.23.

A first application of corollary 1.24 concerns the computation of the Picard group of complete intersections. Recall (see vI.4.3, vI.11.3) that if X is a projective variety, then Pic X is the group of isomorphism classes of algebraic line bundles, or equivalently, of isomorphism classes of holomorphic line bundles, or in the smooth case, of divisors modulo rational equivalence. The second interpretation gives an identification (see vI.4.3) $\mathrm{Pic}\, X \cong H^1(X, \mathcal{O}^*_{X,\mathrm{an}})$, and the exponential exact sequence gives the long exact sequence

$$H^1(X, \mathbb{Z}) \to H^1(X, \mathcal{O}_X) \to \mathrm{Pic}(X) \xrightarrow{c_1} H^2(X, \mathbb{Z}).$$

If X is now a smooth complete intersection in $\mathbb{P}^n$ such that $\dim X \geq 3$, then by corollary 1.24, we have $H^1(X,\mathbb{Z})=0$, so $H^1(X,\mathcal{O}_X)=0$ by Hodge theory, which shows that $H^1(X,\mathcal{O}_X)$ is a quotient of $H^1(X,\mathbb{C})$. Moreover, we have an isomorphism

$$\operatorname{Pic}\mathbb{P}^n \overset{c_1}{\cong} H^2(\mathbb{P}^n,\mathbb{Z})=\mathbb{Z}h,$$

and corollary 1.24 gives a restriction isomorphism $H^2(\mathbb{P}^n,\mathbb{Z})\cong H^2(X,\mathbb{Z})$. The next corollary follows immediately.

Corollary 1.27 *If X is a smooth complete intersection of dimension ≥ 3 in $\mathbb{P}^n$, then* $\operatorname{Pic}X=\mathbb{Z}\mathcal{O}_X(1)$.

In fact, using the arguments of the following section, we can show that under the hypothesis $\dim X\geq 2$, the vanishing property $H^1(X,\mathcal{O}_X)=0$ holds even when X is not smooth, so that also corollary 1.27 holds even when X is not smooth.

1.3 Vanishing theorems and Lefschetz' theorem

As observed by Kodaira and Spencer (see Shiffman & Sommese 1985), it is possible to give a more 'algebraic' proof of the Lefschetz theorem 1.23, at least for cohomology with rational coefficients.

For this, we use the following vanishing theorem due to Akizuki, Kodaira, and Nakano (see Demailly 1996; Griffiths & Harris 1978). Let X be a complex variety, and let L be a holomorphic line bundle on X. Recall that L is said to be positive if L can be equipped with a Hermitian metric whose associated Chern form is positive (i.e. is a Kähler form). By the Kodaira embedding theorem (see vI.7.1.3), this is equivalent to the fact that L is ample, i.e. that the holomorphic sections of $L^{\otimes N}$ for sufficiently large N give an embedding

$$\Phi_{NL}:X\hookrightarrow\mathbb{P}^r.$$

Theorem 1.28 *Let $L\to X$ be a positive line bundle, where X is compact. Then for $p+q>n:=\dim X$, we have*

$$H^q\big(X,\Omega_X^p(L)\big)=0.$$

Applying Serre duality (see vI.5.3.2) and noting that by the exterior product,

$$\big(\Omega_X^p\big)^*\otimes K_X\cong\Omega_X^{n-p},$$

we also obtain the following equivalent statement: *under the same hypotheses, we have*

$$H^q\big(X, \Omega_X^p(-L)\big) = 0 \quad \text{for} \quad p + q < n.$$

Here, the notation $\mathcal{F}(-L)$ means $\mathcal{F} \otimes (L^{-1})$, where $\mathcal{F}$ denotes a coherent sheaf on X.

This theorem yields the following version of the Lefschetz theorem.

Theorem 1.29 *Let X be an n-dimensional compact complex variety, and let $Y \stackrel{j}{\hookrightarrow} X$ be a smooth hypersurface such that the line bundle $\mathcal{O}_X(Y) = (\mathcal{I}_Y)^*$ is positive. Then for $k < n - 1$, the restriction*

$$j^* : H^k(X, \mathbb{Q}) \to H^k(Y, \mathbb{Q})$$

is an isomorphism, and for $k = n - 1$, it is injective.

Remark 1.30 *This statement is weaker than theorem 1.23. Indeed, note that by definition Y is the zero locus of a section σ_Y of $\mathcal{O}_X(Y)$, so NY is the divisor of the section σ_Y^N of $\mathcal{O}_X(NY)$. Under the preceding hypotheses, there exists an embedding Φ_{NY} of X into $\mathbb{P}^r$ such that the pullback of the linear forms gives exactly the sections of $\mathcal{O}_X(NY)$. Thus, under the embedding Φ_{NY}, $Y \subset X$ is the set-theoretic intersection of X with a hyperplane in $\mathbb{P}^r$. Thus, theorem 1.23 implies theorem 1.29. But the latter is strictly weaker, insofar as it deals only with rational cohomology and smooth hypersurfaces.*

Proof of theorem 1.29 By the change of coefficients theorem, we have $H^k(X, \mathbb{C}) = H^k(X, \mathbb{Q}) \otimes_{\mathbb{Q}} \mathbb{C}$, and a similar statement for Y; thus it suffices to prove this theorem for the cohomology with complex coefficients instead of the rational cohomology. Now, under the hypotheses of the theorem, X and Y are Kähler (and even projective), and thus, from the Hodge decomposition for X and Y (see vI.6.1), we have

$$H^k(X, \mathbb{C}) = \bigoplus\nolimits_{p+q=k} H^q\big(X, \Omega_X^p\big)$$

and

$$H^k(Y, \mathbb{C}) = \bigoplus\nolimits_{p+q=k} H^q\big(Y, \Omega_Y^p\big)$$

with $j^* = \bigoplus_{p+q=k} j_{p,q}^*$, where $j_{p,q}^* : H^q(X, \Omega_X^p) \to H^q(Y, \Omega_Y^p)$ is the restriction morphism induced by the morphism

$$j_p^* : \Omega_X^p \to j_*\Omega_Y^p$$

of coherent sheaves. Thus, it suffices to show that $j^*_{p,q}$ is an isomorphism for $p+q < \dim Y$ and is injective for $p+q = \dim Y$. Now, the morphism j^*_p is the composition of the natural morphisms

$$\Omega^p_X \to \Omega^p_X \otimes \mathcal{O}_Y = j_*(\Omega^p_X|_Y) \tag{1.10}$$

and

$$\Omega^p_X|_Y \to \Omega^p_Y \tag{1.11}$$

(the latter should in fact be composed with j_*, which induces an isomorphism in cohomology, and is thus usually omitted). It thus suffices to show that each of the morphisms (1.10) and (1.11) induces an isomorphism on the cohomology of degree q for $p+q < n-1 = \dim Y$, and an injection for $p+q = n-1$.

For this, we apply theorem 1.28 to X and to Y. Let us first consider the case of (1.10). We have the exact sequence

$$0 \to \Omega^p_X(-Y) \to \Omega^p_X \to \Omega^p_X|_Y \to 0.$$

The associated long exact sequence and the vanishing property $H^q(X, \Omega^p_X(-Y)) = 0$ for $p+q < \dim X$ immediately imply that the arrow

$$H^q(X, \Omega^p_X) \to H^q(Y, \Omega^p_X|_Y)$$

induced by (1.10) is an isomorphism for $p+q < \dim Y = \dim X - 1$ and is injective for $p+q = \dim Y$.

Moreover, we have the conormal exact sequence

$$0 \to \mathcal{O}_Y(-Y) \to \Omega_X|_Y \to \Omega_Y \to 0$$

on Y (see vI.3.3.3), where the identification of $\mathcal{O}_Y(-Y) = \mathcal{I}_Y \otimes \mathcal{O}_Y$ with the conormal bundle $N^*_{Y/X}$ is induced by the differential $d : \mathcal{I}_Y \to \Omega_X$. Passing to the pth exterior power, this exact sequence induces the exact sequence

$$0 \to \Omega^{p-1}_Y(-Y) \to \Omega^p_X|_Y \to \Omega^p_Y \to 0.$$

The associated long exact sequence of cohomology and theorem 1.28 applied to Y thus show that the morphism

$$H^q(Y, \Omega^p_X|_Y) \to H^q(Y, \Omega^p_Y)$$

induced by (1.11) is an isomorphism for $p+q < \dim Y$, and is injective for $p+q = \dim Y$. □

Exercises

1. *Morse theory and the Euler–Poincaré characteristic.* Let $X_1 \subset X$ be varieties with compact boundaries.
   (a) Using the long exact sequence of relative cohomology, show that

$$\chi_{\text{top}}(X) = \chi_{\text{top}}(X_1) + \chi_{\text{top}}(X, X_1),$$

   where $\chi_{\text{top}}(X) := \sum_i (-1)^i b_i(X)$ and $\chi_{\text{top}}(X, X_1) := \sum_i (-1)^i \dim H^i(X, X_1)$.

   Now let X be a compact differentiable variety, and let $f : X \to \mathbb{R}$ be a Morse function.
   (b) Let $x \in X$ be a critical point of f of index i. Show that for $\epsilon > 0$, the level sets $X_{\leq f(x) \pm \epsilon}$ satisfy

$$\chi_{\text{top}}\left(X_{\leq f(x)+\epsilon}\right) = \chi_{\text{top}}\left(X_{\leq f(x)-\epsilon}\right) + (-1)^i.$$

   (c) Deduce the formula

$$\chi_{\text{top}}(X) = \sum_i (-1)^i N_i,$$

   where N_i is the number of critical points of index i.
2. *Subvarieties with ample normal bundle and Lefschetz theorems.* Let X be an n-dimensional smooth projective variety and E a holomorphic vector bundle of rank r on X. We say that E is ample if the invertible bundle $\mathcal{O}_{\mathbb{P}(E^*)}(1)$ is ample on the projective bundle $\mathbb{P}(E^*) \xrightarrow{\pi} X$. Thus, $h := c_1(\mathcal{O}_{\mathbb{P}(E^*)}(1))$ is a Kähler class on $\mathbb{P}(E^*)$. Recall (see vI.11.2) that the Chern classes $c_i(E^*)$ are characterised by the relation

$$h^r + \sum_{0<i\leq r} \pi^* c_i(E^*) h^{r-i} = 0 \text{ in } H^{2r}(\mathbb{P}(E^*), \mathbb{Z}).$$

   (a) Deduce from the hard Lefschetz theorem applied to $\mathbb{P}(E^*)$ that if E is ample, then the map

$$\cup h : H^{k+r-2}(\mathbb{P}(E^*), \mathbb{Q}) \to H^{k+r}(\mathbb{P}(E^*), \mathbb{Q})$$

   is injective for $k \leq n$.
   (b) Under the same hypothesis, deduce from the decomposition of $H^*(\mathbb{P}(E^*), \mathbb{Q})$ (see vI.7.3.3) that the map

$$\cup c_r(E) : H^k(X, \mathbb{Q}) \to H^{k+2r}(X, \mathbb{Q})$$

   is injective for $k \leq n - r$.

(c) Show that the conclusion of (b) still holds if E^* is assumed to be ample.

Let X be an $(n+r)$-dimensional complex variety, and let $Y \overset{j}{\hookrightarrow} X$ be an n-dimensional compact complex subvariety of X.

(d) Show that the map

$$j^* \circ j_* : H^k(Y, \mathbb{Z}) \to H^{k+2r}(Y, \mathbb{Z})$$

is equal to $\cup c_r(N_{Y/X})$ (see vI.1.2.2, and vI, chapter 11, exercise 3). Deduce that if the normal bundle $N_{Y/X}$ or its dual is ample, then the map

$$j_* : H^k(Y, \mathbb{Q}) \to H^{k+2r}(X, \mathbb{Q})$$

is injective for $k \leq n - r$.

2
Lefschetz Pencils

In this chapter, we propose an alternative, more holomorphic approach to the results of the preceding section. This approach has the advantage of providing a precise qualitative description of the generators of the relative homology (or cohomology) of the pair (X, Y) in degree n and of the generators of the vanishing homology in degree $n-1$, where $j : Y \hookrightarrow X$ is a smooth hyperplane section of an n-dimensional smooth projective variety, and the vanishing homology is defined by

$$H_*(Y, \mathbb{Z})_{\mathrm{van}} := \mathrm{Ker}\,(j_* : H_*(Y, \mathbb{Z}) \to H_*(X, \mathbb{Z})).$$

This description is obtained by considering a Lefschetz pencil of hyperplane sections of X passing through Y. Such a pencil gives a morphism

$$f : \widetilde{X} \to \mathbb{P}^1,$$

where $\widetilde{X}$ is the blowup of X along the base locus of the pencil.

The holomorphic Morse lemma and the local analysis of the topology of a Lefschetz degeneration then enable us to show that the variety $\widetilde{X} - X_\infty$ has the homotopy type of the union of the variety Y with n-dimensional balls glued onto Y along 'vanishing' spheres which contract to a point when Y degenerates to one of the singular fibres of the pencil. Via an argument using induction on the dimension, one can then deduce another proof of the Lefschetz theorem 1.23 from this description, as well as a proof of the following result.

Theorem 2.1 *The relative homology of the pair (X, Y) in degree n is generated by the classes of cones on the vanishing spheres, and the vanishing homology of Y in degree $n-1$ is generated by the classes of vanishing spheres.*

The local description of the topology of a Lefschetz degeneration will play a fundamental role in the following chapter, where we will describe the local monodromy on the cohomology of the fibres of a Lefschetz degeneration (Picard–Lefschetz formula). Theorem 2.1 will be essential in the proof of the irreducibility theorem for the monodromy action.

This chapter begins with the proof of the existence of Lefschetz pencils. We then study the topology of an ordinary singularity, and establish the following result, which is an easy consequence of real Morse theory.

Theorem 2.2 *Let $f : X \to \Delta$ be a proper holomorphic map, smooth over $\Delta - 0$, such that the fibre X_0 has an ordinary double point as its only singularity. Then X has the homotopy type of the union of a smooth fibre X_t with a ball B^n, where $n = \dim X$, glued onto X_t along a sphere S^{n-1}.*

This sphere $S^{n-1} \subset X_t$ is called a vanishing sphere, and the ball $B^n \subset X$ is called a cone over the vanishing cycle.

In the third and last section of this chapter, we explain how to deduce from this another proof of the Lefschetz theorem, and also clarify the relation between the vanishing cohomology and the primitive cohomology considered in vI.6.2.3. The main reference we used for this section is Andreotti & Frankel (1969).

2.1 Lefschetz pencils

2.1.1 Existence

Let X a be complex variety, and let $f : X \to \mathbb{C}$ be a holomorphic map.

Definition 2.3 *The point $0 \in X$ is a non-degenerate critical point of f if $df(0) = 0$, and the holomorphic Hessian* $\mathrm{Hess}_0 f$ *is a non-degenerate (complex) quadratic form on the tangent space $T^{1,0}_{X,0}$.*

Here, the Hessian is the $\mathbb{C}$-bilinear form on $T^{1,0}_{X,0}$ defined as follows. If $z_1, \ldots, z_n$ is a system of holomorphic coordinates on X in the neighbourhood of 0, then $\mathrm{Hess}_0 f$ is defined in the (complex) basis $\frac{\partial}{\partial z_i}$ of $T^{1,0}_{X,0}$ by

$$\mathrm{Hess}_0 f\left(\frac{\partial}{\partial z_i}, \frac{\partial}{\partial z_j}\right) = \frac{\partial^2 f}{\partial z_i \partial z_j}.$$

Exactly as in the differentiable case, we show that $\mathrm{Hess}_0 f$ does not depend on the choice of holomorphic coordinates as long as $df(0) = 0$.

We will write X_t for the hypersurface $f^{-1}(t) \subset X$ with equation $f - t = 0$.

Definition 2.4 *For $t = f(0)$, we will say that X_t has an ordinary double point at 0 if 0 is a non-degenerate critical point of f.*

Indeed, this definition is independent of the (reduced) equation chosen for the hypersurface X_t: another equation would have to be of the form $f' = g(f - t)$ with $g(0) \neq 0$, and we would then have $df' = gdf + (f - t)dg$. As $df(0) = 0 = f(0) - t$, it follows that $df' = gdf$ modulo a differential form which vanishes to order 2 at 0. Applying Leibniz' rule to the derivatives $\frac{\partial f'}{\partial z_i}$, and using the fact that $df(0) = 0$, we immediately see that

$$\mathrm{Hess}_0 f' = g(0)\mathrm{Hess}_0 f.$$

Definition 2.5 *A pencil of hypersurfaces on a variety X is a projective line $\mathbb{P}^1 \subset \mathbb{P}(H^0(X, L)) =: |L|$, where L is a holomorphic line bundle on X.*

Every element $t \in \mathbb{P}^1$ of this pencil thus gives a section σ_t of L, which is non-zero and well-defined up to a coefficient. If $X_t \subset X$ denotes the hypersurface defined by σ_t, then we write $(X_t)_{t\in\mathbb{P}^1}$ for the pencil. The equation σ_t is of the form $\sigma_0 + t\sigma_\infty$, $t \in \mathbb{C} \subset \mathbb{P}^1$. Assume now that $X \subset \mathbb{P}^N$ is a projective subvariety, and that $L = \mathcal{O}_X(1)$. If the restriction map

$$H^0(\mathbb{P}^N, \mathcal{O}_{\mathbb{P}^N}(1)) \to H^0(X, \mathcal{O}_X(1)) \tag{2.1}$$

is an isomorphism, a pencil in $|L|$ is a projective line in

$$\mathbb{P}(\mathbb{P}^N, H^0(\mathcal{O}_{\mathbb{P}^N}(1))) = (\mathbb{P}^N)^*,$$

where $(\mathbb{P}^N)^*$ is the dual projective space parametrising the hyperplanes of $\mathbb{P}^N$.

If $(X_t)_{t\in\mathbb{P}^1}$ is such a pencil, we write $B := \bigcap_{t\in\mathbb{P}^1} X_t \subset X$ for its base locus. Clearly, with notation as above, B is defined by the equations $\sigma_0 = \sigma_\infty = 0$, so it is a complete intersection of codimension 2 in X if the hypersurfaces X_0 and X_∞ have no common component.

Definition 2.6 *A Lefschetz pencil $(X_t)_{t\in\mathbb{P}^1}$ is a pencil of hypersurfaces satisfying the following conditions:*

(i) The base locus B is smooth of codimension 2 in X. In particular, the hypersurfaces of the pencil are smooth along B. Equivalently, the differentials of the equations σ_0 and σ_∞ are independent along B.

(ii) Every hypersurface X_t has at most one ordinary double point as singularity.

The variety X is contained in $\mathbb{P}^N$, and we will assume that X is non-degenerate, i.e. that the restriction (2.1) is injective. Let $Z \subset X \times (\mathbb{P}^N)^*$ be the algebraic subset defined by

$$Z = \{(x, H) \in X \times (\mathbb{P}^N)^* \mid X_H := X \cap H \text{ is singular at } x\}. \tag{2.2}$$

Write $\mathcal{D}_X = \mathrm{pr}_2(Z) \subset (\mathbb{P}^N)^*$ for the discriminant variety of X. This is the set of singular hyperplane sections of X. We easily see that Z is smooth and that $\dim Z = N - 1$. Indeed, the first projection $Z \to X$ makes Z into a $\mathbb{P}^{N-n-1}$-bundle. However, it can happen that $\dim \mathcal{D}_X < N - 1$. The following result shows that $\dim\mathcal{D}_X = N - 1$ if there exist hyperplane sections of X having an ordinary double point.

Lemma 2.7 *Let $(x, H) \in Z$. Then X_H has an ordinary double point at X if and only if $\mathrm{pr}_2 : Z \to (\mathbb{P}^N)^*$ is a immersion at the point (x, H).*

Proof The argument is exactly the same as in the proof of lemma 1.18. Taking a local chart $\mathbb{C}^N \subset \mathbb{P}^N$ containing x, and taking normalised affine representatives for the homogeneous forms of degree 1 on $\mathbb{P}^N$ in the neighbourhood of H, we can replace $\mathbb{P}^N$ in the statement by $\mathbb{C}^N$, and $(\mathbb{P}^N)^*$ by the set K of affine forms $\sum_{1\leq i\leq N} \alpha_i x_i + \beta$ on $\mathbb{C}^N$.

Then, in the neighbourhood of (x, H), $Z \subset X \times K$ is defined by

$$Z = \{(y, k) \in X \times K \mid k(y) = 0,\ dk(y) = 0\},$$

where k is viewed as the restriction to X of the affine function k on $\mathbb{C}^N$. Thus, in local coordinates $z_1, \ldots, z_n$ on X, Z is defined by the equations

$$k(y) = \frac{\partial k}{\partial z_i}(y) = 0, \quad 1 \leq i \leq n.$$

It follows by differentiation that

$$T_{Z,(x,H)} = \left\{(u, k) \in T_{X,x} \times T_{K,H} \;\middle|\; d_u H + k(x) = 0 \quad \text{and} \right.$$
$$\left. d_u\left(\frac{\partial H}{\partial z_i}\right) + \frac{\partial k}{\partial z_i}(x) = 0 \text{ for } 1 \leq i \leq n\right\},$$

so that

$$\mathrm{Ker}\left((\mathrm{pr}_2)_* : T_{Z,(x,H)} \to T_{K,H}\right) = T_{Z,(x,H)} \cap (T_{X,x} \times \{0\})$$

can be identified with the kernel of the quadratic form $\mathrm{Hess}_0 f$, since $dH_x = 0$. □

The preceding computation also yields the following result.

Corollary 2.8 *If* $\mathrm{Hess}_x f$ *is non-degenerate, then the image* $\mathrm{pr}_{2*}(T_{Z,(x,H)})$ *can be identified with the hyperplane of* $\mathbb{C}^N = T_{K,H}$ *consisting of the functions which vanish at the point* x.

This fact and lemma 2.7 show in particular that if there exists a hyperplane section X_H of X having an ordinary double point, then the generic point H of $\mathcal{D}_X$ corresponds to a hyperplane section of X having a single ordinary double point, and that this happens if and only if $\dim \mathcal{D}_X = N - 1$.

Indeed, Sard's theorem and lemma 2.7 show that the set of hyperplane sections of X having a degenerate singular point is of dimension at most $N - 2$, while if there exists a hyperplane section having an ordinary double point, then the image of Z by pr_2 is of dimension $N - 1$. Finally, corollary 2.8 shows that the set of hyperplane sections having several ordinary double points is of dimension at most $N - 2$.

If $\mathcal{D}_X^0$ denotes the subset of $\mathcal{D}_X$ parametrising the hyperplanes H such that X_H has at most one ordinary double point as singularity, then we have the following conclusion: if $\dim \mathcal{D}_X = N - 1$, then $\mathcal{D}_X^0$ is non-empty and thus dense, since it is clearly a Zariski open set of $\mathcal{D}_X$. Moreover, $\mathcal{D}_X^0$ is smooth, since by lemma 2.7, pr_2 is an isomorphism over $\mathcal{D}_X^0$.

Proposition 2.9 *Let* X *be a smooth subvariety of* $\mathbb{P}^N$. *Then a pencil of hyperplane sections* $(X_t)_{t\in\mathbb{P}^1}$ *is a Lefschetz pencil if and only if one of the following two conditions is satisfied.*

(i) $\mathcal{D}_X$ *is a hypersurface, and the corresponding line* $\Delta \subset (\mathbb{P}^N)^*$ *meets the discriminant hypersurface* $\mathcal{D}_X$ *transversally in the open dense set* $\mathcal{D}_X^0$.
(ii) $\dim \mathcal{D}_X \leq N - 2$ *and the corresponding line* $\Delta \subset (\mathbb{P}^N)^*$ *does not meet* $\mathcal{D}_X$.

Proof Lemma 2.7 shows that in situation (ii), the singular hyperplane sections of X have degenerate singularities. Thus, a Lefschetz pencil cannot meet $\mathcal{D}_X$.

Moreover, in situation (i), in order for the hypersurfaces of the pencil to have at most one ordinary double point, we see from the definition that the pencil must meet $\mathcal{D}_X$ only in $\mathcal{D}_X^0$.

Finally, corollary 2.8 shows that the pencil meets $\mathcal{D}_X$ transversally at a point $H \in \mathcal{D}_X^0$ if and only if the ordinary double point x of X_H does not lie on the base locus of the pencil. Indeed, by the corollary, the projective hyperplane of $(\mathbb{P}^N)^*$ tangent to $\mathcal{D}_X$ at H is the set of $L \in (\mathbb{P}^N)^*$ passing through x. The pencil is thus tangent to $\mathcal{D}_X$ at H if and only if it is contained in this last hyperplane, i.e. if x is contained in B. □

Corollary 2.10 *If $X \subset \mathbb{P}^N$ is a smooth projective complex variety, then a generic pencil $(X_t)_{t\in\mathbb{P}^1}$ of hyperplane sections of X is a Lefschetz pencil.*

2.1.2 The holomorphic Morse lemma

Let X be a complex variety, and let $f : X \to \mathbb{C}$ be a holomorphic map. Let $0 \in X$ be a non-degenerate critical point of f. We then have the following analogue of proposition 1.8.

Lemma 2.11 *There exist holomorphic coordinates $z_1, \ldots, z_n$ on X, centred at 0, such that f can be written in these coordinates as*

$$f(z) = f(0) + \sum_{i=1}^{n} z_i^2.$$

Proof The proof is the same as in the differentiable case. We proceed by induction on the dimension. We locally choose a complex hypersurface $Y \subset X$ passing through 0, of equation $t = 0$, where t is a holomorphic function with non-zero differential at 0. We apply the induction hypothesis to Y, which gives holomorphic coordinates $z_1, \ldots, z_{n-1}$ on Y satisfying the conclusion of the lemma. Extending these functions to holomorphic functions on X, we obtain

$$f - f(0) - \sum_{i=1}^{n-1} z_i^2 = 0$$

on Y. Then we have

$$f - f(0) - \sum_{i=1}^{n-1} z_i^2 = tg,$$

where g is holomorphic and vanishes at 0, since $df(0) = 0 = g(0)dt$. We can then write

$$g = 2\sum_i \alpha_i z_i + t\phi,$$

where ϕ is holomorphic and defined in the neighbourhood of 0. (The latter fact can be seen by noting that the z_i and t give a system of holomorphic coordinates, and by expanding g as a power series.) Setting $z_i' = z_i + \alpha_i t$ for $i \leq n-1$, we thus obtain

$$f = f(0) + \sum_{i \leq n-1} z_i'^2 + t^2 \psi,$$

where ψ is holomorphic and defined in the neighbourhood of 0. As 0 is a non-degenerate critical point of f, we have $\psi(0) \neq 0$, so $\psi = \chi^2$, where χ is holomorphic and defined in the neighbourhood of 0. Then if we set $z_n = t\chi$, f has the desired form in the coordinates $z_1', \ldots, z_{n-1}', z_n$. □

2.2 Lefschetz degeneration

2.2.1 Vanishing spheres

Consider the function

$$f(z) = \sum_i z_i^2$$

defined on $\mathbb{C}^n$. The point 0 is obviously a non-degenerate critical point of this function. Let B be a ball of radius r centred at 0, and write $B_t := B \cap f^{-1}(t)$. The map f restricted to B then has values in the disk Δ of radius r^2. The central fibre B_0 has an ordinary double point at 0 as singularity, whereas the fibres B_t for t near 0 are smooth. Such a map $f : B \to \Delta$ is called a Lefschetz degeneration.

For every point $t = se^{i\theta} \in \Delta^*$ such that $s = |t| \leq r^2$, the fibre B_t contains the sphere S_t^{n-1} defined by

$$S_t^{n-1} = \left\{ z = (z_1, \ldots, z_n) \in B \mid z_i = \sqrt{s} e^{\frac{i}{2}\theta} x_i, \;\; x_i \in \mathbb{R}, \;\; \sum_{1 \leq i \leq n} x_i^2 = 1 \right\} \tag{2.3}$$

and the set $B_{\leq s} = \{z \in B \mid |f(x)| \leq s\}$ contains the ball

$$B_t^n = \left\{ z = (z_1, \ldots, z_n) \in B \mid z_i = \sqrt{s} e^{\frac{i}{2}\theta} x_i, \;\; x_i \in \mathbb{R}, \;\; \sum_{1 \leq i \leq n} x_i^2 \leq 1 \right\}. \tag{2.4}$$

The sphere (2.3) is called a vanishing sphere of the family $(B_t)_{t \in \Delta}$. Indeed, when t tends to 0, the sphere S_t^{n-1} tends to vanish, i.e. to contract to a point. The ball B_t^n is called the cone on the vanishing sphere S_t^{n-1}.

The sphere defined above depends on the choice of coordinates and does not have any privileged orientation. However, it is easy to see that its homology class $\delta \in H_{n-1}(B_t, \mathbb{Z})$, defined by the choice of an orientation, is well-defined up to sign; it is a generator of $H_{n-1}(B_t, \mathbb{Z})$.

Definition 2.12 *The class δ is called the vanishing cycle of the Lefschetz degeneration $f : B \to \Delta$.*

2.2.2 An application of Morse theory

Let $f : B \to \mathbb{C}$, and let $f(z) = \sum_{1 \le i \le n} z_i^2$ be defined on a ball B of $\mathbb{C}^n$ centred at 0. For a disk Δ centred at 0, set $B_\Delta := f^{-1}(\Delta)$ and $S_\Delta = B_\Delta \cap S$, $S = \partial B$. Consider the function $g = \Re f$. Setting $z_i = x_i + iy_i$, we have

$$g(x, y) = \sum_{1 \le i \le n} x_i^2 - y_i^2,$$

so that in the real coordinates x_i, y_i, g has the form studied in the preceding chapter, where 0 is a critical point of index n. Thus, for sufficiently small ϵ, we see by proposition 1.12 that there exists a retraction by deformation of the level set $B_{-\epsilon \le g \le \epsilon}$ onto the union of $B_{g=-\epsilon}$ with the ball B^n defined by

$$x_i = 0 \quad \text{for all} \quad i = 1, \ldots, n, \ \sum_i y_i^2 \le \epsilon,$$

glued to $B_{g=-\epsilon}$ along its boundary, the sphere S^{n-1} defined by

$$x_i = 0 \quad \text{for all} \quad i = 1, \ldots, n, \ \sum_i y_i^2 = \epsilon.$$

Note that this sphere is contained in the fibre $B_{-\epsilon}$, and is in fact the vanishing sphere of (2.3). Similarly, the ball B^n is the ball defined in (2.4).

Lemma 2.13 *For Δ of small radius with respect to the radius of B, the map $f_{|S}$ is of maximal rank along S_Δ.*

Proof It suffices to see that $f_{|S}$ is of maximal rank along $f^{-1}(0) \cap S$. Let $z = (z_1, \ldots, z_n) \in S$. Setting $z_i = x_i + iy_i$, we can write $z = (x, y)$ with $x = (x_1, \ldots, x_n)$, $y = (y_1, \ldots, y_n)$. Then

$$f(z) = (\|x\|^2 - \|y\|^2, 2\langle x, y\rangle),$$

where the norm and the scalar product are the Euclidean norm and scalar product in $\mathbb{R}^n$. If $f(z) = 0$, we thus have $\|x\|^2 = \|y\|^2 \neq 0$, $\langle x, y\rangle = 0$. The differentials of the equations $f_1 = \|x\|^2$, $f_2 = \|y\|^2$ and $f_3 = \langle x, y\rangle$ can be written

$$df_1(u, v) = 2\langle x, u\rangle, \ df_2(u, v) = 2\langle y, v\rangle, \ df_3(u, v) = \langle x, v\rangle + \langle y, u\rangle.$$

A linear relation between df_1, df_2 and df_3 would thus imply that x and y are proportional, which contradicts the fact that $\langle x, y\rangle = 0$, $x \neq 0$, $y \neq 0$. Thus,

$f_1 - f_2$ and f_3 have independent differentials on the tangent space to the sphere, defined by $df_1 + df_2 = 0$. □

By Ehresmann's theorem, there exists a trivialisation of the fibration $f_{|S_\Delta} \to \Delta$, i.e. a diffeomorphism

$$C = (C_0, f) : S_\Delta \cong S_s \times \Delta,$$

where s is a point of Δ.

Such a trivialisation and the choice of a retraction by deformation of the disk onto the point s then induce a retraction by deformation $(R_{S,t})_{t\in[0,1]}$ of S_Δ onto the fibre S_s.

Proposition 2.14 *For Δ of small radius with respect to the radius of B, and $s \in \Delta^*$, there exists a retraction by deformation $(H'_t)_{t\in[0,1]}$ of B_Δ onto the union of the fibre B_s with the ball B^n_s of (2.4). Moreover, this retraction by deformation can be chosen so as to preserve S_Δ and to be induced on S_Δ by a retraction $(R_{S,t})_{t\in[0,1]}$ as above.*

A retraction by deformation satisfying this last property will be said to be compatible with the trivialisation C.

Remark 2.15 *All the fibres B_t for $t \in \Delta^*$ are diffeomorphic, since the punctured disk is arcwise connected and f is a fibration of manifolds with boundary over Δ^*. In what follows, we will set $s = -\epsilon$, $\epsilon \in \mathbb{R}^{*+}$.*

Proof By lemma 2.13, B_Δ is a variety with corners, i.e. its boundary ∂B_Δ consists of real hypersurfaces S_Δ and $B_{\partial\Delta} := f^{-1}(\partial\Delta)$ which are manifolds with boundary meeting transversally along their boundaries. Moreover, for ϵ strictly less than the radius of Δ, the map

$$g = \Re f : B_\Delta \to \mathbb{R}$$

is a submersion of varieties with corners over $[-\epsilon, \epsilon]$, outside of 0. This means that g is submersive outside of 0, and remains submersive on the components of the boundary and on their intersections. The last fact also follows from lemma 2.13.

We can thus apply theorem 1.15, or rather, a version adapted to varieties with corners. As the critical point of g is not located on the boundary of B_Δ, this adaptation is immediate. Writing

$$\Delta_{[-\epsilon,\epsilon]} = \{z \in \Delta \mid -\epsilon \le \Re z \le \epsilon\},$$

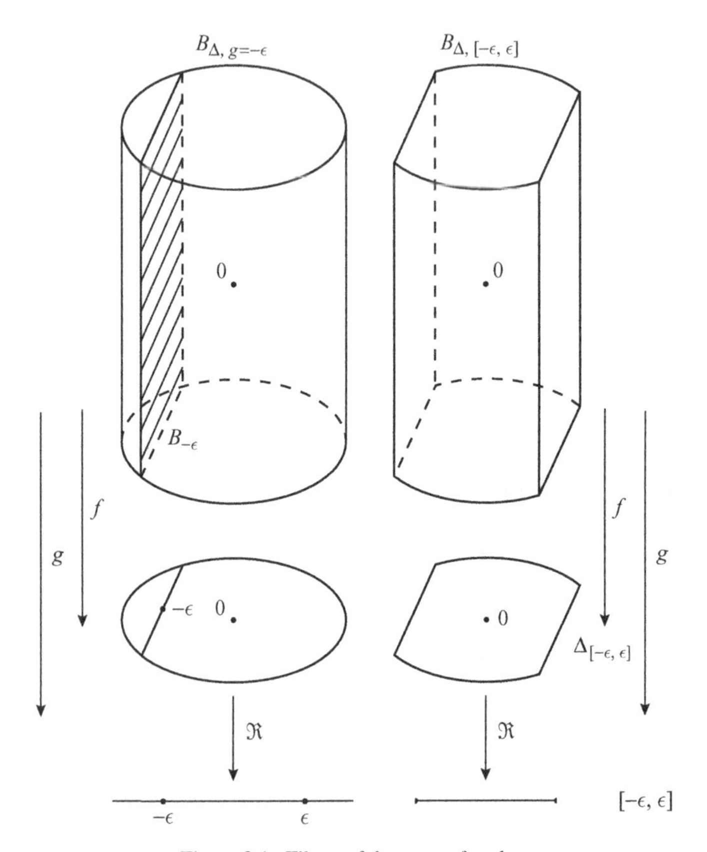

Figure 2.1: Fibres of the maps f and g.

this gives a retraction by deformation $(H_t)_{t\in[0,1]}$ of the set

$$B_{\Delta,[-\epsilon,\epsilon]} := B_\Delta \cap g^{-1}([-\epsilon, \epsilon]) = f^{-1}\big(\Delta_{[-\epsilon,\epsilon]}\big)$$

onto the union of the fibre

$$B_{\Delta,g=-\epsilon} := g^{-1}(-\epsilon) \cap B_\Delta = f^{-1}(I)$$

with the ball B^n defined above (see figure 2.1). Here, $I \subset \Delta_{[-\epsilon,\epsilon]}$ is the segment

$$I = \{z \in \Delta \mid \Re z = -\epsilon\}.$$

This retraction by deformation can moreover be chosen compatible with a trivialisation of the fibration

$$g_{|\partial B_{\Delta,[-\epsilon,\epsilon]}} : \partial B_{\Delta,[-\epsilon,\epsilon]} := \partial B_\Delta \cap g^{-1}([-\epsilon, \epsilon]) \to [-\epsilon, \epsilon].$$

Now, along

$$S_{\Delta,[-\epsilon,\epsilon]} := S \cap f^{-1}\big(\Delta_{[-\epsilon,\epsilon]}\big) \subset \partial B_{\Delta,[-\epsilon,\epsilon]},$$

we obtain such a trivialisation

$$S_{\Delta,[-\epsilon,\epsilon]} \cong S_{\Delta,g=-\epsilon} \times [-\epsilon,\epsilon]$$

by composing the trivialisation

$$C = (C_0, f) : S_{\Delta,[-\epsilon,\epsilon]} \cong S_{-\epsilon} \times \Delta_{[-\epsilon,\epsilon]}$$

with a trivialisation

$$C' = (C'_0, \Re) : \Delta_{[-\epsilon,\epsilon]} \cong I \times [-\epsilon,\epsilon].$$

Having retracted $B_{\Delta,[-\epsilon,\epsilon]}$ onto the union of $B_{\Delta,g=-\epsilon}$ with a ball B^n whose boundary is contained in the fibre $B_{-\epsilon}$, we then obtain a retraction by deformation $(H'_t)_{t\in[0,1]}$ of $B_{\Delta,[-\epsilon,\epsilon]}$ onto the union $B_{-\epsilon} \cup B^n$, noting that the map f restricted to $B_{\Delta,g=-\epsilon} = f^{-1}(I)$ is a fibration in varieties with boundary over the segment I. Thus, by Ehresmann's theorem, there exists a trivialisation

$$T = (T_0, f) : B_{\Delta,g=-\epsilon} \cong B_{-\epsilon} \times I.$$

We may furthermore assume that T_0 coincides with C_0 on $S_{\Delta,g=-\epsilon}$. This trivialisation provides a retraction by deformation of $B_{\Delta,g=-\epsilon}$ onto $B_{-\epsilon}$. Combining this retraction with the previous one, we get the desired retraction of $B_{\Delta,[-\epsilon,\epsilon]}$ onto $B_{-\epsilon} \cup B^n$.

We conclude the proof by noting that by Ehresmann's theorem, B_Δ admits a retraction by deformation onto $B_{\Delta,[-\epsilon,\epsilon]}$, compatible with the trivialisation C of $f : S_\Delta \to \Delta$, since Δ retracts by deformation onto $\Delta_{[-\epsilon,\epsilon]}$, and f is a fibration of varieties with boundary above Δ^*. □

Now consider a proper holomorphic map from an n-dimensional complex variety X to a disk. Assume that f is a submersion over the punctured disk Δ^*, and that over $0 \in \Delta$, f has a non-degenerate critical point x_0. We have the following global version of proposition 2.14.

Theorem 2.16 *There exists a retraction by deformation of X onto the union of X_t with an n-dimensional ball glued to X_t along a vanishing sphere $S_t^{n-1} \subset X_t$, where $t \in \Delta^*$.*

Proof As f is a proper submersion over Δ^*, it suffices by Ehresmann's theorem to show the result for an arbitrarily small disk centred at 0.

By the holomorphic Morse lemma, we may assume that there exists a ball B in X, and holomorphic coordinates $z_1, \ldots, z_n$ centred at x_0, such that

$$f(z) = \sum_i z_i^2$$

in B. Proposition 2.14 then shows that for a sufficiently small disk Δ, the set B_Δ admits a retraction by deformation onto $B_t \cup B^n$, which preserves the boundary S_Δ and is induced on S_Δ by a trivialisation of the fibration

$$f_{|S_\Delta} : S_\Delta \to \Delta \tag{2.5}$$

and by a retraction of the disk Δ onto one of its points $t \neq 0$.

But writing $X_\Delta := f^{-1}(\Delta)$ and $B^0_\Delta = B^0 \cap B_\Delta$, where B^0 denotes the interior of B, the restricted map

$$f : X_\Delta - B^0_\Delta \to \Delta \tag{2.6}$$

is a fibration of varieties with boundary, of boundary $f : S_\Delta \to \Delta$. Ehresmann's theorem then shows that the trivialisation of the fibration (2.5) extends to a trivialisation of the fibration (2.6). Thus, the retraction by deformation of S_Δ onto S_t extends to a retraction by deformation of $X_\Delta - B^0_\Delta$ onto $X_t - B^0_t$. This glues together with the retraction by deformation already constructed on B_Δ to give the desired result. □

Corollary 2.17 *Let i be the inclusion $X_t \hookrightarrow X_\Delta$. Then $i_* : H_k(X_t, \mathbb{Z}) \to H_k(X_\Delta, \mathbb{Z})$ is an isomorphism for $k < n - 1$, and is surjective for $k = n - 1$. Moreover, the kernel of i_* is generated by the class of 'the' vanishing sphere S_t^{n-1} of X_t for $k = n - 1$.*

Proof By theorem 2.16, it suffices to show this with X_Δ replaced by $X_t \cup_{S_t^{n-1}} B_t^n$. The first statement is then equivalent to the vanishing of the relative homology $H_k(X_t \cup_{S_t^{n-1}} B_t^n, X_t, \mathbb{Z})$ for $k \leq n - 1$. This follows from the excision theorem, which shows that

$$H_*\big(X_t \cup_{S_t} B_t^n, X_t, \mathbb{Z}\big) = H_*\big(B_t^n, S_t^{n-1}, \mathbb{Z}\big).$$

As to the second statement, it also follows from the excision theorem, which shows that the first vertical arrow is an isomorphism in the following commutative and exact diagram.

$$\begin{array}{ccccccc} \cdots \longrightarrow & H_n(B_t^n, S_t^{n-1}, \mathbb{Z}) & \xrightarrow{\sim} & H_{n-1}(S_t^{n-1}, \mathbb{Z}) & \longrightarrow & 0 \\ & \downarrow & & \downarrow & & \downarrow \\ \cdots \longrightarrow & H_n(X_\Delta, X_t, \mathbb{Z}) & \longrightarrow & H_{n-1}(X_t, \mathbb{Z}) & \xrightarrow{i_*} & H_{n-1}(X_\Delta, \mathbb{Z}). \end{array}$$

□

2.3 Application to Lefschetz pencils

2.3.1 Blowup of the base locus

Let X be an n-dimensional compact complex variety, and let $(X_t)_{t\in\mathbb{P}^1}$ be a Lefschetz pencil. Let us introduce the variety

$$\widetilde{X} = \{(x, t) \in X \times \mathbb{P}^1 \mid x \in X_t\}.$$

Let $\tau = \mathrm{pr}_1|_{\widetilde{X}}$. It is clear by construction of the blowups (see vI.3.3.3) that $\widetilde{X} \xrightarrow{\tau} X$ can be identified with the blowup of X along the base locus B of the pencil. Moreover, each hypersurface X_t can be naturally identified with the fibre $f^{-1}(t) \subset \widetilde{X}$.

Now set

$$f := \mathrm{pr}_2|_{\widetilde{X}} : \widetilde{X} \to \mathbb{P}^1.$$

As $(X_t)_{t\in\mathbb{P}^1}$ is a Lefschetz pencil, the base locus B is smooth and thus $\widetilde{X}$ is smooth; furthermore, each fibre X_t of f has at most one ordinary double point as singularity. Thus, in the neighbourhood of each critical value of f, we can locally apply theorem 2.16. Let $0_i \in \mathbb{P}^1$, $i = 1, \ldots, M$ be the critical values of f. For each i, let Δ_i be a small disk of $\mathbb{P}^1$ centred at 0_i. Then $\widetilde{X}_{\Delta_i}$ retracts by deformation onto the union of X_{t_i} with an n-dimensional ball n glued to X_{t_i} along a vanishing sphere, where $t_i \in \Delta_i^*$.

Assume that ∞ is not a critical value of f. Let $0 \in \mathbb{C} = \mathbb{P}^1 - \infty$ be a regular value, and let γ_i, $i = 1, \ldots, N$ be paths in $\mathbb{C}$ joining 0 to t_i, not passing through the 0_j, and meeting only at the point 0 (see figure 3.2 in next chapter). It is clear that $\mathbb{C}$ admits a retraction by deformation onto the union of the disks Δ_i with the paths γ_i. As f is a proper fibration above $\mathbb{C} - \{0_1, \ldots, 0_M\}$, it follows from Ehresmann's theorem that $\widetilde{X} - X_\infty$ admits a retraction by deformation onto $\bigcup_i \widetilde{X}_{\gamma_i} \cup \widetilde{X}_{\Delta_i}$, where $\widetilde{X}_{\gamma_i} := f^{-1}(\gamma_i)$. Finally, as f is a fibration above γ_i, each $\widetilde{X}_{\gamma_i}$ admits a trivialisation

$$\widetilde{X}_{\gamma_i} \cong X_{t_i} \times \gamma_i$$

above γ_i, and correspondingly, a retraction by deformation onto X_{t_i}. Moreover, this trivialisation also gives a diffeomorphism from X_{t_i} to X_0.

Combining these different retractions, we have proved the following result.

Theorem 2.18 *The variety $\widetilde{X} - X_\infty$ has the homotopy type of the union of X_0 with n-dimensional balls glued to X_0 along $(n-1)$-dimensional spheres.*

Remark 2.19 *These spheres are images under the identifications $X_{t_i} \cong X_0$ of the vanishing spheres of the X_{t_i}. Even though the homology class of the vanishing sphere of X_{t_i}, $t_i \in \Delta_i^*$ is well-defined up to sign as a generator of the kernel of the arrow $H_{n-1}(X_{t_i}, \mathbb{Z}) \to H_*(\widetilde{X}_{\Delta_i}, \mathbb{Z})$, the class of the corresponding sphere (which we also call vanishing) in $H_{n-1}(X_0, \mathbb{Z})$ depends on the choice of the path γ_i.*

Corollary 2.20 *For $0 \in \mathbb{P}^1 - \infty$ such that X_0 is smooth, the inclusion $i'_0 : X_0 \hookrightarrow \widetilde{X} - X_\infty$ induces an isomorphism*

$$i'_{0*} : H_k(X_0, \mathbb{Z}) \to H_k(\widetilde{X} - X_\infty, \mathbb{Z})$$

for $k < n-1$. For $k = n-1$, i'_{0*} is surjective and the kernel of i_* is generated by the classes of vanishing spheres.

Proof By theorem 2.18, we can replace $\widetilde{X} - X_\infty$ by the union of X_0 with balls B^n glued by their boundaries to the vanishing spheres. The result thus follows, as in corollary 2.17, from repeated applications of the excision theorem and from the corresponding statement for the inclusion of the sphere S^{n-1} into the ball B^n. □

2.3.2 The Lefschetz theorem

We can use Lefschetz pencils in order to give another proof of the Lefschetz theorem 1.23, valid for pairs (X, Y) where Y is a smooth hyperplane section of $X \subset \mathbb{P}^N$. Note that if (X, Y) is such a pair, a refinement of corollary 2.10 shows that there exists a Lefschetz pencil of hyperplane sections of X of which Y is one member X_t. Assume that t is different from the point $\infty \in \mathbb{P}^1$, and that ∞ is also a regular value of the pencil. Let $\widetilde{X}$ be the blowup of X along the base locus of the pencil, and let

$$i_t : X_t \subset X, \quad i'_t : X_t \subset \widetilde{X} - X_\infty, \quad j_t : X_t \subset \widetilde{X}$$

be the inclusions. By the dual version of corollary 2.20, we have the following.

Corollary 2.21 *The restriction arrow*

$$i'^*_t : H^k(\widetilde{X} - X_\infty, \mathbb{Z}) \to H^k(X_t, \mathbb{Z})$$

is an isomorphism for $k < n-1$ and is injective for $k = n-1$, where $n = \dim X$.

Using induction on the dimension, we will use this to deduce another proof of theorem 1.23 (see Andreotti & Frankel 1969):

The restriction arrow

$$i^*_t : H^k(X, \mathbb{Z}) \to H^k(X_t, \mathbb{Z})$$

is an isomorphism for $k < n - 1$ and is injective for $k = n - 1$.

Let us first prove the surjectivity of i^*_t in degree strictly less than $n - 1$.

Lemma 2.22 *The Gysin morphism*

$$(j_\infty)_* : H^{k-1}(X_\infty, \mathbb{Z}) \to H^{k+1}(\widetilde{X}, \mathbb{Z})$$

is injective for $k \leq n - 1$.

Proof Recall (see vI.7.3.3) that we have a canonical isomorphism

$$H^{k+1}(\widetilde{X}, \mathbb{Z}) \simeq H^{k+1}(X, \mathbb{Z}) \oplus H^{k-1}(B, \mathbb{Z}), \tag{2.7}$$

such that the second component of $(j_\infty)_*$ in this decomposition can be identified with l^*, where l is the inclusion of B into X_∞. The injectivity of $(j_\infty)_*$ in degree $k - 1 \leq n - 2$ thus follows from the injectivity of l^* in degree $\leq n - 2$. But this holds by the induction hypothesis, since $B \subset X_\infty$ is a hyperplane section. □

Corollary 2.23 *The restriction map*

$$H^k(\widetilde{X}, \mathbb{Z}) \to H^k(\widetilde{X} - X_\infty, \mathbb{Z})$$

is surjective for $k \leq n - 1$.

Proof This follows from lemma 2.22 and the long exact sequence (see vI.7.2.1)

$$H^k(\widetilde{X}, \mathbb{Z}) \to H^k(\widetilde{X} - X_\infty, \mathbb{Z}) \stackrel{\text{res}}{\to} H^{k-1}(X_\infty, \mathbb{Z}) \stackrel{(j_\infty)_*}{\to} H^{k+1}(\widetilde{X}, \mathbb{Z}) \quad (2.8)$$

obtained using the Thom isomorphism

$$H^{k+1}(\widetilde{X}, \widetilde{X} - X_\infty, \mathbb{Z}) \cong H^{k-1}(X_\infty, \mathbb{Z}).$$

□

From this corollary, together with the surjectivity of

$$i'^*_t : H^k(\widetilde{X} - X_\infty, \mathbb{Z}) \to H^k(X_t, \mathbb{Z})$$

given by corollary 2.21, we deduce the surjectivity of

$$j_t^* : H^k(\widetilde{X}, \mathbb{Z}) \to H^k(X_t, \mathbb{Z})$$

in degree $k < n-1$. It remains to deduce the surjectivity of i_t^*. In the decomposition (2.7) of $H^k(\widetilde{X}, \mathbb{Z})$, the restriction of j_t^* to $H^{k-2}(B, \mathbb{Z})$ is equal to l_*, where l is now the inclusion of B into X_t. To conclude that $i_t^* : H^k(X, \mathbb{Z}) \to H^k(X_t, \mathbb{Z})$ is surjective, it thus suffices to show that we have

$$\operatorname{Im} l_* \subset \operatorname{Im} i_t^*$$

in degree $k < n - 1$. As B is a hyperplane section of X_t, the map

$$l^* : H^m(X_t, \mathbb{Z}) \to H^m(B, \mathbb{Z})$$

is surjective for $m < n - 2$ by the induction hypothesis. Thus, if $k < n - 1$, $H^{k-2}(B, \mathbb{Z}) = \operatorname{Im} l^*$ and the image of $l_* : H^{k-2}(B, \mathbb{Z}) \to H^k(X_t, \mathbb{Z})$ is equal to the image of the composition

$$l_* \circ l^* : H^{k-2}(X_t, \mathbb{Z}) \to H^k(X_t, \mathbb{Z}).$$

But

$$l_* \circ l^* = i_t^* \circ (i_t)_* = h\cup$$

on $H^{k-2}(X_t, \mathbb{Z})$ (see equation (1.6)). Thus, $\operatorname{Im} l_* = \operatorname{Im} l_* \circ l^* \subset \operatorname{Im} i_t^*$ in degree $k < n - 1$. Hence i_t^* is surjective in degree $< n - 1$.

Let us now prove the injectivity of i_t^* in degree $k \leq n - 1$. Let $\alpha \in H^k(X, \mathbb{Z})$ be such that $i_t^*\alpha = 0$ in $H^k(X_t, \mathbb{Z})$. Then, by corollary 2.21, we must have

$$\tau^*\alpha_{|\widetilde{X} - X_\infty} = 0.$$

But then the exact sequence (2.8) shows that we have

$$\tau^*\alpha = (j_\infty)_*\beta, \quad \beta \in H^{k-2}(X_\infty, \mathbb{Z}).$$

Applying τ_* to this equality, we find that $\alpha = (i_\infty)_*\beta$. Thus, the class $\beta \in H^{k-2}(X_\infty, \mathbb{Z})$ satisfies the condition

$$\tau^*(i_\infty)_*\beta = (j_\infty)_*\beta \in H^k(\widetilde{X}, \mathbb{Z}). \tag{2.9}$$

In the decomposition (2.7) in degree k, the first term is equal to $\operatorname{Im}\tau^*$ and the second component of $(j_\infty)_*\beta$ is equal to $l^*\beta \in H^{k-2}(B, \mathbb{Z})$. Thus, the equality (2.9) implies that $l^*\beta = 0$. The Lefschetz theorem for the pair $B \subset X_\infty$ and the fact that $d^0\beta = k - 2 \leq n - 3$ then imply that $\beta = 0$, so $\alpha = 0$. □

Remark 2.24 *This proof of the Lefschetz theorem is not very different from the preceding proof, as the essential point is the computation of the index of particular Morse functions. It is important, however, to note that the function used in the second case is the function $\Re f$, which is not a exhaustion function. Moreover, as the function f is algebraic, the proof given here adapts to the purely algebraic context of étale cohomology (see Katz 1973).*

2.3.3 Vanishing cohomology and primitive cohomology

If Y is a compact Kähler variety and $[\omega] \in H^2(Y, \mathbb{R})$ is a Kähler class, then we have the Lefschetz operator

$$L = [\omega]\cup : H^k(Y, \mathbb{R}) \to H^{k+2}(Y, \mathbb{R}).$$

The primitive cohomology $H^k(Y, \mathbb{R})_{\text{prim}}$ was defined in vI.6.2 by

$$H^k(Y, \mathbb{R})_{\text{prim}} := \operatorname{Ker}(L^{n-k+1} : H^k(Y, \mathbb{R}) \to H^{2n-k+2}(Y, \mathbb{R})),$$

where $n = \dim Y$.

In particular, for $k = n = \dim Y$, we have

$$H^k(Y, \mathbb{R})_{\text{prim}} := \operatorname{Ker}(L : H^k(Y, \mathbb{R}) \to H^{k+2}(Y, \mathbb{R})). \tag{2.10}$$

Assume now that $Y \overset{j}{\hookrightarrow} X$ is a hyperplane section of a projective variety. Then we can take $[\omega] = c_1(\mathcal{O}_Y(1)) = h_Y$, and the equality (1.5) says that the corresponding Lefschetz operator L satisfies

$$L = j^* \circ j_* : H^k(Y, \mathbb{R}) \overset{j_*}{\to} H^{k+2}(X, \mathbb{R}) \overset{j^*}{\to} H^{k+2}(Y, \mathbb{R}). \tag{2.11}$$

Note that in this case, the class of L comes from an integral class, and the above equality holds in the integral cohomology.

For every coefficient ring A, we now define the vanishing cohomology $H^k(Y, A)_{\mathrm{van}}$ by

$$H^k(Y, A)_{\mathrm{van}} = \mathrm{Ker}\,(j_* : H^k(Y, A) \to H^{k+2}(X, A)).$$

A first consequence of the Lefschetz theorem 1.23 and the hard Lefschetz theorem (see vI.6.2.3) is the following result.

Corollary 2.25 *The vanishing cohomology $H^k(Y, \mathbb{Q})_{\mathrm{van}}$ reduces to 0 for $k \neq \dim Y$. Furthermore, it is contained in the primitive cohomology $H^k(Y, \mathbb{Q})_{\mathrm{prim}}$ for $k = \dim Y$.*

Proof For $k < \dim Y$, the equality (2.11) shows that j_* is injective if L is. Now, we know that L is injective on the cohomology of degree $k < \dim Y$. If moreover $k > \dim Y$, the injectivity of $j_* : H^k(Y, \mathbb{Q}) \to H^{k+2}(X, \mathbb{Q})$ follows by Poincaré duality from the surjectivity of $j^* : H^m(X, \mathbb{Q}) \to H^m(Y, \mathbb{Q})$ for $m < \dim Y$, i.e. from the Lefschetz theorem 1.23. The last assertion follows immediately from the equalities (2.11) and (2.10). □

Note also that for $k = \dim Y = n - 1$, the terminology 'vanishing cohomology' is justified by the following result.

Lemma 2.26 *The vanishing cohomology $H^{n-1}(Y, \mathbb{Z})_{\mathrm{van}}$ is generated by the classes of vanishing spheres of a Lefschetz pencil passing through Y.*

Proof By definition, the Gysin morphism

$$j_* : H^{n-1}(Y, \mathbb{Z}) \to H^{n+1}(X, \mathbb{Z})$$

can be identified via Poincaré duality with the morphism

$$j_* : H_{n-1}(Y, \mathbb{Z}) \to H_{n-1}(X, \mathbb{Z}).$$

Thus, the lemma follows from corollary 2.20 and the fact that

$$\mathrm{Ker}\,(j_* : H_{n-1}(Y, \mathbb{Z}) \to H_{n-1}(X, \mathbb{Z})) = \\ \mathrm{Ker}\,(j_* : H_{n-1}(Y, \mathbb{Z}) \to H_{n-1}(\widetilde{X} - X_\infty, \mathbb{Z})),$$

where the notation is that of section 2.3.1. This last point follows from corollary 2.23. □

To conclude, thanks to the hard Lefschetz theorem, we can give a precise comparison of the primitive cohomology with the vanishing cohomology.

Proposition 2.27

(i) We have a decomposition as an orthogonal direct sum (relative to the intersection form on $H^{n-1}(Y,\mathbb{Q})$)

$$H^{n-1}(Y,\mathbb{Q}) = H^{n-1}(Y,\mathbb{Q})_{\mathrm{van}} \oplus j^*H^{n-1}(X,\mathbb{Q}).$$

(ii) Similarly, we have a decomposition as an orthogonal direct sum

$$H^{n-1}(Y,\mathbb{Q})_{\mathrm{prim}} = H^{n-1}(Y,\mathbb{Q})_{\mathrm{van}} \oplus j^*H^{n-1}(X,\mathbb{Q})_{\mathrm{prim}}.$$

Proof (i) Note first that the dimensions coincide since by definition of the Gysin morphism j_*, $H^{n-1}(Y,\mathbb{Q})_{\mathrm{van}}$ is also the subspace orthogonal to $j^*H^{n-1}(X,\mathbb{Q})$ for the intersection form on $H^{n-1}(Y,\mathbb{Q})$. Thus, it suffices to show that

$$H^{n-1}(Y,\mathbb{Q})_{\mathrm{van}} \cap j^*H^{n-1}(X,\mathbb{Q}) = \{0\}.$$

Now, if $\alpha = j^*\beta$ satisfies $j_*\alpha = 0$, then $j_* \circ j^*\beta = 0$ in $H^{n+1}(X,\mathbb{Q})$. But if L_X denotes the Lefschetz operator of X corresponding to the polarisation $h = c_1(\mathcal{O}_X(1))$, then by (1.6), we have the equality $j_* \circ j^* = L_X$, and we know that

$$L_X : H^k(X,\mathbb{Q}) \to H^{k+2}(X,\mathbb{Q})$$

is injective for $k < n = \dim X$. Thus $\beta = 0 = \alpha$.
(ii) As we know that $H^{n-1}(Y,\mathbb{Q})_{\mathrm{van}}$ is contained in $H^{n-1}(Y,\mathbb{Q})_{\mathrm{prim}}$, the preceding decomposition induces a decomposition as an orthogonal direct sum

$$H^{n-1}(Y,\mathbb{Q})_{\mathrm{prim}} = H^{n-1}(Y,\mathbb{Q})_{\mathrm{van}} \oplus j^*H^{n-1}(X,\mathbb{Q}) \cap H^{n-1}(Y,\mathbb{Q})_{\mathrm{prim}}.$$

It thus suffices to prove the equality

$$j^*H^{n-1}(X,\mathbb{Q}) \cap H^{n-1}(Y,\mathbb{Q})_{\mathrm{prim}} = j^*H^{n-1}(X,\mathbb{Q})_{\mathrm{prim}}. \tag{2.12}$$

But if $\alpha = j^*\beta \in H^{n-1}(Y,\mathbb{Q})_{\mathrm{prim}}$, then $L_Y\alpha = 0$, so

$$j_*L_Y\alpha = j_*L_Y j^*\beta = 0 = j_*j^*j_*j^*\beta = L_X^2\beta$$

in $H^{n+3}(X,\mathbb{Z})$. Thus, β is primitive, and we have proved the inclusion $\subset$ in (2.12).

The inclusion in the other direction follows from the equality of the dimensions:

$$\dim H^{n-1}(Y, \mathbb{Q})_{\mathrm{prim}} = \dim H^{n-1}(Y, \mathbb{Q})_{\mathrm{van}} \oplus j^* H^{n-1}(X, \mathbb{Q})_{\mathrm{prim}}. \qquad (2.13)$$

To see this, note that the codimension of $H^{n-1}(Y, \mathbb{Q})_{\mathrm{prim}}$ in $H^{n-1}(Y, \mathbb{Q})$ is equal to $\dim H^{n+1}(Y, \mathbb{Q})$, since $L : H^{n-1}(Y, \mathbb{Q}) \to H^{n+1}(Y, \mathbb{Q})$ is surjective.

Moreover, the codimension of

$$H^{n-1}(Y, \mathbb{Q})_{\mathrm{van}} \oplus j^* H^{n-1}(X, \mathbb{Q})_{\mathrm{prim}}$$

in

$$H^{n-1}(Y, \mathbb{Q})_{\mathrm{van}} \oplus j^* H^{n-1}(X, \mathbb{Q}) = H^{n-1}(Y, \mathbb{Q})$$

is equal to $\dim H^{n+3}(X, \mathbb{Q})$.

Indeed, j^* is injective, and the codimension of $H^{n-1}(X, \mathbb{Q})_{\mathrm{prim}}$ in $H^{n-1}(X, \mathbb{Q})$ is equal to $\dim H^{n+3}(X, \mathbb{Q})$, since the operator

$$L_X^2 : H^{n-1}(X, \mathbb{Q}) \to H^{n+3}(X, \mathbb{Q})$$

is surjective. But we also have $\dim H^{n+1}(Y, \mathbb{Q}) = \dim H^{n+3}(X, \mathbb{Q})$, since

$$j_* : H^{n+1}(Y, \mathbb{Q}) \to H^{n+3}(X, \mathbb{Q})$$

is an isomorphism by the Lefschetz theorem 1.23 and Poincaré duality. The equality (2.13) is thus proved. □

2.3.4 Cones over vanishing cycles

Theorem 2.18 also allows us to describe the relative homology $H_n(X, Y, \mathbb{Z})$, where $Y = X_0$ is a smooth hypersurface of X which is a member of a Lefschetz pencil $(X_t)_{t \in \mathbb{P}^1}$.

Indeed, for every critical value $0_i \in \mathbb{P}^1$, we introduced the cone B_i^n on the vanishing sphere $S_i^{n-1} \subset X_{t_i}$, where $t_i \in \Delta_i^*$ and the disk Δ_i is a small disk centred at 0_i.

Introducing the paths γ_i as in subsection 2.3.1 (see figure 3.2 in next chapter), we obtain a diffeomorphism $\widetilde{X}_{\gamma_i} \cong X_0 \times [0, 1]$, and $B'^n_i := B_i^n \cup S_i^{n-1} \times [0, 1]$ can then be viewed as a cone over the vanishing sphere

$$S'^{n-1}_i := S_i^{n-1} \times 1 \subset X_0.$$

An orientation of this cone then enables us to assign to it a class

$$\Gamma_i \in H_n(\widetilde{X} - X_\infty, X_0, \mathbb{Z}),$$

also called a cone over the vanishing cycle δ_i.

Theorem 2.18 then says that $\widetilde{X} - X_\infty$ has the homotopy type of $X_0 \cup_i B'^n_i$. The following result can be deduced directly from this.

Proposition 2.28 *The relative homology $H_n(\widetilde{X} - X_\infty, X_0, \mathbb{Z})$ is generated by the cones over the vanishing cycles.*

Proof By theorem 2.18, we can replace $\widetilde{X} - X_\infty$ by the union of X_0 with the cones B'^n_i over the vanishing spheres S'^{n-1}_i. The result is then a consequence of the excision theorem. □

In fact, we can then deduce the following result as in section 2.3.2.

Proposition 2.29 *The relative homology $H_n(X, X_0, \mathbb{Z})$ is generated by the images in X under τ of the cones on the vanishing cycles.*

Proof Consider the continuous map

$$\tau : \widetilde{X} - X_\infty \to X$$

sending $X_0 \subset \widetilde{X}$ to $X_0 \subset X$. In view of proposition 2.28, proposition 2.29 simply says that

$$\tau_* : H_n(\widetilde{X} - X_\infty, X_0, \mathbb{Z}) \to H_n(X, X_0, \mathbb{Z})$$

is surjective.

Consider the diagram of exact sequences of relative homology for the pairs $(\widetilde{X} - X_\infty, X_0)$ and (X, X_0):

$$\begin{array}{ccccccc}
H_n(\widetilde{X}-X_\infty, \mathbb{Z}) & \longrightarrow & H_n(\widetilde{X}-X_\infty, X_0, \mathbb{Z}) & \longrightarrow & H_{n-1}(X_0, \mathbb{Z}) & \xrightarrow{i'_{0*}} & H_{n-1}(\widetilde{X}-X_\infty, \mathbb{Z}) \\
\downarrow & & \downarrow & & \downarrow & & \downarrow \\
H_n(X, \mathbb{Z}) & \longrightarrow & H_n(X, X_0, \mathbb{Z}) & \longrightarrow & H_{n-1}(X_0, \mathbb{Z}) & \xrightarrow{i_{0*}} & H_{n-1}(X, \mathbb{Z}),
\end{array}$$

where the vertical arrows are induced by τ_*. We have already seen that the kernels of

$$i'_{0*} : H_{n-1}(X_0, \mathbb{Z}) \to H_{n-1}(\widetilde{X} - X_\infty, \mathbb{Z})$$

and

$$i_{0*} : H_{n-1}(X_0, \mathbb{Z}) \to H_{n-1}(X, \mathbb{Z})$$

are equal. It thus suffices to show that the arrow

$$\tau_* : H_n(\widetilde{X} - X_\infty, \mathbb{Z}) \to H_n(X, \mathbb{Z})$$

is surjective. But this follows from the dual exact sequence of (2.8):

$$H_n(\widetilde{X} - X_\infty, \mathbb{Z}) \to H_n(\widetilde{X}, \mathbb{Z}) \stackrel{\eta}{\to} H_{n-2}(X_\infty, \mathbb{Z}), \tag{2.14}$$

where the last arrow η is given by the intersection (which can be identified by Poincaré duality with the restriction). Indeed, we also have the decomposition (2.7) in homology

$$H_n(\widetilde{X}, \mathbb{Z}) \cong H_n(X, \mathbb{Z}) \oplus H_{n-2}(B, \mathbb{Z}),$$

where the first projection is given by τ_*. In particular, this decomposition gives an inclusion $H_{n-2}(B, \mathbb{Z}) \to H_n(\widetilde{X}, \mathbb{Z})$, and the composition

$$H_{n-2}(B, \mathbb{Z}) \to H_n(\widetilde{X}, \mathbb{Z}) \to H_{n-2}(X_\infty, \mathbb{Z}),$$

where the second arrow is the map η of (2.14), is simply given by l_*, where l is the inclusion of B into X_∞.

By the Lefschetz theorem applied to the pair (B, X_∞), this composition is surjective, and it follows immediately by the exact sequence (2.14) that

$$\tau_* : H_n(\widetilde{X} - X_\infty, \mathbb{Z}) \to H_n(X, \mathbb{Z})$$

is surjective. □

Exercises

1. *Bertini's lemma.* Let X be a smooth complex variety, $\mathcal{L} \in \operatorname{Pic} X$, and let Y, Y' be two hypersurfaces defined by equations

$$\sigma, \sigma' \in H^0(X, \mathcal{L}).$$

   Show that the (possibly empty) hypersurface X_t, $t = (\alpha, \beta) \in \mathbb{P}^1$ defined by the equation $\alpha\sigma + \beta\sigma'$ is smooth outside the base locus $B = Y \cap Y'$, for a general point t of $\mathbb{P}^1$. (Apply Sard's theorem to the holomorphic map $\frac{\sigma}{\sigma'} : X - B \to \mathbb{P}^1$.)

2. *Quadrics of even dimension and vanishing cycles.* Let $X_0, \ldots, X_{2m+1}$ be a system of homogeneous coordinates for the projective space $\mathbb{P}^{2m+1}$, and let $Q \subset \mathbb{P}^{2m+1}$ be the hypersurface defined by the equation

$$q(X) := \sum_{0 \le i \le m} X_i X_{m+1+i} = 0.$$

   Such a hypersurface is called a projective quadric.

(a) Show that Q is smooth.

(b) Let $P,\ P' \subset Q$ be the m-dimensional projective subspaces defined by

$$P = \{X \in \mathbb{P}^{2m+1} \mid X_i = 0 \quad \text{for all} \quad i \leq m\}$$
$$P' = \{X \in \mathbb{P}^{2m+1} \mid X_i = 0 \quad \text{for all} \quad i \leq m-1, \quad X_{2m+1} = 0\}.$$

Show that $P \cup P'$ is the proper intersection of Q with the projective subspace $K \subset \mathbb{P}^{2m+1}$ of codimension m defined by

$$K = \{X \in \mathbb{P}^{2m+1} \mid X_i = 0 \ \text{for all} \ i \leq m-1\}.$$

Deduce that the cohomology classes $[P],\ [P']$ satisfy

$$[P] + [P'] = h^m \ \text{in} \ H^{2m}(Q, \mathbb{Z}),$$

where $h := c_1(\mathcal{O}_Q(1))$.

(c) Deduce that

$$\langle [P], [P] + [P'] \rangle = 1 \tag{2.15}$$

where $\langle\ ,\ \rangle$ is the intersection form on $H^{2m}(Q, \mathbb{Z})$.

(d) Recall that the self-intersection $\langle [P], [P] \rangle$ is equal to the Euler class of the normal bundle $N_{P/Q}$, or to its Chern class of maximal degree (see volume I, chapter 11, exercise 3). Deduce from the exact sequence

$$0 \to N_{P/Q} \to N_{P/\mathbb{P}^{2m+1}} \stackrel{dq}{\to} \mathcal{O}_P(2) \to 0$$

together with Whitney's formula that

$$\langle [P], [P] \rangle = 0 \ \text{if} \ m \ \text{is odd},$$
$$\langle [P], [P] \rangle = 1 \ \text{if} \ m \ \text{is even}.$$

(e) Using the fact that $\langle [P] - [P'], h^m \rangle = 0$ deduce that

$$\langle [P] - [P'], [P] - [P'] \rangle = -2 \ \text{if} \ m \ \text{is odd},$$
$$\langle [P] - [P'], [P] - [P'] \rangle = 2 \ \text{if} \ m \ \text{is even}.$$

We can show that $[P] - [P']$ is the vanishing cycle δ of any Lefschetz degeneration of Q. Thus, the last result is in agreement with the computation of the self-intersection

$$\langle \delta, \delta \rangle = (-1)^{\frac{n(n-1)}{2}} 2, \quad n = 2m + 1$$

(cf. remark 3.21).

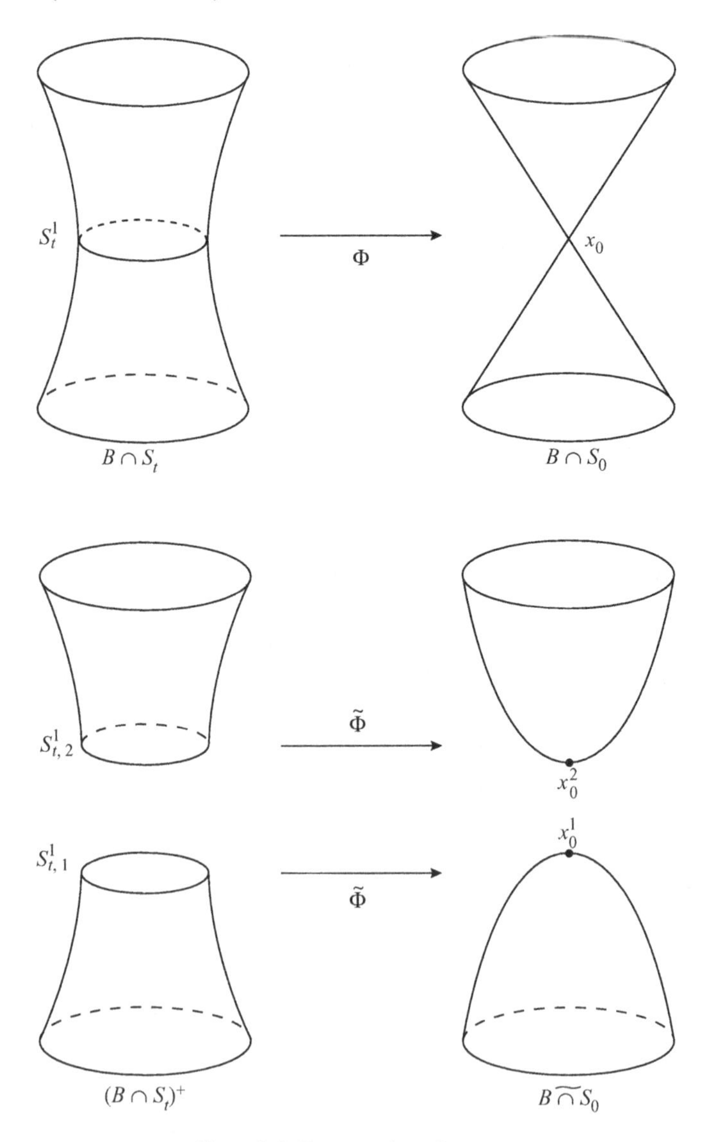

Figure 2.2: Degeneration of a curve.

3. *Ordinary singularities of curves* (see figure 2.2). Let $S \xrightarrow{f} \Delta$ be a proper holomorphic map from a smooth complex surface to the unit disk, such

that f is submersive over Δ^* and f has a non-degenerate critical point x_0 over 0. The curve $S_0 := f^{-1}(0)$ thus admits local coordinates $x,\ y$ on S centred at x_0 given by the equation

$$xy = 0.$$

(a) *Normalisation.* Let $\tau : \widetilde{S} \to S$ be the blowup of S at x_0. Show that the proper transform $\widetilde{S}_0$ of S_0 is smooth. (This proper transform is the curve $\tau^{-1}(S_0)$, from which we subtract the exceptional divisor, which appears in $\tau^{-1}(S_0)$ with multiplicity 2.) Show that $\tau : \widetilde{S}_0 \to S_0$ is an isomorphism outside $\tau^{-1}(x_0)$, and that $\tau^{-1}(x_0)$ consists of two points $x_0^1,\ x_0^2$, which can be naturally identified with the two 'branches' $x = 0,\ y = 0$ of S_0 at x_0. $\widetilde{S}_0$ is called the normalisation of S_0.

(b) In the local coordinates $x,\ y$ as above, let $B \subset S$ be a ball of radius 1, and for small ϵ and $t = \epsilon^2$, let S_t^1 be the circle with equations

$$y = \frac{\epsilon^2}{x}, \quad |x| = \epsilon$$

contained in $S_t \cap B$, where $S_t = f^{-1}(t)$. Show that S_t^1 is the vanishing sphere of the fibre S_t. Show that the continuous map ϕ defined by

$$\phi(x, y) = (x(|x| - |y|), 0) \text{ if } |x| \geq |y|,$$
$$\phi(x, y) = (0, y(|y| - |x|)) \text{ if } |x| \leq |y|$$

defines a homeomorphism from $B \cap S_t - S_t^1$ to its image (which is an open set of S_0). Show that ϕ contracts the circle S_t^1 onto the point x_0.

(c) Show that ϕ can be extended to a continuous map

$$\Phi : S_t \to S_0$$

which induces a homeomorphism from $S_t - S_t^1$ to $S_0 - x_0$ and contracts S_t^1 onto x_0.

(d) Show that Φ induces a continuous map $\tilde{\Phi}$ of S_t^+ onto $\widetilde{S}_0$, where S_t^+ is the surface obtained by 'cutting S_t along S_t^1': a neighbourhood U of S_t^1 in S_t is homeomorphic to $S^1 \times]-1, 1[$. The surface S_t^+ is by definition the union of $S_t - S_t^1$ with $S^1 \times [0, 1)$, $S^1 \times (-1, 0]$, glued respectively along $S^1 \times (0, 1) \subset S_t - S_t^1$ and $S^1 \times (-1, 0) \subset S_t - S_t^1$.

Show that $\tilde{\Phi}$ topologically identifies $\widetilde{S}_0$ with the surface obtained by contracting each circle $S_{t,1}^1,\ S_{t,2}^1$ forming the boundary of S_t^+ onto the point $x_0^1,\ x_0^2$ of $\tau^{-1}(0)$ which corresponds to it naturally. (Via $\tilde{\Phi}$, the two components of $U - S_t^1$ correspond bijectively to the branches of S_0 near x_0, i.e. to the components of $\widetilde{S}_0 \cap \tau^{-1}(S_0 \cap B)$, or to the points of $\tau^{-1}(x_0)$.)

(e) Show that the surface with boundary S_t^+ has the same number of connected components as S_t if and only if the vanishing cycle $\delta \in H^1(S_t, \mathbb{Z})$ (which is the cohomology class of the circle S_t^1) is non-zero. Show that if $\delta = 0$, the curve S_t^+ has one more connected component than S_t.

(f) Deduce from (d) that the same result holds when S_t^+ is replaced by $\widetilde{S}_0$.

3
Monodromy

In this chapter, we study the monodromy action on the cohomology of the fibres of a projective morphism. The monodromy representation is an essential ingredient of the theory of variations of Hodge structure, partly because of the following result, which can be proved using curvature properties of the domains of polarised periods (see Griffiths *et al.* 1984).

Theorem 3.1 *A polarised variation of Hodge structures parametrised by a quasi-projective base is trivial if the corresponding monodromy representation is trivial.*

Here, the monodromy representation is actually associated to the local system underlying the variation of Hodge structure.

In the case of a variation of Hodge structure coming from geometry, this theorem is a consequence of the following theorem, called Deligne's invariant cycles theorem, which will be proved in the following chapter, and which closely relates Hodge theory and monodromy.

Theorem 3.2 *Let $\phi : X \to Y$ be a smooth projective morphism between smooth quasi-projective varieties. Then for $y \in Y$, the space of invariants under the monodromy action $\rho : \pi_1(Y, y) \to \mathrm{Aut}\, H^k(X_y, \mathbb{Q})$ is equal to the image of the restriction $H^k(\overline{X}, \mathbb{Q}) \to H^k(X_y, \mathbb{Q})$ for any smooth projective compactification $\overline{X}$ of X. In particular, it is a constant Hodge substructure of $H^k(X_y, \mathbb{Z})$.*

In this text, we restrict ourselves to studying the monodromy of Lefschetz pencils, for which one can show by very simple methods that the monodromy groups are necessarily quite large whenever the fibres possess vanishing cohomology. More precisely, the analysis of the local monodromy for Lefschetz degenerations (Picard–Lefschetz formula) gives the generators for the monodromy group explicitly.

Proposition 3.3 *Let $f : X \to \Delta$ be a smooth proper holomorphic map over 0 such that the fibre X_0 has an ordinary double point as its only singularity. Then the monodromy action T of the generator of $\pi_1(\partial\Delta, t) = \mathbb{Z}$ on the cohomology of the smooth fibre $H^{n-1}(X_t, \mathbb{Z})$ is given by*

$$T(\alpha) = \alpha \pm \langle \alpha, \delta \rangle \delta,$$

where $\delta \in H^{n-1}(X_t, \mathbb{Z})$ is the cohomology class of the vanishing sphere $S_t^{n-1} \subset X_t$.

Together with the fact that the vanishing cycles generate the vanishing cohomology, and also using Zariski's theorem, this proposition implies the following result.

Let $X \subset \mathbb{P}^N$ be a smooth connected projective n-dimensional variety, and let $U \subset (\mathbb{P}^N)^*$ be the open set parametrising the smooth hyperplane sections.

Theorem 3.4 *For $y \in U$ parametrising a smooth hyperplane section X_y of X, the monodromy representation*

$$\rho : \pi_1(U, y) \to \mathrm{Aut}\, H^{n-1}(X_y, \mathbb{Q})_{\mathrm{van}}$$

is irreducible.

This theorem says that there exists no non-trivial local subsystem of the local system with stalk $H^{n-1}(X_y, \mathbb{Q})_{\mathrm{van}}$. The most famous application of this result is the Noether–Lefschetz theorem, which concerns the algebraic cycles of a surface of degree ≥ 4 in $\mathbb{P}^3$. (In fact, the same reasoning gives an analogous statement for sufficiently ample surfaces in any smooth 3-dimensional variety X.)

Theorem 3.5 *Let $S \subset \mathbb{P}^3$ be a general surface of degree ≥ 4. Then* $\mathrm{Pic}\, S = \mathrm{Pic}\, \mathbb{P}^3 = \mathbb{Z}$.

This chapter begins with generalities on the equivalence of categories between local systems and monodromy representations. The second section is devoted to the monodromy of Lefschetz pencils. We prove the Picard–Lefschetz formula there, as well as Zariski's theorem and the irreducibility theorem for the monodromy action. The last section is devoted to the Noether–Lefschetz locus and the Noether–Lefschetz theorem.

3.1 The monodromy action

3.1.1 Local systems and representations of π_1

Let X be a locally connected topological space, and G an abelian group. Recall (see vI.9.2) that a local system of stalk G is a sheaf which is locally isomorphic to the constant sheaf of stalk G. More generally, if A is a ring and G is an A-module, a local system of A-modules of stalk G is a sheaf of A-modules which is locally isomorphic as a sheaf of A-modules to the constant sheaf of stalk G.

If $i : Y \to X$ is a continuous map and $\mathcal{F}$ is a local system of stalk G on X, then $i^{-1}(\mathcal{F})$ is a local system of stalk G on Y. This follows from the fact that the inverse image of a constant sheaf is constant. When i is the inclusion of a subset of X, we write $i^{-1}(\mathcal{F}) := \mathcal{F}_{|Y}$.

Lemma 3.6 *Let $\mathcal{F}$ be a local system of stalk G on $X \times [0, 1]$. Then*

$$\mathcal{F} \cong \mathrm{pr}_1^{-1}(\mathcal{F}_{|X\times 0}).$$

Proof Let $\mathcal{G}$ denote the right-hand term. The two sheaves considered here are both locally constant sheaves of stalk G which canonically have the same restriction to $X \times 0$. Let

$$\sigma \in \Gamma(X \times 0, \mathrm{Hom}(\mathcal{F}_{|X\times 0}, \mathcal{G}_{|X\times 0}))$$

be the canonical section. The lemma then follows from the two following lemmas.

Lemma 3.7 *Let $\mathcal{F}$ be a local system on $X \times [0, 1]$, and let σ be a section of $\mathcal{F}$ on $X \times 0$. Then σ extends to a unique section of $\mathcal{F}$.*

Lemma 3.8 *Let Z be a locally connected topological space, and let $\phi : \mathcal{F} \to \mathcal{G}$ be a morphism of local systems which is an isomorphism in the neighbourhood of $Y \subset Z$. Then ϕ is an isomorphism on the connected component of Y in Z.*

To prove lemma 3.6, we apply lemma 3.7 to the section σ of $\mathrm{Hom}(\mathcal{F}_{|X\times 0}, \mathcal{G}_{|X\times 0})$. This gives a uniquely determined section $\tilde{\sigma}$ of $\mathrm{Hom}(\mathcal{F}, \mathcal{G})$ on $X \times [0, 1]$. As σ is an isomorphism on $X \times 0$, $\tilde{\sigma}$ is an isomorphism by lemma 3.8. □

Proof of lemma 3.7 By the compactness of $[0, 1]$, we can find an open covering of X by connected open sets U_i, and real numbers $0 \leq \epsilon_{i,j} < \epsilon_{i,j+1}$, $1 \leq j \leq N_i$, such that $\epsilon_{i,1} = 0$, $\epsilon_{i,N_i} = 1$, and $\mathcal{F}$ is constant on $U_i \times [\epsilon_{i,j}, \epsilon_{i,j+1}]$. Then, by the restriction isomorphisms, we have

$$\Gamma(U_i \times [\epsilon_{i,j}, \epsilon_{i,j+1}], \mathcal{F}) = \Gamma\big(U_i, \mathcal{F}_{|U_i \times \epsilon_{i,j}}\big) = \Gamma(U_i \times [\epsilon_{i,j-1}, \epsilon_{i,j}], \mathcal{F}).$$

Thus, $\sigma_{|U_i}$ extends uniquely to σ_i on $U_i \times [0, 1]$, and the extensions coincide on the intersections $U_i \times [0, 1] \cap U_j \times [0, 1]$, since they coincide on $U_i \cap U_j$ and the sheaf is locally constant. By gluing the σ_i, we have thus obtained the desired section $\tilde{\sigma}$. The uniqueness follows from the uniqueness of the σ_i. □

Proof of lemma 3.8 Using the fact that ϕ is a morphism of local systems, we show that the set of points $x \in X$ such that $\phi_x : \mathcal{F}_x \to G_x$ is an isomorphism is both open and closed in X. For this, thanks to the local triviality, we can reduce to the case of a locally constant morphism of constant sheaves, for which the result is obvious. □

Proposition 3.9 *If X is connected, locally arcwise connected and simply connected, then every local system $\mathcal{G}$ of stalk G is trivial on X, i.e. isomorphic to the constant sheaf G.*

Proof Fix $x \in X$, and for every $y \in X$, let $\gamma : [0, 1] \to X$ be a path from x to y. By lemma 3.6, the inverse image $\gamma^{-1}\mathcal{G}$ is canonically isomorphic to the constant sheaf of stalk $\mathcal{G}_x$, as well as to the constant sheaf of stalk $\mathcal{G}_y$. We thus have restriction isomorphisms of

$$\Gamma([0, 1], \gamma^{-1}\mathcal{G}) \cong \mathcal{G}_x, \quad \Gamma([0, 1], \gamma^{-1}\mathcal{G}) \cong \mathcal{G}_y,$$

and in particular an isomorphism of abelian groups $\gamma^* : \mathcal{G}_x \cong \mathcal{G}_y$. This isomorphism does not depend on the choice of γ. Indeed, if γ' is another path from x to y, there exists a homotopy of paths

$$H : [0, 1]^2 \to X$$

between γ and γ' satisfying

$$H(s, 0) = \gamma(s), \quad H(s, 1) = \gamma'(s), \quad H(0, t) = x, \quad H(1, t) = y.$$

By lemma 3.6, the local system $M := H^{-1}(\mathcal{G})$ is constant on $[0, 1]^2$. We thus obtain canonical isomorphisms, as all the stalks are isomorphic to $\Gamma([0, 1]^2, M)$:

$$M_{(0,0)} \cong M_{(1,0)} \cong M_{(1,1)} \cong M_{(0,1)} \cong M_{(0,0)},$$

with composition equal to the identity, or

$$\mathcal{G}_x \cong \mathcal{G}_y \cong \mathcal{G}_y \cong \mathcal{G}_x \cong \mathcal{G}_x.$$

But by the definition of the isomorphisms γ^* and γ'^*, the first of these isomorphisms is equal to γ^*, being induced by the path $H(t, 0)$, the second is equal to the identity, being induced by the constant path $H(1, s)$, and the third is $(\gamma'^*)^{-1}$, being induced by the path $H(1-t, 1)$. Finally, the last isomorphism is equal to the identity, since it is induced by the constant path $H(0, 1-t)$. Thus, we have $\gamma^* = \gamma'^*$.

We thus have a canonical isomorphism $\gamma_{xy} : \mathcal{G}_x \cong \mathcal{G}_y$ for $x, y \in X$. Let us show that these isomorphisms are locally constant. Let P be the space of paths on X, i.e. of continuous maps $\phi : [0, 1] \to X$, equipped with the topology of uniform convergence. We have a continuous map

$$\Phi : P \times [0, 1] \to X$$

sending (ϕ, t) on $\phi(t)$. By lemma 3.6, we have an isomorphism

$$\Phi^{-1}\mathcal{G} \cong \mathrm{pr}_1^{-1}\Phi_0^{-1}\mathcal{G}$$

where $\Phi_0 = \Phi_{|P\times\{0\}}$, which gives an isomorphism of local systems

$$\gamma : \Phi_0^{-1}\mathcal{G} \cong \Phi_1^{-1}\mathcal{G}$$

where $\Phi_1 = \Phi_{|P\times\{1\}}$. By construction, the induced isomorphism

$$\gamma_\phi : \mathcal{G}_{\phi(0)} \cong \mathcal{G}_{\phi(1)}$$

between the stalks of $\Phi_0^{-1}\mathcal{G}$ and $\Phi_1^{-1}\mathcal{G}$ at $\phi \in P$ is equal to the map $\gamma_{\phi(0),\phi(1)}$. Now, the maps Φ_0 and Φ_1 are open, as X is locally arcwise connected, so the fact that γ is a morphism of local systems implies that the $\gamma_{xx'}$ are locally constant.

In conclusion, the isomorphisms γ_{xy} form a section of the local system

$$\mathrm{Hom}\,(\mathcal{G}_x, \mathcal{G})$$

which is an isomorphism at the point x (here the left-hand term is the constant sheaf of stalk $\mathcal{G}_x \cong G$). Thus, it is an isomorphism everywhere by lemma 3.8. □

Corollary 3.10 *If X is arcwise connected and locally simply connected and x is a point of X, we have a natural bijection between the set of isomorphism classes of local systems of stalk G and the set of representations*

$$\pi_1(X, x) \to \mathrm{Aut}\, G,$$

modulo the action of $\mathrm{Aut}\, G$ *by conjugation.*

Remark 3.11 *We can also formulate this bijection by saying that there exists a bijection between the set of representations* $\pi_1(X, x) \to \operatorname{Aut} G$ *and the set of pairs* $(\mathcal{G}, \alpha)$ *consisting of a local system* $\mathcal{G}$ *on* X *and an isomorphism* $\alpha : \mathcal{G}_x \cong G$*. The bijection statement in the corollary is then obtained by passing to the quotient by the action of* G *on these two sets.*

Remark 3.12 *Everything above was formulated in the context of local systems of abelian groups. If we work more generally with local systems of* A*-modules, then all the local trivialisations used are morphisms of* A*-modules, and the representation of* π_1 *obtained above has values in the subgroup of* $\operatorname{Aut} G$ *consisting of automorphisms of* A*-modules.*

Proof of Corollary 3.10 Let $\mathcal{G}$ be a local system of stalk G. Let $\pi : \widetilde{X} \to X$ be the universal cover of X. By proposition 3.9, the local system $\pi^{-1}\mathcal{G}$ is constant on $\widetilde{X}$, with stalk G. If we are given an isomorphism $\alpha : \mathcal{G}_x \cong G$ and a point $\tilde{x} \in \widetilde{X}$ over x, we even have a unique isomorphism of locally constant sheaves

$$\beta : \pi^{-1}(\mathcal{G}) \cong G,$$

determined by the condition that the induced isomorphism

$$\pi^{-1}(\mathcal{G})_{\tilde{x}} \cong G$$

is equal to the composition

$$\pi^{-1}(\mathcal{G})_{\tilde{x}} \cong \mathcal{G}_x \overset{\alpha}{\cong} G.$$

Now let $\gamma \in \pi_1(X, x)$. We associate to γ the element $\rho(\gamma) \in \operatorname{Aut} G$ defined in the following way. Let $y = \gamma \cdot \tilde{x} \in \widetilde{X}$; then $\rho(\gamma)$ is equal to the composition

$$\alpha \circ \mu_y \circ \beta_y^{-1} : G \to G,$$

where $\mu_y : \pi^{-1}(\mathcal{G})_y \cong \mathcal{G}_x$ is the natural isomorphism, and $\beta_y : \mathcal{G}_y \cong G$ is the isomorphism of stalks induced by β.

It remains to show that ρ is a representation, i.e. a morphism of groups. For this, it suffices to note that $\rho(\gamma)$ can also be identified with

$$\alpha \circ \gamma^* \circ \alpha^{-1},$$

where $\gamma^* : \mathcal{G}_x \cong \mathcal{G}_x$ is the isomorphism induced by any path from x to x of homotopy class γ. This follows from the proof of proposition 3.9, which constructs the trivialisation β using trivialisations of the locally constant system $\pi^{-1}(\mathcal{G})$ along paths in the simply connected variety $\widetilde{X}$.

We now construct the inverse correspondence. Let ρ be a representation of $\pi_1(X, x)$. Note that we can view X as the quotient of $\widetilde{X}$ by the action of $\pi_1(X, x)$. Let us associate to ρ the sheaf $\mathcal{G}$ on X whose sections over an open set U are the $\pi_1(X)$-equivariant locally constant functions from $\pi^{-1}(U)$ to G:

$$\begin{aligned}\Gamma(U, \mathcal{G}) = \{\sigma : \pi^{-1}(U) \to G \mid \sigma(\gamma \cdot u) \\ = \rho(\gamma) \cdot \sigma(u),\ \ \forall u \in \pi^{-1}(U),\ \gamma \in \pi_1(X, x)\}.\end{aligned} \tag{3.1}$$

The fact that $\pi : \widetilde{X} \to X$ is a trivial covering over the simply connected open sets of X immediately implies that the sheaf defined above is locally constant. Moreover, by formula (3.1), the choice of a point $\tilde{x} \in \widetilde{X}$ over x gives an isomorphism $\alpha : \mathcal{G}_x \cong G$, obtained as the limit over the open sets U containing x of the maps $\Gamma(U, \mathcal{G}) \to G$, $\sigma \mapsto \sigma(\tilde{x})$, in the notation of (3.1). We easily check that these two correspondences are inverses of each other. □

Definition 3.13 *The representation $\rho : \pi_1(X, x) \to \operatorname{Aut} \mathcal{G}_x = \operatorname{Aut} G$ corresponding to a local system (see corollary 3.10) is called the monodromy representation.*

3.1.2 Local systems associated to a fibration

Let $\phi : Y \to X$ be a fibration of topological spaces: this means that locally on X, there exists a trivialisation of ϕ, i.e. a homeomorphism

$$C = (C_0, \phi) : Y_U \cong Y_0 \times U,$$

over X, where U is an open neighbourhood of $0 \in X$ and $Y_U := \phi^{-1}(U)$.

By Ehresmann's theorem, this is the case if Y and X are differentiable varieties and ϕ is submersive and proper.

If we assume moreover that X is locally contractible, then for sufficiently small U, the open sets Y_U as above have the same homotopy type as the fibre Y_u, $u \in U$. Using the invariance under homotopy, i.e. the fact that for a contractible space U, $H^l(Y_u \times U, A) = H^l(Y_u, A)$ for all $l \geq 0$, we deduce that for every ring of coefficients A, the sheaves $R^k\phi_*A$ are locally constant sheaves. Here, $R^k\phi_*$ is the kth derived functor of the functor ϕ_* from the category of sheaves on Y to the category of sheaves on X.

For local systems obtained as above using a fibration, the monodromy representation

$$\rho : \pi_1(X, x) \to \operatorname{Aut} H^k(Y_x, A)$$

is in fact induced by homeomorphisms of the fibre Y_x. Indeed, for $\gamma \in \pi_1(X, x)$

represented by a loop $\gamma : [0,1] \to X$ based at x, by gluing local trivialisations above segments $[\epsilon_i, \epsilon_{i+1}]$, we can trivialise the fibration $Y_\gamma \to [0,1]$ defined as the fibred product

$$\begin{array}{ccc} Y_\gamma & \longrightarrow & Y \\ {\scriptstyle \phi_\gamma}\downarrow & & \downarrow{\scriptstyle \phi} \\ [0,1] & \longrightarrow & X. \end{array}$$

From such a trivialisation

$$T = (T_0, \phi_\gamma) : Y_\gamma \cong \phi_\gamma^{-1}(0) \times [0,1],$$

we deduce a homeomorphism $\psi = T_0 : \phi_\gamma^{-1}(1) \cong \phi_\gamma^{-1}(0)$. These two spaces are canonically homeomorphic to the fibre Y_x. It is clear by construction of the monodromy representation that $\rho(\gamma)$ is induced by this homeomorphism of Y_x, i.e. that

$$\rho(\gamma)(\eta) = \psi^*\eta = T_0^*\eta, \quad \eta \in H^k(Y_x, A).$$

In particular, it follows that the monodromy representation on the cohomology of the fibre $H^*(Y_x, A)$ is compatible with the cup-product on $H^*(Y_x, A)$, in the sense that each $\rho(\gamma) \in \operatorname{Aut} H^*(Y_x, A)$ is a ring automorphism for the ring structure given by the cup-product. This fact is also a direct consequence of the construction of the local system $R^*\phi_* A$, which shows that this local system is equipped with the structure given by the cup-product.

In the case of a proper differentiable fibration whose fibres are oriented and $2n$-dimensional, the local system $R^{2n}\phi_*\mathbb{Z}$ is trivial, and we deduce from the above that by Poincaré duality, we have an isomorphism

$$R^n\phi_*\mathbb{Z} \cong (R^n\phi_*\mathbb{Z})^*$$

of local systems. Equivalently, writing $\langle\, , \rangle$ for the corresponding intersection form on the stalk

$$H^n(Y_x, \mathbb{Z}) = R^n\phi_*\mathbb{Z}_x,$$

the monodromy representation takes values in

$$\operatorname{Aut}(H^n(Y_x, \mathbb{Z}), \langle\, , \rangle).$$

3.1.3 Monodromy and variation of Hodge structure

Let $\phi : Y \to X$ be a holomorphic, submersive and projective morphism between complex varieties, where 'projective' means that there exists a holomorphic immersion

$$i : Y \hookrightarrow X \times \mathbb{P}^N$$

such that $\mathrm{pr}_1 \circ i = \phi$. We then have a monodromy representation

$$\rho : \pi_1(X, x) \to \mathrm{Aut}\, H^k(Y_x, \mathbb{Z})$$

as above for every k. It is a remarkable fact that Hodge theory imposes constraints on the monodromy representation ρ. Hodge theory says that every group $H^k(Y_x, \mathbb{Z})$ is equipped with a Hodge structure, i.e. with a decomposition

$$H^k(Y_x, \mathbb{C}) = H^k(Y_x, \mathbb{Z}) \otimes \mathbb{C} = \bigoplus\nolimits_{p+q=k} H^{p,q}(Y_x)$$

into complex subspaces. Recall that a Hodge substructure of $H^k(Y_x, \mathbb{Z})$ is a sublattice $L \subset H^k(Y_x, \mathbb{Z})$ having an induced decomposition as a direct sum

$$L_{\mathbb{C}} := L \otimes \mathbb{C} = \bigoplus\nolimits_{p+q=k} L_{\mathbb{C}} \cap H^{p,q}(Y_x).$$

The first constraint imposed on ρ, given in the following proposition, is a consequence of the global invariant cycles theorem, which will be proved in the following chapter (corollary 4.25).

Proposition 3.14 *If X is quasi-projective, then the space*

$$H^k(Y_x, \mathbb{Z})^\rho := \{\alpha \in H^k(Y_x, \mathbb{Z}) \mid \rho(\gamma)(\alpha) = \alpha, \ \ \forall \gamma \in \pi_1(X, x)\}$$

of invariants under ρ is a Hodge substructure of $H^k(Y_x, \mathbb{Z})$.

Another consequence of Hodge theory, the quasi-unipotence theorem, concerns the local monodromy. Let us assume here that X is a punctured disk Δ^*, so that $\pi_1(X, x) = \mathbb{Z}$ and the monodromy group $\mathrm{Im}\, \rho$ is generated by a single element T.

Theorem 3.15 *T is quasi-unipotent, i.e. there exist integers N and M such that*

$$(T^N - 1)^M = 0.$$

In fact, we can even take $M \leq k + 1$.

Apart from the final statement, which uses the theory of mixed Hodge structures, this theorem whose proof sketched below is due to Borel says that the eigenvalues of the operator T are roots of unity. We restrict ourselves to sketching the main lines of the proof.

As these eigenvalues are also algebraic integers, Kronecker's theorem shows that it suffices to prove that they are of modulus equal to 1. An essential ingredient of the proof is the period map $\mathcal{P}$, constructed on the universal cover $e : \mathbb{H} \to \Delta^*$ of Δ^*, with values in the polarised period domain $\mathcal{D}$ (see vI.10.1.3). By definition, $\mathcal{P}(\tilde{z})$ is the point of $\mathcal{D}$ parametrising the Hodge decomposition on $H^k(Y_z, \mathbb{Z})$, where $z = e(\tilde{z}) \in \Delta^*$. Here, as $\mathbb{H}$ is simply connected, we identity $H^k(Y_z, \mathbb{Z})$ with a constant lattice K (the local system $e^{-1}(R^k\phi_*\mathbb{Z})$ is constant on $\mathbb{H}$ by proposition 3.9).

The group $\mathrm{Aut}\,(K, \langle\, , \rangle)$ has an obvious action on $\mathcal{D}$, where $\langle\, , \rangle$ is the intersection form

$$\langle \alpha, \beta \rangle = \int_{Y_z} L^{n-k} \alpha \cup \beta,$$

where L is the Lefschetz cup-product operator with class

$$h = c_1((\mathrm{pr}_2 \circ j)^* \mathcal{O}_{\mathbb{P}^N}(1))$$

on $H^k(Y_z, \mathbb{Z})$. (By Poincaré duality, we may of course assume that $k \leq n$.) Clearly the period map satisfies

$$\mathcal{P}(\tilde{z} + 1) = T\mathcal{P}(\tilde{z}),$$

where $\tilde{z} \mapsto \tilde{z} + 1$ is the action of the generator of $\pi_1(\Delta^*, x)$ $\mathbb{H}$ and $T \in \mathrm{Aut}\,(K, \langle\, , \rangle)$.

To prove theorem 3.15, we may assume that the cohomology $H^k(Y_z, \mathbb{Z})$ is primitive (see vI.6.2.3), the general case then being obtained using the Lefschetz decomposition relative to the polarisation h. One of the steps then consists in proving that $\mathcal{P}$ decreases distances for the hyperbolic metric on $\mathbb{H}$ and for a translation-invariant metric on $\mathcal{D}$, which is a homogeneous space under the group $G := \mathrm{Aut}\,(K_{\mathbb{R}}, \langle\, , \rangle)$. Here, the metric on $\mathcal{D}$ is constructed using the intersection form $\langle\, , \rangle$, or more precisely its Hermitian version

$$H_k(\alpha, \beta) = i^k \int_{Y_z} L^{n-k} \alpha \wedge \overline{\beta}.$$

The transversality property of the period map (see vI.10.2.2) is used in a crucial manner, as well as the fact that the Hodge decomposition is orthogonal for H_k at every point of $\mathcal{D}$, and that the Hermitian form H_k is definite of sign $(-1)^{\frac{k(k-1)}{2}} i^{k-p+q}$ on each space $K^{p,q}$ at every point of $\mathcal{D}$. This last fact implies (see Griffiths *et al.* 1984) that the holomorphic sectional curvature of $\mathcal{D}$ in the directions satisfying the transversality condition is negative. We then apply Ahlfors' lemma (see Kobayashi 1970), to conclude that the period map $\mathcal{P}$ decreases distances up to a coefficient when $\mathbb{H}$ is equipped with the hyperbolic metric (see Kobayashi 1970), i.e. satisfies the equality

$$d_{\mathcal{D}}(\mathcal{P}(x), \mathcal{P}(y)) \leq C d_{\mathbb{H}}(x, y)$$

for a constant $C > 0$.

Thus, for every $x \in \mathbb{H}$, we have

$$d_{\mathcal{D}}(\mathcal{P}(x+1), \mathcal{P}(x)) \leq C d_{\mathbb{H}}(x, x+1),$$

i.e.

$$d_{\mathcal{D}}(T\mathcal{P}(x), \mathcal{P}(x)) \leq C d_{\mathbb{H}}(x, x+1).$$

Now let $x_0 \in \mathbb{H}$ be a given point, and let $x_n \in \mathbb{H}$ be a sequence of points such that $\lim_{n\to\infty} \mathcal{I} x_n = 0$. Then $\lim_{n\to\infty} d_{\mathbb{H}}(x_n, x_n + 1) = 0$. Let us write

$$\mathcal{P}(x_n) = g_n \mathcal{P}(x_0)$$

with $g_n \in \mathrm{Aut}\,(K_{\mathbb{R}}, \langle\, ,\, \rangle)$. Then we get

$$d_{\mathcal{D}}(T\mathcal{P}(x_n), \mathcal{P}(x_n)) = d_{\mathcal{D}}(T g_n \mathcal{P}(x_0), g_n \mathcal{P}(x_0)) \leq C d_{\mathbb{H}}(x_n, x_n + 1).$$

But as $d_{\mathcal{D}}$ is translation-invariant, we have

$$d_{\mathcal{D}}(T g_n \mathcal{P}(x_0), g_n \mathcal{P}(x_0)) = d_{\mathcal{D}}\big(g_n^{-1} T g_n \mathcal{P}(x_0), \mathcal{P}(x_0)\big).$$

Thus, we finally obtain

$$\lim_{n\to\infty} d_{\mathcal{D}}\big(g_n^{-1} T g_n \mathcal{P}(x_0), \mathcal{P}(x_0)\big) = 0.$$

This means that the sequence $g_n^{-1} T g_n \in \mathrm{Aut}\,(K_{\mathbb{R}}, \langle\, ,\, \rangle)$ is adherent to the stabiliser of the point $\mathcal{P}(x_0) \in \mathcal{D}$. To conclude, we use the fact that this stabiliser is a compact group, which follows from the fact that the Hermitian form H_k is of definite sign on each of the $H^{p,q}$ at the point $\mathcal{P}(x_0)$, and conclude from this that the eigenvalues of T are of modulus 1. □

3.2 The case of Lefschetz pencils

3.2.1 The Picard–Lefschetz formula

Let X be a smooth n-dimensional complex variety, and $f : X \to \Delta$ a Lefschetz degeneration. Recall that this means that f is proper with non-zero differential over the punctured disk Δ^*, and that the fibre X_0 has an ordinary double point as its only singularity. Let $t \in \Delta^*$. As $\pi_1(\Delta^*, t) = \mathbb{Z}$, the monodromy representation ρ on the cohomology of the fibre $H^{n-1}(Y_t, \mathbb{Z})$ is determined by the image of the natural generator of $\pi_1(\Delta^*, t)$, which we denote by T as above.

Note also that by corollary 2.17, the monodromy acts trivially on $H^k(X_t, \mathbb{Z})$ for $k < n-1$, and thus also for $k > n-1$ by Poincaré duality.

Recall that X_t contains vanishing spheres S_t^{n-1}, contained in the intersection of X_t with a ball centred at the singular point of X_0. The cohomology class of such a sphere defined by an orientation is written δ. It is a generator of

$$\operatorname{Ker}(H^{n-1}(X_t, \mathbb{Z}) \cong H_{n-1}(X_t, \mathbb{Z}) \to H_{n-1}(X, \mathbb{Z}))$$

(cf. corollary 2.17).

Finally, the fibre X_t is a real oriented $(2n-2)$-dimensional variety, so we have the intersection form $\langle\, , \rangle$ on $H^{n-1}(X_t, \mathbb{Z})$.

Theorem 3.16 *(Picard–Lefschetz formula)* *For every $\alpha \in H^{n-1}(X_t, \mathbb{Z})$, we have*

$$T(\alpha) = \alpha + \epsilon_n \langle \alpha, \delta \rangle \delta, \tag{3.2}$$

where $\epsilon_n = \pm 1$ according to the value of n.

We have the following corollary of the Picard–Lefschetz formula.

Corollary 3.17 *If $n-1$ is even, the monodromy operator is an involution. If $n-1$ is odd and $\delta \in H^{n-1}(X_t, \mathbb{Z})$ is not of torsion, the monodromy operator is of infinite order.*

Proof If $n-1$ is odd, the intersection form $\langle\, , \rangle$ is antisymmetric, and thus $\langle \delta, \delta \rangle = 0$. Iterating (3.2), we then find that

$$T^k(\alpha) = \alpha + k\epsilon_n \langle \alpha, \delta \rangle \delta.$$

If $n-1$ is even, then since the monodromy operator must preserve the symmetric intersection form $\langle\, , \rangle$, we find that

$$\begin{aligned} \langle T(\alpha), T(\beta) \rangle &= \langle \alpha, \beta \rangle \\ &= \langle \alpha, \beta \rangle + \epsilon_n \langle \alpha, \delta \rangle \langle \beta, \delta \rangle + \epsilon_n \langle \beta, \delta \rangle \langle \alpha, \delta \rangle + \langle \alpha, \delta \rangle \langle \beta, \delta \rangle \langle \delta, \delta \rangle. \end{aligned}$$

This gives

$$2\epsilon_n + \langle \delta, \delta \rangle = 0 \tag{3.3}$$

whenever $\delta \neq 0$. We will show below (remark 3.21) that $\delta \neq 0$ in the case where $n-1$ is even. But then

$$\begin{aligned}T^2(\alpha) &= \alpha + \epsilon_n\langle\alpha, \delta\rangle\delta + \epsilon_n\langle\alpha + \epsilon_n\langle\alpha, \delta\rangle\delta, \delta\rangle\delta \\ &= \alpha + (2\epsilon_n + \langle\delta, \delta\rangle)\langle\alpha, \delta\rangle\delta = \alpha.\end{aligned}$$

□

Remark 3.18 *The sign ϵ_n can be made explicit. It suffices to take the orientations carefully into account in the computations below. In the case where $n-1$ is even, the sign is easier to compute, thanks to formula (3.3); we find*

$$\epsilon_n = -(-1)^{\frac{n(n-1)}{2}}.$$

The Picard–Lefschetz formula is a consequence of the local Picard–Lefschetz formula, which we now present.

Let B be a ball of $\mathbb{C}^n$ centred at 0, and let $f : B \to \mathbb{C}$, $f(z) = \sum_{i=1}^n z_i^2$. If Δ is a sufficiently small disk, we know that with the notation $S = \partial B$, $B_\Delta = f^{-1}(\Delta)$, $S_\Delta = S \cap B_\Delta$ and $B_{\partial\Delta} = f^{-1}(\partial\Delta)$, $f : S_\Delta \to \Delta$ is a fibration, while $f : B_{\partial\Delta} \to \partial\Delta$ is a fibration of varieties with boundary.

As the disk Δ is contractible, it follows that we can find a trivialisation

$$C = (C_0, f) : S_\Delta \cong S_t \times \Delta, \quad t \in \partial\Delta.$$

Moreover, letting

$$\gamma : [0, 1] \to \partial\Delta, \quad \gamma(0) = \gamma(1) = t$$

be a parametrisation of the boundary of Δ, we can find a trivialisation of the induced fibration

$$f_\gamma : B^\gamma_{\partial\Delta} \to [0, 1],$$

i.e. a diffeomorphism

$$C' = (C'_0, f) : B^\gamma_{\partial\Delta} \cong B_t \times [0, 1].$$

Moreover, we may assume that $C'_0 = C_0 \circ \gamma$ on the boundary $S^\gamma_{\partial\Delta}$ of $B^\gamma_{\partial\Delta}$. But then the trivialisation C' induces a diffeomorphism

$$\psi = \left(C'_{0|f_\gamma^{-1}(1)}\right)^{-1} : f_\gamma^{-1}(0) \cong f_\gamma^{-1}(1), \tag{3.4}$$

i.e. a diffeomorphism ψ from B_t to B_t, which is the identity on the boundary S_t of B_t since the trivialisation C' restricted to $S^\gamma_{\partial\Delta}$ comes from the trivialisation

$$C : S_{\partial\Delta} \cong S_t \times \partial\Delta.$$

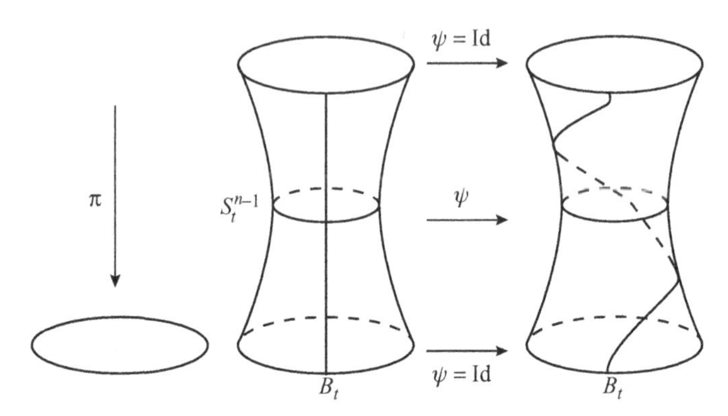

Figure 3.1: The Dehn twist.

Remark 3.19 *The diffeomorphism ψ is called a 'Dehn twist' (see figure 3.1). Its effect is particularly obvious in dimension $n-1=1$. Then the fibre B_t is a cylinder, and ψ twists the generators of the cylinder once around the cylinder. We will shortly give a general description of this diffeomorphism.*

This diffeomorphism enables us to construct a refined version of the monodromy on the homology of the pair (B_t, S_t) as follows. If $\alpha \in H_*(B_t, S_t)$ is a relative cycle, i.e. a chain with boundary $\partial\alpha$ contained in S_t, then the chain $\psi_*(\alpha) - \alpha$ satisfies $\partial(\psi_*(\alpha) - \alpha) = 0$ since $\psi = \text{Id}$ on S_t. This gives a map

$$\begin{aligned} \phi : H_k(B_t, S_t, \mathbb{Z}) &\to H_k(B_t, \mathbb{Z}) = H_k\big(B_t^0, \mathbb{Z}\big), \\ \alpha &\mapsto \psi_*(\alpha) - \alpha. \end{aligned}$$

Note also that as the variety B_t^0 is smooth, $(2n-2)$-dimensional and oriented, we have an intersection form

$$H_k(B_t, S_t, \mathbb{Z}) \times H_{2n-2-k}\big(B_t^0, \mathbb{Z}\big) \to \mathbb{Z} \tag{3.5}$$

given by Poincaré duality (1.8). The local Picard–Lefschetz formula is then as follows.

Proposition 3.20 *For $\alpha \in H_{n-1}(B_t, S_t, \mathbb{Z})$, we have*

$$\phi(\alpha) = \epsilon_n \langle \alpha, \delta \rangle \delta, \tag{3.6}$$

where $\epsilon_n = \pm 1$ according to the value of n, $\delta \in H_{n-1}(B_t, \mathbb{Z})$ is the class of the vanishing sphere (2.3), and $\langle\, , \rangle$ is the pairing of (3.5).

Let us first show how proposition 3.20 implies theorem 3.16.

Proof of theorem 3.16 We may assume that the disk Δ is sufficiently small, so that by Morse's lemma, in the neighbourhood of the critical point $0 \in X_0$, X_Δ contains a set of the form B_Δ in which f is as above.

As the restriction of f to $X_\Delta - B^0_\Delta$ is a fibration into varieties with boundary, the trivialisation

$$C = (C_0, f) : S_\Delta \cong S_t \times \Delta$$

extends to a trivialisation

$$C = (C_0, f) : X_\Delta - B^0_\Delta \cong \left(X_t - B^0_t\right) \times \Delta.$$

For a parametrisation $\gamma : [0, 1] \to \partial\Delta$ of the boundary of the disk with $\gamma(0) = \gamma(1) = t$, the trivialisation

$$C' = (C'_0, f_\gamma) : B^\gamma_{\partial\Delta} \cong B_t \times [0, 1]$$

of the fibration $f_\gamma : B^\gamma_{\partial\Delta} \to [0, 1]$ introduced above then extends to a trivialisation

$$C'' = (C''_0, f_\gamma) : X^\gamma \to X_t \times [0, 1],$$

where X^γ is the fibred product

$$\begin{array}{cccl} & X^\gamma & \longrightarrow & X_{\partial\Delta} \qquad := f^{-1}(\partial\Delta) \\ & \downarrow & & \downarrow \\ \gamma : & [0, 1] & \longrightarrow & \partial\Delta. \end{array}$$

Indeed, it suffices to set $C''_0 = C_0$ in $X^\gamma - B^\gamma_\Delta$ and $C''_0 = C'_0$ in B^γ_Δ.

The diffeomorphism

$$\Psi = \left(C''_{0\,|f_\gamma^{-1}(1)}\right)^{-1} : X_t = X^\gamma_0 \cong X^\gamma_1 = X_t$$

thus obtained is clearly equal to the identity on $X_t - B^0_t$, since the trivialisation C'' of $X^\gamma - B^\gamma_\Delta \to [0, 1]$ comes from a trivialisation of $f : X_{\partial\Delta} - B_{\partial\Delta} \to \partial\Delta$. Moreover, $\Psi = \psi$ on B_t.

Now let $\alpha \in H^{n-1}(X_t, \mathbb{Z}) = H_{n-1}(X_t, \mathbb{Z})$. We know that $T(\alpha) = \Psi_*(\alpha)$, so $T(\alpha) - \alpha = \Psi_*\alpha - \alpha$ is represented by a chain with support in B_t since $\Psi = \mathrm{Id}$ outside B_t.

In fact, by the excision theorem, we have a natural map

$$a : H_{n-1}(X_t, \mathbb{Z}) \to H_{n-1}(B_t, S_t, \mathbb{Z}),$$

since the last group can be identified with $H_{n-1}(X_t, X_t - B_t, \mathbb{Z})$. This map sends a cycle of X_t to its intersection with B_t, whose boundary is contained in ∂B_t.

Since $\Psi = \mathrm{Id}$ on $X_t - B_t$, it is now clear that

$$T(\alpha) - \alpha = \Psi_*(\alpha) - \alpha = j_*(\psi_*(a(\alpha)) - a(\alpha)) = j_*\phi(a(\alpha)) \in H_{n-1}(X_t, \mathbb{Z}),$$

where j is the inclusion of B_t into X_t. Thus, formula (3.2) follows from formula (3.6), by noting firstly that the vanishing cycle of X_t is equal to $j_*\delta$ by definition, and secondly that for $\alpha \in H_{n-1}(X_t, \mathbb{Z})$, we have

$$\langle \alpha, j_*\delta \rangle_{X_t} = \langle a(\alpha), \delta \rangle_{B_t},$$

where the left-hand $\langle\, ,\rangle$ is the intersection in $H_{n-1}(X_t, \mathbb{Z}) = H^{n-1}(X_t, \mathbb{Z})$ while the right-hand one denotes the pairing (3.5). □

Proof of proposition 3.20 Note first that B_t has a very simple topology. Indeed, for $t \in \mathbb{R}^+$, B_t is the set

$$\{z = (z_1, \ldots, z_n) = (x_1, \ldots, x_n, y_1, \ldots, y_n) \mid$$
$$\|x\|^2 + \|y\|^2 \le r^2,\ \|x\|^2 - \|y\|^2 = t,\ \langle x, y\rangle = 0\},$$

where $x = (x_1, \ldots, x_n)$, $y = y_1, \ldots, y_n$ and $\langle\, ,\rangle$ is the scalar product in $\mathbb{R}^n$. Clearly, for $0 < t < r^2$, this is a bundle of $(n-1)$-dimensional balls on the sphere S^{n-1} via the map

$$\pi : B_t \to S^{n-1}, \quad z = (x, y) \mapsto \frac{x}{\|x\|}. \tag{3.7}$$

More precisely, it is a disk bundle in the tangent bundle of the vanishing sphere $S_t^{n-1} \subset B_t$, which is defined by $y = 0$ in the notation above.

It follows that B_t retracts by deformation onto S_t^{n-1}, and that the homology $H_{n-1}(B_t, \mathbb{Z})$ is isomorphic to $\mathbb{Z}$, generated by $\delta = \left[S_t^{n-1}\right]$. Let $\delta' \in H_{n-1}(B_t, S_t, \mathbb{Z})$ be the class of the ball $Z_x := \pi^{-1}(x)$, $x \in S^{n-1}$. As we have $\langle \delta, \delta' \rangle = \pm 1$ (depending on the choice of the orientations of the cycles), the relative homology $H_{n-1}(B_t, S_t, \mathbb{Z})$ is isomorphic to $\mathbb{Z}$, generated by the class δ'. Hence, we necessarily have a formula of the type

$$\phi(\delta') = \epsilon_n \langle \delta', \delta \rangle \delta$$

for a certain $\epsilon_n \in \mathbb{Z}$, which gives more generally

$$\phi(\alpha) = \epsilon_n \langle \alpha, \delta \rangle \delta$$

for every $\alpha \in H_{n-1}(B_t, S_t, \mathbb{Z})$.

Thus, proposition 3.20 simply says that $\epsilon_n = \pm 1$. Consider the diffeomorphism ψ of 3.4. The chain $\psi_*(Z_x) - Z_x$ of B_t has no boundary, and its class in $H_{n-1}(B_t, \mathbb{Z})$ is equal to $\phi(\delta')$. The number ϵ_n can be determined up to sign; it is the intersection number $\langle \psi(Z_x) - Z_x, \delta' \rangle$. The map π contracts Z_x and also the boundary of $\psi(Z_x)$ onto a point, since ψ is the identity on ∂B. Thus, it gives a map

$$\pi' = \pi \circ \psi : Z'_x \cong S^{n-1} \to S^{n-1}, \tag{3.8}$$

where Z'_x is the sphere S^{n-1} obtained by contracting the boundary of the ball $B^{n-1} = Z_x$ to a point. Then we clearly have

$$\langle \psi(Z_x) - Z_x, \delta' \rangle = \pm d^0 \pi',$$

since δ' is the class of $\pi^{-1}(y)$, $y \in S^{n-1}$. It remains only to show that $d^0 \pi' = \pm 1$.

For a parametrisation γ of $\partial \Delta$, let us take the map defined by $\gamma(u) = t \exp(2i\pi u)$. Points of $B^{\gamma}_{\partial\Delta}$ are then written as (z, u), $z \in B_{\partial\Delta}$, $f(z) = e^{2i\pi u}$. We then have a trivialisation

$$D = (D_0, f) : B^{\gamma}_{\Delta} \cong B_t \times [0, 1]$$

of the fibration $f_\gamma : B^{\gamma}_{\Delta} \to [0, 1]$, given by

$$D_0(z, u) = \exp(-i\pi u) z. \tag{3.9}$$

This trivialisation does not have the desired property that its restriction to the boundary S^{γ}_{Δ} comes from a trivialisation $C = (C_0, f)$ of the fibration $f : S_\Delta \to \Delta$ or better of the fibration $S^{\eta}_{\Delta} \to \Delta$, where S^{η}_{Δ} is a neighbourhood of S_Δ in B_Δ. But since D_0 and C_0 are the identity on the fibre $S^{\eta}_t = f_\gamma^{-1}(0)$, they are homotopic and even naturally homotopic, via the homotopy $(H_s)_{s \in [0,1]}$ given by

$$H_s(z) = C_0(D^{-1}(D_0(z), s f_\gamma(z))).$$

For $s = 1$, we have $H_1 = C_0$, and for $s = 0$, we have $H_0 = D_0$.

It then suffices to take a function μ which takes the value 1 in a neighbourhood of S_Δ and 0 in the neighbourhood of B'_Δ, where the ball B' has radius strictly smaller than the radius r of B, and to define

$$C'_0(z, u) = H_{\mu(z)}(z, u), \quad z \in S^{\eta}_{\Delta}, \quad C'_0(z, u) = D_0(z, u), z \in B'_\Delta, \tag{3.10}$$

in order to obtain a trivialisation having the desired properties. Finally, note that for C_0, we can take the map defined by

$$C_0(z) = (\alpha(z) x, \beta(z) y'), \quad z = (x, y) \in S^{\eta}_{\Delta}, \tag{3.11}$$

where y' is the projection of y onto the hyperplane orthogonal to x, and the real positive functions α, β are defined by the equations

$$\alpha^2(z)\|x\|^2 - \beta(z)^2\|y'\|^2 = t, \quad \alpha^2(z)\|x\|^2 + \beta(z)^2\|y'\|^2 = \|z\|^2.$$

Let us now return to the ball

$$Z_x = \pi^{-1}(x) = \{(\lambda x, y) \mid \langle x, y\rangle = 0, \quad \lambda^2 + \|y\|^2 \leq r^2, \ \lambda^2 - \|y\|^2 = t\}.$$

We can assume that the function μ is the function $\frac{2}{r^2-t}|y|^2$ on Z_x. Then we have $\pi \circ \psi^{-1}(z) = \pi \circ C_0'(z, 1)$, $z \in Z_x$, i.e.

$$\pi \circ \psi^{-1}(z) = \pi \circ C_0\left(D^{-1}\left(D_0(z, 1), \frac{2}{r^2 - t}|y|^2\right)\right).$$

Thus, by (3.7), (3.9), (3.10) and (3.11), noting that $D_0(z, 1) = -z$, we now have

$$\pi \circ \psi^{-1}(z) = -\pi \circ C_0\left(\exp\left(\frac{2i\pi|y|^2}{r^2 - t}\right) z\right) = \frac{u}{|u|},$$

with

$$u(z) = \lambda \cos\left(\frac{2i\pi|y|^2}{r^2 - t}\right) x - \sin\left(\frac{2i\pi|y|^2}{r^2 - t}\right) y, \quad z = (\lambda x, y),$$
$$\langle x, y\rangle = 0, \quad \lambda^2 - \|y\|^2 = t.$$

It is now clear that the map from the sphere Z_x' to S^{n-1} induced by $\pi \circ \psi^{-1}$ is of degree 1. It follows immediately that the map π' of (3.8) induced by $\pi \circ \psi$ is also of degree 1. □

Remark 3.21 *The explicit description of the fibre B_t given at the beginning of the proof shows that the normal bundle of the vanishing sphere S_t^{n-1} in B_t is isomorphic to its tangent bundle, and thus the self-intersection $[S_t^{n-1}]^2 = \delta_t^2$, which is equal to the Euler class of this normal bundle up to sign, is equal by Hopf's theorem to the Euler–Poincaré characteristic of the sphere, i.e. to 2 if $n-1$ is even and 0 if $n-1$ is odd. In fact, the sign which appears is due to the change of orientation that occurs when passing from the coordinate system $(x_1, y_1, \ldots, x_n, y_n)$, which gives the complex orientation, to the coordinate system $(x_1 \ldots, x_n, y_1, \ldots, y_n)$, which arises when we represent B_t as a bundle in balls on the tangent bundle of S_t^{n-1}. Clearly, the sign of the Jacobian of this change of coordinates is equal to $(-1)^{\frac{n(n-1)}{2}}$, which for $n-1$ even gives the formula*

$$\langle\delta, \delta\rangle = (-1)^{\frac{n(n-1)}{2}} 2.$$

3.2.2 Zariski's theorem

Assume that $X \subset \mathbb{P}^N$ is a smooth projective connected non-degenerate variety. The discriminant variety $\mathcal{D}_X \subset (\mathbb{P}^N)^*$ was defined above as the set of singular hyperplane sections of X. Let $U := (\mathbb{P}^N)^* - \mathcal{D}_X$ be its complement. U parametrises the universal hypersurface $\phi : \mathcal{X}_U \to U$ defined by

$$\mathcal{X}_U = \{(x, H) \in X \times U \mid x \in X_H\}.$$

By the definition of U, the map $\phi = \mathrm{pr}_2$ is a submersion with smooth fibre X_H over H.

We saw that $\mathcal{D}_X$ is irreducible, as the image in $(\mathbb{P}^N)^*$ of the smooth irreducible variety Z of (2.2). We also saw that a Lefschetz pencil is a pencil $\Delta \subset (\mathbb{P}^N)^*$ which meets $\mathcal{D}_X$ transversally in its smooth locus, or else does not meet $\mathcal{D}_X$ at all in the case where $\dim \mathcal{D}_X \leq N-2$. In the latter case we have $\pi_1(U, 0) = 1$, so there is no monodromy action associated to the fibration ϕ.

In general, the following result due to Zariski shows that the monodromy $\rho : \pi_1(U, 0) \to \mathrm{Aut}\,(H^k(X_0, \mathbb{Z}))$ can be computed by restricting to a Lefschetz pencil.

Theorem 3.22 *Let $\mathcal{Y} \subset \mathbb{P}^r$ be a hypersurface, and let $U = \mathbb{P}^r - \mathcal{Y}$ be its complement. Then for $0 \in U$ and for every projective line $\Delta \subset \mathbb{P}^r$ passing through 0 which meets $\mathcal{Y}$ transversally in its smooth locus, the natural map*

$$\pi_1(\Delta - \Delta \cap \mathcal{Y}, 0) \to \pi_1(U, 0)$$

is surjective.

Proof Let W be the set of lines passing through 0 which meet $\mathcal{Y}$ transversally in its smooth locus. We have a $\mathbb{P}^1$-bundle $\pi = \mathrm{pr}_2 : P \to W$ on W given by

$$P = \{(x, t) \in \mathbb{P}^r \times W \mid x \in \Delta_t\}.$$

Consider the first projection $\mathrm{pr}_1 : P \to \mathbb{P}^r$, and let

$$U' := \mathrm{pr}_1^{-1}(U) = P - \mathrm{pr}_1^{-1}(\mathcal{Y}) \subset P.$$

By the definition of W, the restriction of π to U' is a fibration. Moreover, as all the lines parametrised by W pass through 0, $\pi' := \pi_{|U'}$ admits a natural section $W' \subset U'$ defined by $W' = \{(0, t) \mid t \in W\}$.

Let $0' = (0, o) \in W'$, and let $\Delta = \Delta_o$. Noting that the fibre $\pi'^{-1}(t)$ for $t \in W$ can be identified with $\Delta_t - \Delta_t \cap \mathcal{Y}$, we have the homotopy exact sequence

$$\pi_1(\Delta - \Delta \cap \mathcal{Y}, 0) \to \pi_1(U', 0') \to \pi_1(W, o) \to 1.$$

We can now apply pr_{1*} to this exact sequence. As $W' \subset U'$ is a section of π', $\pi_1(W', 0') \subset \pi_1(U', 0')$ is a section of this exact sequence. Now, pr_1 contracts W' to 0, and we conclude that

$$\mathrm{pr}_{1*}(\pi_1(\Delta - \Delta \cap \mathcal{Y}, 0)) = \mathrm{pr}_{1*}(\pi_1(U', 0')).$$

It thus remains simply to show that the arrow $\mathrm{pr}_{1*} : \pi_1(U', 0') \to \pi_1(U, 0)$ is surjective. For this, it suffices to show that

$$\mathrm{pr}_{1*} : \pi_1(U', 0'_1) \to \pi_1(U, 0_1)$$

is surjective, for $0'_1$ generic and $0_1 = pr_1(0'_1)$. But this follows from the fact that pr_1 is birational, so there exists a Zariski open set U'' of U' containing $0'_1$, isomorphic to a Zariski open set U'' of U via pr_1. Then the morphism $\pi_1(U'', 0_1) \to \pi_1(U, 0_1)$ is surjective since $U - U''$ is of real codimension ≥ 2 in U, and the commutative diagram

$$\begin{array}{ccc} \pi_1(U'', 0_1) & \longrightarrow & \pi_1(U', 0'_1) \\ \downarrow & & \downarrow \mathrm{pr}_{1*} \\ \pi_1(U'', 0_1) & \longrightarrow & \pi_1(U, 0_1) \end{array}$$

where the bottom horizontal arrow is surjective shows that $\mathrm{pr}_{1*} : \pi_1(U', 0'_1) \to \pi_1(U, 0_1)$ is also surjective. □

Let us return to the discriminant hypersurface $\mathcal{D}_X$, and its open set $\mathcal{D}_X^0$ parametrising the hypersurfaces X_H having exactly one ordinary double point. Fix $0 \in U$ as above. We have the monodromy representation

$$\rho : \pi_1(U, 0) \to \mathrm{Aut}\, H^{n-1}(X_0, \mathbb{Z}), \quad n = \dim X,$$

associated to the fibration ϕ. Moreover, for every $y \in \mathcal{D}_X^0$, let $y' \in U$ be near y, contained in a disk D_y which meets $\mathcal{D}_X^0$ transversally at y, and such that $D_y - \{y\} \subset U$. Then we have a vanishing cycle $\delta_y \subset H^{n-1}(X_{y'}, \mathbb{Z}) = H_{n-1}(X_{y'}, \mathbb{Z})$, well-defined up to sign as a generator of the kernel of the map

$$H_{n-1}(X_{y'}, \mathbb{Z}) \to H_{n-1}(X_{D_y}, \mathbb{Z}),$$

where $X_{y'} := \phi^{-1}(y')$, $X_{D_y} = \phi^{-1}(D_y)$.

Let us now choose a path γ from y' to 0 contained in U; then, by trivialising the fibration ϕ over γ, we can construct a diffeomorphism $\psi : X_{y'} \cong X_0$, well-defined up to homotopy. Thus, we have a vanishing cycle

$$\delta_\gamma = \psi_*(\delta_y) \in H_{n-1}(X_0, \mathbb{Z}) = H^{n-1}(X_0, \mathbb{Z}).$$

We can then deduce the following result from the irreducibility of the discriminant hypersurface.

Proposition 3.23 *All the vanishing cycles constructed above (and defined up to sign) are conjugate (up to sign) under the monodromy action ρ.*

Proof Clearly, by definition of the monodromy action, if we change the path γ above by composing it with a loop γ' based at 0, the morphism ψ_* becomes $\rho(\gamma') \circ \psi_*$, so that

$$\delta_{\gamma' \cdot \gamma} = \rho(\gamma')(\delta_\gamma).$$

It thus suffices to check what happens when we change the point y. But as $\mathcal{D}_X$ is irreducible, its smooth locus is connected, so it is arcwise connected, as is $\mathcal{D}_X^0$. If y_1 is another point of $\mathcal{D}_X^0$, we can choose a path l from y to y_1 in $\mathcal{D}_X^0$ and lift it to a path l' from y' to y_1' contained in the boundary of a tubular neighbourhood of $\mathcal{D}_X^0$ in $(\mathbb{P}^N)^*$. Obviously, a trivialisation of ϕ over l' transports the vanishing cycle $\delta_y \in H_{n-1}(X_{y'}, \mathbb{Z})$ to the vanishing cycle $\delta_{y_1} \in H_{n-1}(X_{y_1'}, \mathbb{Z})$. If γ is a path from 0 to y' and γ' is a path from 0 to y_1', then the loop $\gamma'' := \gamma'^{-1} \cdot l \cdot \gamma$ based at 0 satisfies

$$\rho(\gamma'')(\delta_\gamma) = \delta_{\gamma'}.$$

∎

Zariski's theorem 3.22 and the proposition above imply the following result.

Corollary 3.24 *Let $(X_t)_{t \in \mathbb{P}^1}$ be a Lefschetz pencil of hyperplane sections of X, 0_i, $i = 1, \ldots, M$ the critical values, and $0 \in \mathbb{P}^1$ a regular value. Then all the vanishing cycles $\delta_i \in H^{n-1}(X_0, \mathbb{Z})$ of the pencil are conjugate under the monodromy action of $\rho : \pi_1(\mathbb{P}^1 - \{0_1, \ldots, 0_M\}, 0) \to \mathrm{Aut}\, H^{n-1}(X_0, \mathbb{Z})$.*

Remark 3.25 *These vanishing cycles are not well-defined without specifying the choice of paths γ_i from 0 to 0_i. However, as we saw above, changing the path comes down to letting a loop act via the monodromy action.*

3.2.3 Irreducibility of the monodromy action

From the arguments above, we will deduce the irreducibility theorem for the monodromy action, theorem 3.27, which has very important applications. Recall that if $X_0 \overset{j}{\hookrightarrow} X$ is a smooth hyperplane section, the vanishing cohomology of

X_0 is defined by

$$H^{n-1}(X_0, \mathbb{Q})_{\mathrm{van}} = \mathrm{Ker}\left(j_* : H^{n-1}(X_0, \mathbb{Q}) \to H^{n+1}(X, \mathbb{Q})\right).$$

It is clear that if $U \subset (\mathbb{P}^N)^*$ denotes the open set parametrising the smooth hyperplane sections of X, then the monodromy action

$$\rho : \pi_1(U, 0) \to \mathrm{Aut}\, H^{n-1}(X_0, \mathbb{Q})$$

leaves $H^{n-1}(X_0, \mathbb{Q})_{\mathrm{van}}$ stable. Indeed, we have an inclusion

$$J : \mathcal{X}_U \to U \times X$$

of fibrations over U which gives a morphism of local systems

$$J_* : R^{n-1} f_* \mathbb{Q} \to R^{n+1} \mathrm{pr}_{1*} \mathbb{Q}$$

whose value on the stalk at the point 0 is the map j_*. Thus, we have a local subsystem $\mathrm{Ker}\, J_*$ whose stalk at the point 0 is $\mathrm{Ker}\, j_*$. The monodromy ρ preserves the stalk of this local subsystem, i.e. leaves $\mathrm{Ker}\, j_*$ stable.

Definition 3.26 *The action of a group G on a vector space E is said to be irreducible if every vector subspace $F \subset E$ stable under G is equal to $\{0\}$ or E.*

Theorem 3.27 *Let the notation and hypotheses be as in corollary 3.24. Then the monodromy action*

$$\rho : \pi_1(U, 0) \to \mathrm{Aut}\, H^{n-1}(X_0, \mathbb{Q})_{\mathrm{van}}$$

is irreducible.

Proof Obviously, it suffices to prove the irreducibility of the monodromy action

$$\pi_1(\mathbb{P}^1 - \{0_1, \ldots, 0_M\}, 0) \to \mathrm{Aut}\, H^{n-1}(X_0, \mathbb{Q})_{\mathrm{van}},$$

where $(X_t)_{t \in \mathbb{P}^1}$ is a Lefschetz pencil.

By lemma 2.26, the vanishing cohomology is generated by the vanishing cycles δ_i of the pencil.

By proposition 2.27, the restriction of the intersection form $\langle\, ,\rangle$ is non-degenerate on $H^{n-1}(X_0, \mathbb{Q})_{\mathrm{van}}$.

Let $F \subset H^{n-1}(X_0, \mathbb{Q})_{\mathrm{van}}$ be a non-trivial vector subspace which is stable under the monodromy action ρ. For $i \in \{1, \ldots, M\}$, let $\tilde{\gamma}_i$ be the loop in B based at 0 which is equal to γ_i until t_i, winds around the disk Δ_i once in the positive

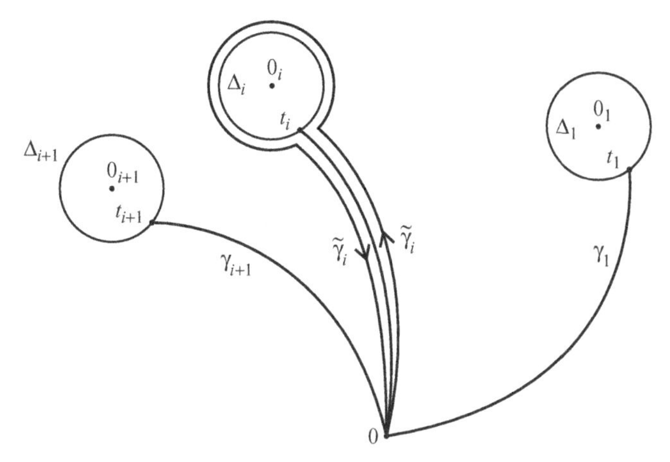

Figure 3.2: The loops $\tilde{\gamma}_i$ generate $\pi_1(\mathbb{P}^1 - \{0_1, \ldots, 0_M\}, 0)$.

direction, and then returns to 0 via γ_i^{-1} (see figure 3.2). The Picard–Lefschetz formula (3.2) shows that

$$\rho(\tilde{\gamma}_i)(\alpha) = \alpha \pm \langle \alpha, \delta_i \rangle \delta_i, \quad \forall \alpha \in H^{n-1}(X_0, \mathbb{Q}).$$

Now let $0 \neq \alpha \in F$. As $\langle \, , \rangle$ is non-degenerate on $H^{n-1}(X_0, \mathbb{Q})_{\text{van}}$, which is generated by the vanishing cycles, there exists $i \in \{1, \ldots, M\}$ such that $\langle \alpha, \delta_i \rangle \neq 0$. As $\rho(\tilde{\gamma}_i)(\alpha) - \alpha \in F$, the Picard–Lefschetz formula implies that $\delta_i \in F$. But by corollary 3.24, all the vanishing cycles are conjugate under the monodromy action, so F, which is stable under the monodromy action, must contain all the vanishing cycles. Thus $F = H^{n-1}(X_0, \mathbb{Q})_{\text{van}}$. □

To conclude, note that the equivalence between representations of the π_1 and local systems (see corollary 3.10) yields the following consequence of theorem 3.27.

Corollary 3.28 *Let $\phi : \mathcal{X}_U \to U$ be the universal smooth hypersurface. Then there exists no non-trivial local subsystem of $R^{n-1}\phi_*\mathbb{Q}_{\text{van}}$, where $n = \dim X$.*

3.3 Application: the Noether–Lefschetz theorem

3.3.1 The Noether–Lefschetz locus

Let $X \subset \mathbb{P}^N$ be a 3-dimensional projective variety, and let $U \subset (\mathbb{P}^N)^*$ be the complement of the discriminant variety. U parametrises the universal smooth hypersurface

$$\mathcal{X}_U = \{(x, H) \in X \times U \mid x \in X_H\}.$$

We know (see Kollár 1996) that there exists a countable set of projective schemes H_i, called Hilbert schemes, indexed by the Hilbert polynomial, i.e. essentially by the degree and the genus, parametrising the 1-dimensional subschemes S of X. (The idea of the construction of these schemes, due to Grothendieck, is as follows. Observe that by Riemann–Roch, thanks to a vanishing theorem for $H^1(S, \mathcal{O}_S(n))$, $n \geq n_0$ (for an n_0 depending only on the degree and the genus), the number

$$h^0(S, \mathcal{O}_S(n)) := \dim H^0(S, \mathcal{O}_S(n))$$

is independent of S for $n \geq n_0$. Moreover, the restriction map

$$H^0(X, \mathcal{O}_X(n)) \to H^0(S, \mathcal{O}_S(n))$$

is surjective. Finally, there exists some sufficiently large n, independent of S, such that $S \subset X$ is defined by equations of degree n. Thus, we have an injective map of the set of subschemes S of given degree and genus in the Grassmannian of subspaces of $H^0(X, \mathcal{O}_X(n))$, which to S associates

$$I_S(n) := \mathrm{Ker}\,(H^0(X, \mathcal{O}_X(n)) \to H^0(S, \mathcal{O}_S(n))).$$

One then produces explicit algebraic equations for the image of this injection, which equip this set with the structure of a closed algebraic subscheme of the Grassmannian.)

It follows from the definition of the Hilbert scheme that for $H \in U$, the subset

$$H_{i,H} := \{Z \in H_i \mid Z \subset X_H\}$$

is a closed algebraic subset of H_i. More generally, the subset

$$H_{i,U} := \{(Z, H) \in H_i \times U \mid Z \subset X_H\}$$

is a closed algebraic subset of $H_i \times U$, called the relative Hilbert scheme. In what follows, we are only interested in relative Hilbert schemes parametrising locally complete intersection schemes of dimension 1, i.e. divisors in the surfaces X_H.

As H_i is projective, the image of $H_{i,U}$ under the second projection onto U is a closed algebraic subset of U.

By construction, every pair (X_H, C) consisting of a curve C contained in a smooth hyperplane section X_H of X is parametrised by one of the Hilbert schemes $H_{i,U}$.

In the case where $X = \mathbb{P}^3$ embedded in $\mathbb{P}^N$ via the linear system $|\mathcal{O}_{\mathbb{P}^3}(d)|$, the corresponding surfaces X_H are the surfaces of degree d in $\mathbb{P}^3$.

Lemma 3.29 *A curve $Z \subset X_H$ (i.e. a subscheme of pure dimension 1) satisfies the condition*

$$[Z] = c_1(\mathcal{O}_{X_H}(k)) \quad in \quad H^2(X_H, \mathbb{Z})$$

if and only if there exists a hypersurface Y of degree k in $\mathbb{P}^3$ such that $Z = Y \cap X_H$.

Proof By the Lefschetz theorem on hyperplane sections, we know that X_H satisfies $H^1(X_H, \mathbb{Z}) = 0$, so we also have $H^1(X_H, \mathcal{O}_{X_H}) = 0$. It then follows from the exponential exact sequence that $[Z] = c_1(\mathcal{O}_{X_H}(k))$ if and only if we have the equality

$$\mathcal{O}_{X_H}(Z) = \mathcal{O}_{X_H}(k)$$

in the Picard group. Under this hypothesis, Z is thus the zero locus of a section of $\mathcal{O}_{X_H}(k)$. It then suffices to show that the restriction map

$$H^0(\mathcal{O}_{\mathbb{P}^3}(k)) \to H^0(\mathcal{O}_{X_H}(k))$$

is surjective. But this follows from the exact sequence

$$0 \to \mathcal{I}_{X_H}(k) \to \mathcal{O}_{\mathbb{P}^3}(k) \to \mathcal{O}_{X_H}(k) \to 0,$$

with $\mathcal{I}_{X_H}(k) = \mathcal{O}_{\mathbb{P}^3}(k-d)$, together with the fact that

$$H^1(\mathbb{P}^3, \mathcal{O}_{\mathbb{P}^3}(l)) = 0 \text{ for all } l \in \mathbb{Z}.$$

This last fact is part of Bott's vanishing theorem (see Griffiths & Harris 1978), and can also be obtained as a consequence of Kodaira's vanishing theorem for negative l, and Kodaira's theorem together with Serre duality for positive l, since $K_{\mathbb{P}^3} = \mathcal{O}_{\mathbb{P}^3}(-4)$. □

Lemma 3.30 *Consider the set C_U^k consisting of the pairs (Z, H) such that $Z = X_H \cap Y$, where $H \in U$ and Y is a hypersurface of $\mathbb{P}^3$ of degree k. C_U^k has an obvious scheme structure: it is a projective bundle above U, with fibre $|\mathcal{O}_{X_H}(k)|$ over $H \in U$. Then C_U^k is open in the component of the relative Hilbert scheme $H_{i,U}$ containing it.*

Proof It suffices to show that the image $\mathrm{pr}_1(C_U^k)$ in H_i is open. For this, it suffices to see that $\mathrm{pr}_1(C_U^k)$ and H_i have the same Zariski tangent space at Z. But

$$T_{H_i,Z} = \mathrm{Hom}_{\mathcal{O}_{\mathbb{P}^3}}(\mathcal{I}_Z, \mathcal{O}_Z) = H^0(Z, N_{Z/\mathbb{P}^3}),$$

and as Z is a complete intersection of two hypersurfaces of degree d and k, its normal bundle is equal to $\mathcal{O}_Z(d) \oplus \mathcal{O}_Z(k)$.

Moreover, we have a natural map

$$H^0(\mathbb{P}^3, \mathcal{O}_{\mathbb{P}^3}(d)) \oplus H^0(\mathbb{P}^3, \mathcal{O}_{\mathbb{P}^3}(k)) \to T_{C_U^k,(Z,H)},$$

which to (G, K) (considered as a pair of infinitesimal deformations of H and Y respectively) associates the corresponding infinitesimal deformation of the pair $(Y \cap X_H, H)$. Clearly, the composition of this arrow with pr_{1*} is the restriction

$$H^0(\mathbb{P}^3, \mathcal{O}_{\mathbb{P}^3}(d)) \oplus H^0(\mathbb{P}^3, \mathcal{O}_{\mathbb{P}^3}(k)) \to H^0(Z, \mathcal{O}_Z(d)) \oplus H^0(Z, \mathcal{O}_Z(k)).$$

We conclude by showing that for a complete intersection Z of dimension ≥ 1 in $\mathbb{P}^r$, the restriction arrow

$$H^0(\mathbb{P}^r, \mathcal{O}_{\mathbb{P}^r}(l)) \to H^0(Z, \mathcal{O}_Z(l))$$

is surjective for every $l \in \mathbb{Z}$, which follows easily from the vanishing property

$$H^i(\mathbb{P}^r, \mathcal{O}_{\mathbb{P}^r}(l')) = 0, \quad 0 < i < r, \quad l' \in \mathbb{Z}.$$

□

We deduce from this lemma that $H_{i,U} - C_U^k$ is closed in $H_{i,U}$, as is its projection via pr_2 onto U.

Definition 3.31 *The Noether–Lefschetz locus for surfaces of* $\mathbb{P}^3$ *is the countable union of the closed algebraic subsets* $\mathrm{pr}_2(H_{i,U} - C_U^k)$ *if* $C_U^k \subset H_{i,U}$, *and* $\mathrm{pr}_2(H_{i,U})$ *if* $H_{i,U}$ *contains no subset* C_U^k.

In other words, by lemma 3.29, it is the set of smooth surfaces X_H of degree d containing a curve Z which is not of the form $X_H \cap Y$, or whose class $[Z]$ is not proportional to $c_1(\mathcal{O}_{X_H}(1))$.

Every irreducible component of one of the closed algebraic sets $\mathrm{pr}_2(H_{i,U} - C_U^k)$ and $\mathrm{pr}_2(H_{i,U})$ as above is called a component of the Noether–Lefschetz locus.

The Noether–Lefschetz locus defined above has the following specific property in the case of surfaces of $\mathbb{P}^3$: a curve $C \subset S$ is of class proportional to a multiple of the class of a plane section if and only if it is a complete intersection $C = S \cap Y$. Forgetting this last point, and more generally, considering the open set U parametrising the surfaces X_H which are smooth hyperplane sections of a 3-dimensional variety X, we can define the Noether–Lefschetz locus $\mathrm{NL}(U) \subset U$ to be the set of surfaces X_H such that the restriction map $\mathrm{Pic}\, X \to \mathrm{Pic}\, X_H$ is not surjective. Using the Lefschetz theorem on hyperplane sections, the exponential exact sequence, and the fact that the quotient $H^2(X_H, \mathbb{Z})/H^2(X, \mathbb{Z})$ is torsion-free, we see that it is also the set of surfaces X_H such that the restriction map $\mathrm{NS}(X) \otimes \mathbb{Q} \to NS(X_H) \otimes \mathbb{Q}$ is not surjective, where the Néron–Severi group $\mathrm{NS}(Y)$ is defined as the image of the map $c_1 : \mathrm{Pic}\, Y \to H^2(Y, \mathbb{Z})$.

3.3.2 The Noether–Lefschetz theorem

Theorem 3.32 *With notation as in the preceding section, for $d \geq 4$, the components of the Noether–Lefschetz locus are proper algebraic subsets of U.*

A surface S which is general of degree at least 4, i.e. which corresponds to a point in moduli space lying outside a countable union of proper closed algebraic subsets, thus satisfies the property that all of its curves are complete intersections $S \cap Y$. The condition $d \geq 4$ is essential. Indeed, for $d = 3$, a surface S of degree 3 in $\mathbb{P}^3$ contains lines, as is easily shown by dimension-counting. In fact it contains exactly 27 lines (see exercise 1). But such a line $l \subset S$ is not a complete intersection $S \cap Y$. Indeed, otherwise we would have

$$\deg l := [l] \cup c_1(\mathcal{O}_{\mathbb{P}^3}(1)) = [S] \cup [Y] \cup c_1(\mathcal{O}_{\mathbb{P}^3}(1)) \quad \text{in} \quad H^6(\mathbb{P}^3, \mathbb{Z}) = \mathbb{Z},$$

so $\deg l$ would be divisible by 3 while in fact $\deg l = 1$.

The Noether–Lefschetz theorem admits the following generalisation in the case of smooth hyperplane sections $X_H \subset X$ of a smooth connected 3-dimensional variety embedded via a linear system $|L|$.

Theorem 3.33 *If the very ample linear system $|L|$ on X is such that the smooth surfaces $X_H \in |L|$ satisfy $H^{2,0}(X_H) \cap H^2(X_H, \mathbb{C})_{\mathrm{van}} \neq 0$, then the Noether–Lefschetz locus $\mathrm{NL}(U) \subset U$ is a countable union of proper algebraic subsets of U.*

The application to the case of $\mathbb{P}^3$ can be shown by noting that the hypothesis of theorem 3.33 for surfaces of degree d in $\mathbb{P}^3$ is equivalent to $d \geq 4$ by the adjunction formula and the fact that $K_{\mathbb{P}^3} = \mathcal{O}_{\mathbb{P}^3}(-4)$.

Proof of theorem 3.33 We want to show that for every relative Hilbert scheme $K = H_{i,U}$ as above, the projection $\mathrm{pr}_2 : K \to U$ cannot be dominant unless the pairs (C, H), $C \subset X_H$ parametrised by $H_{i,U}$ satisfy $c_1(C) \in \mathrm{NS}(X)_{|X_H} \otimes \mathbb{Q}$. Assume, then, that $\mathrm{pr}_2 : K \to U$ is dominant. Clearly, up to replacing U by a Zariski open set V and K by a closed algebraic subset, we may assume that K is smooth and $\mathrm{pr}_2 : K \to V$ is a proper étale cover of degree s. (We first reduce to the case where pr_2 is generically finite, then to the étale case by Sard's theorem.)

For every $H \in V$, the fibre $\mathrm{pr}_2^{-1}(H)$ parametrises curves $Z_{1,H}, \ldots, Z_{s,H} \subset X_H$. Clearly, the classes

$$\alpha_{1,H}, \ldots, \alpha_{s,H} \in H^2(X_H, \mathbb{Z})$$

of these curves, together with the space $\mathrm{NS}(X)_{|X_H}$, generate over $\mathbb{Q}$ a local subsystem F of the local system $R^2\phi_*\mathbb{Q}_{|V}$, where

$$\phi : \mathcal{X}_U \to U, \quad \mathcal{X}_U \subset X \times U$$

is the universal hypersurface.

Let $F' \subset F$ be the local subsystem consisting of the vanishing classes, or equivalently the intersection $F \cap (R^2\phi_*\mathbb{Q}_{\mathrm{van}})_{|V}$. As V is a Zariski open set of U, the map

$$\pi_1(V, 0) \to \pi_1(U, 0)$$

is surjective and corollary 3.28 remains valid on V. It follows that $F' = 0$ or $F' = (R^2\phi_*\mathbb{Q}_{\mathrm{van}})_{|V}$.

In the second case, we conclude in particular that the cohomology $H^2(X_H, \mathbb{Q})_{\mathrm{van}}$ for $H \in V$ is generated by divisor classes which are classes of type $(1, 1)$ in the Hodge decomposition. We thus have $H^{2,0}(X_H) \cap H^2(X_H, \mathbb{C})_{\mathrm{van}} = 0$, which contradicts the hypothesis of the theorem. In the first case, we claim that the classes $\alpha_{i,H}$ lie in $\mathrm{NS}(X)_{|X_H}$, which is what we wanted to prove.

Indeed, it follows from the Lefschetz theorem on $(1, 1)$-classes that the orthogonal complement of $\mathrm{NS}(X)_{|X_H}$ in $\langle \alpha_{i,H}, \mathrm{NS}(X)_{|X_H} \rangle$ with respect to the intersection pairing on $H^2(X_H, \mathbb{Q})$ is contained in $H^2(X_H)_{\mathrm{van}}$, hence in F'. Now this orthogonal complement is 0 if and only if the $\alpha_{i,H}$ belong to $\mathrm{NS}(X)_{|X_H}$. □

Exercises

1. *The 27 lines on a cubic surface.* Let $S \subset \mathbb{P}^3$ be a smooth surface of degree 3 defined by an equation $f = 0$, $f \in H^0(\mathbb{P}^3, \mathcal{O}_{\mathbb{P}^3}(3))$.
   (a) Let $G := \mathrm{Grass}\,(2, 4)$ be the Grassmannian parametrising the vector subspaces of rank 2 of V, $\mathbb{P}^3 = \mathbb{P}(V)$, or the projective lines

$\mathbb{P}^1 \subset \mathbb{P}^3$. Show that the Plücker embedding (see vI.10.1.1)

$$G \to \mathbb{P}\left(\bigwedge^2 V\right) = \mathbb{P}^5, \quad \Delta \mapsto e_1 \wedge e_2,$$

where e_1, e_2 form a basis of Δ, identifies G with a quadric of $\mathbb{P}^5$.

(b) Let $\mathcal{E}$ be the bundle of rank 2 on G which is dual to the tautological subbundle, i.e. whose fibre at Δ is the space Δ^* of linear forms on Δ. Let $\sigma \in H^0(\mathbb{P}^3, \mathcal{O}_{\mathbb{P}^3}(1))$. Show that the locus of zeros of the induced section $\tilde{\sigma}$ of $\mathcal{E}$, defined by

$$\tilde{\sigma}_\Delta = \sigma_{|\Delta},$$

is a projective plane $P_\sigma \subset Q$ (the dual plane of the plane defined by σ).

Recall that if $\mathcal{F}$ is a holomorphic vector bundle of rank r on a variety X, the Chern class $c_r(\mathcal{F})$ is equal to the Euler class of $\mathcal{F}$, so that if τ is a holomorphic section of $\mathcal{F}$ whose zero locus W is of codimension r, then the class of the cycle associated to W (see vI.11.1.2) is equal to $c_r(\mathcal{F})$ in $H^{2r}(X, \mathbb{Z})$. See also Exercise 11.3 of volume 1.

(c) Deduce from this that $c_2(\mathcal{E}) = [P_\sigma]$.

(d) For a complex vector bundle $\mathcal{F}$ of rank 2, express the Chern classes $c_4(\mathrm{Sym}^3\mathcal{F})$ as a polynomial in the Chern classes $c_1(\mathcal{F})$, $c_2(\mathcal{F})$, where $\mathrm{Sym}^3\mathcal{F}$ is the third symmetric product of $\mathcal{F}$. (Use the splitting principle vI.11.2.1.)

(e) Deduce from this that the bundle $\mathcal{E}$ on Q satisfies $c_4(\mathrm{Sym}^3\mathcal{E}) = 27$, where

$$c_4(\mathrm{Sym}^3\mathcal{E}) \in H^8(Q, \mathbb{Z}) \cong \mathbb{Z}.$$

(f) Show that for generic f, the section $\tilde{f}$ of $\mathrm{Sym}^3\mathcal{E}$ given by

$$\tilde{f}_\Delta = f_{|\Delta},$$

where we use the isomorphism $S^3H^0(\mathbb{P}^1, \mathcal{O}_{\mathbb{P}^1}(1)) \cong H^0(\mathbb{P}^1, \mathcal{O}_{\mathbb{P}^1}(3))$, vanishes transversally on Q, i.e. its zero locus is a reduced finite set. (Use Sard's theorem and the fact that the set

$$Z \subset G \times H^0(\mathbb{P}^3, \mathcal{O}_{\mathbb{P}^3}(3))$$
$$Z = \{(\Delta, f) \mid f_{|\Delta} = 0\}$$

is smooth of codimension 4.)

(g) Deduce from this that for generic f, S contains 27 lines.

(h) Using the adjunction formula, the fact that $\deg K_{\mathbb{P}^1} = -2$ and the fact that $K_S = \mathcal{O}_S(1)$, prove that these lines satisfy

$$\langle \Delta, \Delta \rangle = -1.$$

(i) Show that two lines Δ, Δ' have non-proportional cohomology classes in $H^2(S, \mathbb{Z})$.

(j) Deduce from theorem 3.27 that, taken over all pairs of lines Δ, Δ' in S, the classes $[\Delta] - [\Delta']$ generate the vanishing cohomology $H^2(S, \mathbb{Q})_{\mathrm{van}}$.

NB. In fact, these statements hold for every smooth cubic surface.

2. *The monodromy action on the points of a generic hyperplane section of an irreducible curve.* Let $X \subset \mathbb{P}^N$ be an irreducible reduced curve of degree d. For K in a Zariski open set U of $(\mathbb{P}^N)^*$, the set $X_K := K \cap X$ (which does not intersect the singular locus Sing X) then consists of d distinct points. If $0 \in U$, we have a monodromy representation

$$\rho : \pi_1(U, 0) \to S_d,$$

where S_d is the symmetric group acting on the set X_{K_0}. We explain this action by noting that by the definition of U, the set

$$\widetilde{U} := \{(x, K) \in X \times U \mid x \in X_K\}$$

is a topological cover of U, so that the parallel transport of points of X_K along a path in U makes sense.

(a) Let $(K_t)_{t\in\mathbb{P}^1}$ be a pencil of hyperplanes such that $\bigcap_{t\in\mathbb{P}^1} K_t \cap X = \emptyset$. This pencil then induces a map

$$f : X \to \mathbb{P}^1.$$

Show that we can choose this pencil in such a way that for a point $t_0 \in \mathbb{P}^1$, the fibre K_{t_0} is contained in the smooth part of X and consists of $d - 1$ points, of which one (called x_0) has multiplicity 2. Show that for a small disk $\Delta \subset \mathbb{P}^1$, $f^{-1}(\Delta)$ consists of $d - 1$ disks, of which $d - 2$ are diffeomorphic to Δ via f, and that f makes the last one, centred at x_0, into a double cover of Δ. For a suitable coordinate centred at x_0, the map f is then of the form $u \mapsto u^2$ on this last disk.

(b) Deduce that the image of ρ contains a transposition.

(c) Show that the group $\operatorname{Im} \rho$ acts transitively on K_0. (Prove that for $x \in K_0$, the set of $y \in X_{\mathrm{smooth}} := X - \operatorname{Sing} X$ such that there exists a path in U transporting x to y is closed and open in X_{smooth}, and use the fact that the irreducibility of X is equivalent to the connectedness of X_{smooth} (see vI.11.1.1).)

(d) Show similarly that the group $\operatorname{Im}\rho$ acts bitransitively on K_0. (If $d = 1$, there is nothing to show; otherwise, apply the above reasoning to the set $(\mathbb{P}^N)^*_x$ of hyperplanes passing through $x \in K_0$, introducing the monodromy action $\rho' : \pi_1(U', 0) \to S_{d-1}$, where U' is the intersection of U with $(\mathbb{P}^N)^*_x$ and S_{d-1} is the symmetric group acting on $K_0 - x$.)

(e) Show that a subgroup of the symmetric group S_d of permutations of the set of d points K_0 which contains a transposition and acts bitransitively on K_0 is the whole symmetric group.

We have now proved the following result, which is called the *uniform position principle*, and is very useful in practice.

Theorem 3.34 (see Arbarello *et al.* 1985, p. 111; Harris 1979) *The monodromy acts on the points of a generic hyperplane section X_K of an irreducible projective curve X of degree d as the symmetric group S_d of permutations of X_K.*

4
The Leray Spectral Sequence

Except for the second proof of the Lefschetz theorem, which uses the vanishing theorems, we did not make any real use of Hodge theory in the preceding chapters. In fact, one can even consider that the study of the monodromy of Lefschetz pencils and its application to the Noether–Lefschetz theorem constitute something closer to an application of topology to Hodge theory. In the present chapter, however, we will illustrate the consequences of Hodge theory on the topology of families of algebraic varieties.

The Leray spectral sequence shows that the cohomology of a locally trivial fibration $\phi : X \to Y$ can be computed starting from the cohomology of Y with values in the local systems $R^q\phi_*\mathbb{Z}$, in the sense that there exists a filtration L (called the Leray filtration) on $H^n(X, \mathbb{Z})$ such that each graded group $\mathrm{Gr}_L^p H^{p+q}(X, \mathbb{Z})$ is a quotient of a subgroup of $H^p(Y, R^q\phi_*\mathbb{Z})$. However, giving only the groups $H^p(Y, R^q\phi_*\mathbb{Z})$, or even just the local systems $R^q\phi_*\mathbb{Z}$ on Y, does not suffice to determine the cohomology of X. In fact, we have the Leray spectral sequence, whose term $E_2^{p,q}$ is equal to $H^p(Y, R^q\phi_*\mathbb{Z})$ and whose term $E_\infty^{p,q}$ is equal to $\mathrm{Gr}_L^p H^{p+q}(X, \mathbb{Z})$, and the piece of additional data needed to compute the last group is just the differentials d_r of the Leray spectral sequence. The simplest example of a non-trivial fibration, given by the sphere bundles, shows that these differentials d_r can be non-zero for arbitrarily large r.

A first important theorem, which should be compared to the formality theorem, and which gives a remarkable topological property of families of projective varieties, is the following.

Theorem 4.1 (Deligne) *The Leray spectral sequence with rational coefficients of a projective fibration $\phi : X \to Y$ degenerates at E_2.*

The argument is formal and uses only the hard Lefschetz theorem, which gives a relative Lefschetz decomposition, i.e. a decomposition on the local systems

$R^q\phi_*\mathbb{Q}$. The existence of this decomposition, which is compatible (in a sense to be made precise) with the Leray spectral sequence, implies that it suffices to prove that the arrows d_r, $r \geq 2$, vanish on the groups $H^p(Y, R^q\phi_*\mathbb{Q}_{\text{prim}})$, which follows from their compatibility with the relative Lefschetz isomorphisms.

This argument is valid over any base (it is not even necessary to assume that Y is complex). Moreover, the only property used is the existence of the Lefschetz isomorphisms, which makes moderate use of Hodge theory.

When we assume furthermore that ϕ is a morphism of quasi-projective varieties, mixed Hodge theory combined with the degeneracy theorem above gives the following result. Let $\phi : X \to Y$ be a projective morphism between projective varieties, and let U be the open set of Y consisting of the regular values of ϕ. Let

$$\rho : \pi_1(U, y) \to \operatorname{Aut} H^k(X_y, \mathbb{Q}), \quad y \in U$$

be the monodromy representation.

Theorem 4.2 (Deligne) *The image of the restriction map*

$$H^k(X, \mathbb{Q}) \to H^k(X_y, \mathbb{Q})$$

is the invariant cohomology group $H^k(X_y, \mathbb{Q})^{\text{inv}}$.

If we replace X by $X_U := \phi^{-1}(U)$, this follows from the degeneracy at E_2 of the Leray spectral sequence. The additional ingredient needed here is the existence of a mixed Hodge structure on the cohomology of X_U, and also the strictness of the morphisms of mixed Hodge structures, which enable us to show that the images of the restrictions

$$H^k(X, \mathbb{Q}) \to H^k(X_y, \mathbb{Q}), \quad H^k(X_U, \mathbb{Q}) \to H^k(X_y, \mathbb{Q})$$

coincide.

The first section of this chapter is devoted to the construction of the Leray spectral sequence, which is itself a special case of the spectral sequence of a composed functor.

In the second section, we explain the compatibility of the Lefschetz decomposition with the Leray spectral sequence, and prove the degeneracy theorem 4.1. We conclude the chapter with an introduction to mixed Hodge structures and the proof of theorem 4.2.

4.1 Definition of the spectral sequence

4.1.1 The hypercohomology spectral sequence

Let $\mathcal{A}$, $\mathcal{B}$ be abelian categories, and let F be a left exact functor from $\mathcal{A}$ to $\mathcal{B}$. Assume that $\mathcal{A}$ has sufficiently many injective objects, so that for a left bounded complex $(M^\bullet, d)$ of objects of $\mathcal{A}$, we have the derived objects

$$R^i F(M^\bullet) \in \mathrm{Ob}(\mathcal{B}).$$

We will now construct a canonical spectral sequence

$$E_r^{p,q} \Rightarrow R^{p+q} F(M^\bullet)$$

with

$$E_2^{p,q} = R^p F(H^q(M^\bullet)).$$

For this, we need the following result.

Proposition 4.3 *There exists an injective resolution*

$$(M^{\bullet,\bullet}, D_1, D_2),\ i : M^\bullet \hookrightarrow M^{\bullet,0}$$

of the complex $M^\bullet$ having the following properties: for every $l \in \mathbb{Z}$, the induced complexes $(B^{l,\bullet}, D_2)$ form an injective resolution of B^l, the induced complexes $(Z^{l,\bullet}, D_2)$ form an injective resolution of Z^l, and the complexes $(H^l(M^{\bullet,\bullet}, D_1), D_2)$ form an injective resolution of $H^l(M^\bullet)$.

Moreover, such a resolution satisfies the following universal property: for every morphism of complexes $\phi : N^\bullet \to M^\bullet$, and every resolution $(N^\bullet, d_N) \hookrightarrow (N^{\bullet,\bullet}, d_1, d_2)$, there exists a morphism of double complexes $\tilde{\phi} : N^{\bullet,\bullet} \to M^{\bullet,\bullet}$ extending ϕ.

Here, the complexes $(Z^{l,\bullet}, D_2)$ are defined as the kernel $\mathrm{Ker}\, D_1 \subset M^{l,\bullet}$ equipped with the differential induced by D_2, and the complexes $(B^{l,\bullet}, D_2)$ are defined as the image $\mathrm{Im}\, D_1 \subset M^{l,\bullet}$ equipped with the differential induced by D_2. Similarly, $Z^l = \mathrm{Ker}\, d \subset M^l$ and $B^l = \mathrm{Im}\, d \subset M^l$.

Proof Let us first construct the complex of injective objects $(M^{\bullet,0}, D_1)$ and the injective morphism

$$i : (M^\bullet, d) \hookrightarrow (M^{\bullet,0}, D_1).$$

We require that $B^{i,0}$ and $Z^{i,0}$ be injective and that the induced morphisms

$$H^i(M^\bullet, d) = Z^i / B^i \to H^i(M^{\bullet,0}, D_1) = Z^{i,0} / B^{i,0}$$

be injective.

To construct such a complex $(M^{i,0}, D_1)$, we note that given an injection

$$i : B^l \to R^l$$

into an injective object, there exist injective objects L^l and K^l and a commutative diagram

$$\begin{array}{ccccc} B^l & \hookrightarrow & Z^l & \hookrightarrow & M^l \\ {\scriptstyle i}\downarrow & & {\scriptstyle i}\downarrow & & {\scriptstyle i}\downarrow \\ R^l & \hookrightarrow & L^l & \hookrightarrow & K^l, \end{array}$$

in which all the vertical arrows i are injective and satisfy the property:
(*) *The induced morphisms* $Z^l/B^l \to L^l/R^l$ *and* $M^l/Z^l \to K^l/L^l$ *are injective.*

Indeed, we first take an injection

$$j : Z^l \hookrightarrow L_1^l$$

into an injective object. The inclusion $B^l \hookrightarrow Z^l$ then extends to an arrow $t : R^l \to L_1^l$, by the universal property of injective objects. Furthermore, let $k : R^l/B^l \hookrightarrow L_2^l$ be an injection into an injective object, and set

$$L_3^l = L_1^l \oplus L_2^l, \quad i' = (j, 0), \quad s = (t, k \circ \pi) : R^l \to L_3^l,$$

where $\pi : R^l \to R^l/B^l$ is the quotient map. Then clearly s is injective, and the following diagram is commutative:

$$s : \qquad \begin{array}{ccc} B^l & \hookrightarrow & Z^l \\ {\scriptstyle i}\downarrow & & \downarrow{\scriptstyle i'} \\ R^l & \hookrightarrow & L_3^l. \end{array}$$

Finally, take an injection $m : Z^l/B^l \to L_4^l$ into an injective object. Set

$$L^l = L_3^l \oplus L_4^l, \quad i = (i', m \circ \pi) : Z^l \to L^l,$$

where $\pi : Z^l \to Z^l/B^l$ is the quotient map. The injection $s' = (s, 0) : R^l \hookrightarrow L^l$ makes the following diagram commutative:

$$s' : \qquad \begin{array}{ccc} B^l & \hookrightarrow & Z^l \\ {\scriptstyle i}\downarrow & & \downarrow{\scriptstyle i} \\ R^l & \hookrightarrow & L^l, \end{array}$$

and clearly the induced map $Z^l/B^l \to L^l/R^l$ is now injective.

We proceed similarly to construct the inclusion $i : M^l \hookrightarrow K^l$.

The complex $M^{\bullet,0}$ is then constructed inductively. Let p be such that $M^i = 0$, $i < p$. We then set $R^p = 0$ and $M^{p,0} = K^p$, where K^p is constructed as above. By hypothesis, we have an induced inclusion

$$B^{p+1} \cong M^p/Z^p \hookrightarrow K^p/L^p.$$

But this last quotient is injective, as it is a quotient of two injective objects. Setting $R^{p+1} = K^p/L^p$, we construct K^{p+1} as above. We then set $M^{p+1,0} = K^{p+1}$, and

$$D^1 : M^{p,0} \to M^{p+1,0}$$

is the composition

$$K^p \to K^p/L^p = R^{p+1} \hookrightarrow K^{p+1}.$$

Then Ker $D_1 = L^p$ is indeed injective. In general, as the complex $M^{\bullet,0}$ and the injection i are constructed to order k, as well as the injective objects

$$L^i \subset M^{i,0},\ i \leq k, \quad L^i = \operatorname{Ker} D_1,\ i < k$$
$$R^i \subset M^{i,0},\ i \leq k, \quad R^i = \operatorname{Im} D_1,\ i \leq k$$

satisfying the conditions

$$i_l : M^l \hookrightarrow M^{l,0},\ \ i_l : Z^l \hookrightarrow L^l \ \text{ for } l \leq k$$

and

$$i_l : M^l/Z^l \hookrightarrow M^{l,0}/L^l,\ \ i_l : Z^l/B^l \hookrightarrow L^l/R^l \ \text{ for } l \leq k,$$

we set $R^{k+1} = M^{k,0}/L^k$. By hypothesis, the morphism $B_{k+1} = M^k/Z^k \to R^{k+1}$ is injective. We then construct L^{k+1}, K^{k+1} as above and set $M^{k+1,0} = K^{k+1}$, and

$$D^1 : M^{k,0} \to M^{k+1,0}$$

is the composition

$$M^{k,0} \to M^{k,0}/L^k = R^{k+1} \hookrightarrow M^{k+1,0}.$$

The complex of injective objects $(M^{\bullet,0}, D_1)$ and the inclusion i then satisfy the conclusion that $Z^{i,0} = L^i$ is injective, $B^{i,0} = R^i$ is injective and $H^i(M^{\bullet,0})$ is injective, as the quotient of two injective objects. Finally, the map

$$H^i(M^\bullet) \to H^i(M^{\bullet,0})$$

induced by i is injective by the property (*).

To construct the desired resolution, it suffices to apply the preceding construction to the quotient complex

$$M^{\bullet,0}/M^{\bullet}$$

and so on. To see that all the conditions are satisfied, we simply note that this complex satisfies

$$H^k(M^{\bullet,0}/M^{\bullet}) = H^k(M^{\bullet,0})/H^k(M^{\bullet}),$$

which follows from the fact that the arrows

$$H^k(M^{\bullet}, d) \to H^k(M^{\bullet,0}, D_1)$$

are injective.

We leave it to the reader to check the universal property. □

This resolution now allows us to construct a spectral sequence, which is canonical starting from E_2, and abutting to the hypercohomology $R^k F(M^{\bullet})$.

Theorem 4.4 *There exists a spectral sequence*

$$E_r^{p,q} \Rightarrow R^{p+q} F(M^{\bullet}),$$

and thus a filtration F on $R^{p+q}F(M^{\bullet})$, such that $\mathrm{Gr}_F^p R^{p+q} F(M^{\bullet}) = E_{\infty}^{p,q}$. This spectral sequence is canonical starting from $E_2^{p,q}$, and satisfies

$$E_2^{p,q} = R^p F(H^q(M^{\bullet})).$$

Proof Consider the double complex of $\mathcal{B}$ obtained by applying the functor F to the resolution $(M^{\bullet,\bullet}, D_1, D_2)$:

$$(N^{\bullet,\bullet}, \delta_1, \delta_2) = (F(M^{\bullet,\bullet}), F(D_1), F(D_2)).$$

Writing $(S^{\bullet}, D)$ for the associated simple complex, we have

$$R^i F(M^{\bullet}) = H^i(S^{\bullet}, D)$$

by definition (see vI.8.1.2). Furthermore, the complex $S^{\bullet}$ admits the filtration G by the second index

$$S^n = \bigoplus_{p+q=n} N^{p,q}, \quad G^r(S^n) = \bigoplus_{p+q=n, q\geq r} N^{p,q}.$$

We thus have a corresponding spectral sequence

$$E_r^{p,q} \Rightarrow H^{p+q}(S^{\bullet}, D) = R^{p+q} F(M^{\bullet}),$$

and a filtration G on the cohomology $H^i(S^\bullet, D)$, whose associated graded object is

$$\mathrm{Gr}_G^p H^{p+q}(S^\bullet, D) = E_\infty^{p,q} = E_r^{p,q}$$

for sufficiently large r. The computation of the first terms of the spectral sequence of a double complex (see vI.8.3.2) shows that we have

$$E_0^{p,q} = N^{q,p}, \quad d_0 = \delta_1.$$

But we know that the complex $(M^{q,p}, D_1)$ is a complex of injective objects, whose boundaries $B^{p,q}$ and cycles $Z^{p,q}$ are also injective objects. It follows easily that we have

$$H^q(F(M^{\bullet,p}), \delta_1) = F(H^q(M^{\bullet,p}, D_1)),$$

and thus

$$E_1^{p,q} = H^q(N^{\bullet,p}, \delta_1) = H^q(F(M^{\bullet,p}), \delta_1) = F(H^q(M^{\bullet,p}, D_1)).$$

Moreover, we know that $d_1 : E_1^{p,q} \to E_1^{p+1,q}$ is simply induced by δ_2:

$$d_1 = H^q(\delta_2) : H^q(N^{\bullet,p}, \delta_1) \to H^q(N^{\bullet,p+1}, \delta_1).$$

But as noticed above,

$$H^q(N^{\bullet,p}, \delta_1) = F(H^q(M^{\bullet,p}, D_1)),$$

and the differential induced by δ_2 is equal to $F(D_2^q)$, where

$$D_2^q : H^q(M^{\bullet,\bullet}, D_1) \to H^q(M^{\bullet,\bullet+1}, D_1)$$

is induced by D_2. As the complex $H^q(M^{\bullet,\bullet}, D_1)$ equipped with the differential D_2^q is an injective resolution of $H^q(M^\bullet, d)$, we see that by definition of the derived functor $R^i F$, we have

$$E_2^{p,q} = H^p\big(E_1^{\bullet,q}, d_1\big) = H^p\big(F(H^q(M^{\bullet,\bullet}, D_1)), F\big(D_2^q\big)\big) = R^p F(H^q(M^\bullet)).$$

This concludes the computation of the term $E_2^{p,q}$. □

We now show that this filtration and this spectral sequence (starting from the term E_2) depend on the complex $(M^\bullet, d)$ only up to quasi-isomorphism of complexes, in the sense that a quasi-isomorphism

$$\phi : (M^\bullet, d) \to (K^\bullet, d_K)$$

induces a filtered isomorphism

$$R^{p+q} F(M^\bullet) \to R^{p+q} F(K^\bullet)$$

and an isomorphism of spectral sequences

$$\phi_r^{p,q} : \left({}_M E_r^{p,q}, d_r\right) \cong \left({}_K E_r^{p,q}, d_r\right)$$

starting from the term E_2.

Here, an isomorphism between filtered objects (A, F), (B, G) is an isomorphism $\phi : A \cong B$ which is compatible with the filtrations and induces an isomorphism $F^p A \cong G^p B$ for every p.

For this, we first show using the notation above that the morphism ϕ extends to a morphism of double complexes $\phi' : N^{p,q} \to N_K^{p,q}$, which can be done using the universal property of the resolutions of proposition 4.3. (Here and in what follows, the lower index K denotes the objects $N^{\bullet,\bullet}$, $S^\bullet$ constructed for the complex K.) This morphism thus gives a morphism of filtered complexes

$$\phi'' : S \to S_K,$$

and thus a morphism $\phi_r^{p,q}$ between the corresponding spectral sequences. Now, it is clear that at E_2, this morphism

$$\phi_2^{p,q} : R^p F(H^q(M^\bullet, d)) \to R^p F(H^q(K^\bullet, d_K))$$

is simply induced by applying the derived functor $R^q F$ to the morphism $H^q(\phi)$.

In our case, since ϕ is a quasi-isomorphism, $\phi_2^{p,q}$ must be an isomorphism. The following lemma then suffices to conclude the proof of theorem 4.4.

Lemma 4.5 *Let $\psi : S^\bullet \to S_K^\bullet$ be a morphism of filtered complexes bounded below and satisfying the convergence property*

$$G^l S^p = 0 \ \forall l \gg p, \quad G^l S_K^p = 0 \ \forall l \gg p.$$

Then, if the morphism of spectral sequences induced by ψ is an isomorphism at E_2, ψ induces an isomorphism of spectral sequences for $r \geq 2$ and an isomorphism of filtered objects

$$H^q(\psi) : H^q(S^\bullet) \to H^q(S_K^\bullet).$$

□

Proof As each $E_r^{p,q}$ is the cohomology of the preceding complex

$$d_{r-1} : E_{r-1}^{p,q} \to E_{r-1}^{p+r-1,q-r+2},$$

we see immediately that all the $\psi_r^{p,q}$ are isomorphisms for $r \geq 2$, and thus $\psi_\infty^{p,q}$ is an isomorphism. We thus have a morphism of filtered objects

$$H^q(\psi) : H^q(S^\bullet) \to H^q(S_K^\bullet)$$

whose filtrations are bounded and which induces an isomorphism on the associated graded object. It is clear by induction on the length of the filtration that it is an isomorphism of filtered objects. □

Remark 4.6 *To obtain this spectral sequence, we can replace the injective resolution with a resolution $(M^{\bullet,\bullet}, D_1, D_2)$ by F-acyclic objects in proposition 4.3, such that each complex $(Z^{l,\bullet}, D_2)$ is an F-acyclic resolution of $Z^l(M^\bullet)$, each complex $(B^{l,\bullet}, D_2)$ is an F-acyclic resolution of $B^l(M^\bullet)$ and each complex $(H^l(M^\bullet, D_1), D_2)$ is an F-acyclic resolution of $H^l(M^\bullet)$. Indeed, this follows from the universal property of the resolutions in proposition 4.3, together with the fact that derived functors can be computed using F-acyclic resolutions.*

Consider the special case where the M^p are F-acyclic for every p. We know (see vI.8.1.2) that

$$R^n F(M^\bullet) = H^n(F(M^\bullet)).$$

Thus, in particular, we obtain the following result.

Theorem 4.7 *Let $(M^\bullet, d)$ be a complex of F-acyclic objects. Then there exist a natural filtration L on $H^{p+q}(F(M^\bullet), F(d))$, and a spectral sequence $E_r^{p,q} \Rightarrow R^{p+q} F(M^\bullet)$, with*

$$E_2^{p,q} = R^p F(H^q(M^\bullet)), \quad E_\infty^{p,q} = \mathrm{Gr}_L^p R^{p+q} F(M^\bullet).$$

Finally, the universal property of the resolutions of proposition 4.3 implies the functoriality of the spectral sequence given in theorem 4.4.

Proposition 4.8 *If $\phi : M^\bullet \to K^\bullet$ is a morphism of complexes, then ϕ induces a morphism ϕ_r of spectral sequences which is canonical starting from E_2, and the morphism*

$$R^k F(M^\bullet) \to R^k F(K^\bullet)$$

induced by $F(\phi)$ is compatible with the filtrations defined above. The induced morphism at the level of the graded object is equal to the morphism ϕ_∞.

Indeed, with notation as above, ϕ extends to a morphism of filtered complexes

$$N^\bullet \to N_K^\bullet,$$

which gives the desired morphism of spectral sequences. The morphism ϕ_2 : $E_r^{p,q}(M) \to E_r^{p,q}(K)$ is then equal to the morphism

$$R^p F(H^q(M^\bullet)) \to R^p F(H^q(K^\bullet))$$

induced by $H^q(\phi)$, and thus ϕ_2 and ϕ_r for $r \geq 2$ are determined by ϕ. □

4.1.2 Spectral sequence of a composed functor

Theorem 4.7 applies particularly to the computation of the derived functors of a composed functor. We now assume that we have three abelian categories $\mathcal{A}$, $\mathcal{B}$, $\mathcal{C}$, where $\mathcal{A}$ and $\mathcal{B}$ have sufficiently many injective objects, and left exact functors

$$F : \mathcal{A} \to \mathcal{B}, \quad F' : \mathcal{B} \to \mathcal{C}.$$

We assume that the functor F transforms the injective objects of $\mathcal{A}$ into F'-acyclic objects of $\mathcal{B}$. If $M \in \mathrm{Ob}(\mathcal{A})$, then the derived functors $R^i(F' \circ F)(M)$ are by definition computed as follows. Let

$$(M^\bullet, d),\ i : M \hookrightarrow M^0$$

be an injective resolution of M; then

$$R^i(F' \circ F)(M) = H^i(F' \circ F(M^\bullet), F' \circ F(d)).$$

But furthermore, the complex $(F(M^\bullet), F(d))$ is a complex of F'-acyclic objects of $\mathcal{B}$, and we have

$$H^q(F(M^\bullet), F(d)) = R^q F(M).$$

Thus we can apply theorem 4.7 to the complex $(F(M^\bullet), F(d))$, which gives the following result.

Theorem 4.9 *There is a canonical filtration L on the objects $R^i(F' \circ F)(M)$, and a spectral sequence*

$$E_r^{p,q} \Rightarrow R^{p+q}(F' \circ F)(M)$$

with

$$E_2^{p,q} = R^p F'(R^q F(M)), \quad E_\infty^{p,q} = \mathrm{Gr}_L^p R^{p+q}(F' \circ F)(M).$$

We can even add that since the complex $(F(M^\bullet), F(d))$ is well-defined up to quasi-isomorphism of complexes of F'-acyclic objects, this spectral sequence starting from E_2 and the corresponding filtration on $R^i(F' \circ F)(M)$ are canonical.

This spectral sequence is called the spectral sequence of a composed functor.

Example. Consider the category $\mathcal{M}_{X_{an}}$ or $\mathcal{M}_{X_{alg}}$ of sheaves of $\mathcal{O}_X$-modules on an analytic or algebraic variety, i.e. in the analytic case, the sheaves of $\mathcal{O}_{X_{an}}$-modules for the usual topology, and in the second case the sheaves of $\mathcal{O}_{X_{alg}}$-modules for the Zariski topology. We will simply use the notation $\mathcal{O}_X$ and $\mathcal{M}_X$. If $\mathcal{G} \in \mathrm{Ob}\,\mathcal{M}_X$, we have the functor

$$\underline{\mathrm{Hom}}\,(\mathcal{G}, \cdot) : \mathcal{M}_X \to \mathcal{M}_X$$

which to $\mathcal{F} \in \mathrm{Ob}\,\mathcal{M}_X$ associates the sheaf of $\mathcal{O}_X$-modules $\underline{\mathrm{Hom}}\,(\mathcal{G}, \mathcal{F})$, associated to the presheaf

$$U \mapsto \mathrm{Hom}_{\mathcal{O}_X(U)}\,(\mathcal{G}(U), \mathcal{F}(U)).$$

Furthermore, we also have the functor

$$\mathrm{Hom}\,(\mathcal{G}, \cdot) : \mathcal{M}_X \to \mathcal{E},$$

where $\mathcal{E}$ is the category of $\mathbb{C}$-vector spaces, which to $\mathcal{F} \in \mathrm{Ob}\,\mathcal{M}_X$ associates

$$\mathrm{Hom}_{\mathcal{O}_X}(\mathcal{G}, \mathcal{F}) := \Gamma(\underline{\mathrm{Hom}}\,(\mathcal{G}, \mathcal{F})).$$

By definition, the functor $\mathrm{Hom}\,(\mathcal{G}, \cdot)$ is a composed functor, and in order to apply the preceding results, it suffices to show that this functor transforms injective sheaves into acyclic sheaves for the functor Γ. For this, by vI.4.3.1, it suffices to prove the following result.

Lemma 4.10 *If $\mathcal{F}$ is injective, then the sheaf $\underline{\mathrm{Hom}}_{\mathcal{O}_X}(\mathcal{G}, \mathcal{F})$ is flasque.*

Proof Saying that $\mathcal{F}$ is injective is equivalent to saying that for every injective morphism $A \to B$ of sheaves of $\mathcal{O}_X$-modules, the induced arrow

$$\mathrm{Hom}_{\mathcal{O}_X}(B, \mathcal{F}) \to \mathrm{Hom}_{\mathcal{O}_X}(A, \mathcal{F})$$

is surjective.

If $j : U \to X$ is the inclusion of an open set, apply this to the sheaf $A = j_!\mathcal{G}_{|U}$ (the extension by 0 of $\mathcal{G}_{|U}$), which is the sheaf associated to the presheaf

$$V \mapsto 0 \text{ if } V \not\subset U, \quad V \mapsto \mathcal{G}(V) \text{ if } V \subset U.$$

We have a natural inclusion

$$j_!\mathcal{G}_{|U} \subset \mathcal{G}.$$

The injectivity of $\mathcal{F}$ thus gives the surjectivity of the morphism

$$\mathrm{Hom}_{\mathcal{O}_X}(\mathcal{G}, \mathcal{F}) \to \mathrm{Hom}_{\mathcal{O}_X}(j_!\mathcal{G}_{|U}, \mathcal{F}).$$

Now, clearly, this arrow can be identified with the restriction arrow

$$\mathrm{Hom}_{\mathcal{O}_X}(\mathcal{G}, \mathcal{F}) \to \mathrm{Hom}_{\mathcal{O}_U}(\mathcal{G}_{|U}, \mathcal{F}_{|U}),$$

i.e. with the restriction

$$\Gamma\big(X, \underline{\mathrm{Hom}}_{\mathcal{O}_X}(\mathcal{G}, \mathcal{F})\big) \to \Gamma\big(U, \underline{\mathrm{Hom}}_{\mathcal{O}_X}(\mathcal{G}, \mathcal{F})\big).$$

Thus, $\underline{\mathrm{Hom}}_{\mathcal{O}_X}(\mathcal{G}, \mathcal{F})$ is flasque. □

Thus, for every pair of sheaves $\mathcal{G}$, $\mathcal{F} \in \mathcal{M}_X$, we have a 'spectral sequence of Ext', whose $E_2^{p,q}$ term is equal to $H^p(X, \mathcal{E}xt^q_{\mathcal{O}_X}(\mathcal{G}, \mathcal{F}))$, and which converges to $\mathrm{Ext}^{p+q}_{\mathcal{O}_X}(\mathcal{G}, \mathcal{F})$. Here, $\mathcal{E}xt^q_{\mathcal{O}_X}(\mathcal{G}, \cdot)$ denotes the qth derived functor of the functor $\underline{\mathrm{Hom}}_{\mathcal{O}_X}(\mathcal{G}, \cdot)$ and $\mathrm{Ext}^q_{\mathcal{O}_X}(\mathcal{G}, \cdot)$ denotes the qth derived functor of the functor $\mathrm{Hom}_{\mathcal{O}_X}(\mathcal{G}, \cdot)$.

4.1.3 The Leray spectral sequence

The Leray spectral sequence is a special case of the spectral sequence of a composed functor. Consider a continuous map $\phi : X \to Y$ between two topological spaces, and the corresponding functor ϕ_* from the category of sheaves of abelian groups on X to the category of sheaves of abelian groups on Y. It follows from the definition that if $\mathcal{F}$ is flasque on X, then $\phi_*\mathcal{F}$ is flasque on Y. Moreover, the functor $\Gamma(X, \cdot)$ of global sections on X is the composition of the functor ϕ_* with the functor $\Gamma(Y, \cdot)$ of global sections on Y. Thus, we can apply theorem 4.9.

Theorem 4.11 (Leray) *For every sheaf $\mathcal{F}$ on X, there exists a canonical filtration L on $H^q(X, \mathcal{F})$, and a spectral sequence $E_r^{p,q} \Rightarrow H^{p+q}(X, \mathcal{F})$ that is canonical starting from E_2, and satisfies*

$$E_2^{p,q} = H^p(Y, R^q\phi_*\mathcal{F}), \quad E_\infty^{p,q} = \mathrm{Gr}^p_L H^{p+q}(X, \mathcal{F}).$$

This spectral sequence is called the Leray spectral sequence.

Assume now that Y and X are differentiable varieties, and $\phi : X \to Y$ is a proper submersion. Recalling that we can compute the cohomology of X with coefficients in $\mathbb{R}$ using the de Rham complex

$$H^q(X, \mathbb{R}) = \frac{\mathrm{Ker}(d : A^q(X) \to A^{q+1}(X))}{\mathrm{Im}(d : A^{q-1}(X) \to A^q(X))}$$

of X, we will give another construction of the Leray spectral sequence in this case, for the cohomology with coefficients in $\mathbb{R}$. Consider the exact sequence of vector bundles on X

$$0 \to \phi^*\Omega_{Y,\mathbb{R}} \to \Omega_{X,\mathbb{R}} \to \Omega_{X/Y,\mathbb{R}} \to 0,$$

defining the bundle of real relative differential forms $\Omega_{X/Y,\mathbb{R}}$. It gives a decreasing filtration $L^\bullet\Omega^q_{X,\mathbb{R}}$ of the vector bundle $\Omega^q_{X,\mathbb{R}}$ defined by

$$L^p\Omega^q_{X,\mathbb{R}} = \phi^*\Omega^p_{Y,\mathbb{R}} \wedge \Omega^{q-p}_{X,\mathbb{R}},$$

whose associated graded vector bundle is given by

$$\mathrm{Gr}^p_L\Omega^q_{X,\mathbb{R}} = \phi^*\Omega^p_{Y,\mathbb{R}} \otimes \Omega^{q-p}_{X/Y,\mathbb{R}}.$$

We thus have the corresponding filtration on $A^q(X)$, which we denote in the same way; $L^pA^q(X)$ is the space of differentiable sections of the bundle Ω^q_X.

A form α in $L^pA^q(X)$ can be written locally as a combination $\sum_i \phi^*\omega_i \wedge \beta_i$, where ω_i is a differential form of degree $\geq p$ on Y. Leibniz' rule then shows that we also have $d\alpha \in L^pA^{q+1}(X)$. Thus, the complex $(A^\bullet(X), d)$ is filtered by L, and we have a corresponding spectral sequence

$$E_r^{p,q} \Rightarrow H^{p+q}(A^\bullet(X), d) = H^{p+q}(X, \mathbb{R}).$$

We then obtain the following result, whose proof uses certain facts concerning the de Rham complex of a flat bundle, which are proved in vI.9.2.2 and in subsection 5.1.1 of the next chapter.

Theorem 4.12 *Starting at E_2, this spectral sequence coincides with the Leray spectral sequence of the constant sheaf $\mathbb{R}$ on X with respect to the map ϕ.*

Proof The complex of sheaves $(\mathcal{A}^q, d)$ of differential forms on X is a fine and thus acyclic resolution of $\mathbb{R}$, and the sheaves $\phi_*\mathcal{A}^q$ are fine and thus acyclic on Y (see vI.4.3.1). If we take a resolution $\mathcal{A}^{\bullet,\bullet}$ of the complex $(\phi_*\mathcal{A}^q, d)$ by Γ-acyclic sheaves which satisfies the properties of proposition 4.3 (or more precisely those of remark 4.6), then the Leray spectral sequence is by definition induced by the filtration by the second index, which we call H, on the simple complex B associated to the double complex $A^{\bullet,\bullet}$ of global sections of $\mathcal{A}^{\bullet,\bullet}$. Let us take for $\mathcal{A}^{\bullet,\bullet}$ a canonical resolution, for example the Čech resolution, which is the limit of the Čech resolutions on all covers. Then the filtration induced by L on each $\phi_*\mathcal{A}^p$ extends to a filtration L on each $\mathcal{A}^{p,q}$, compatible with the two differentials $D_1 = d,\ D_2 = \delta$. Now consider the following filtration G on $\mathcal{B}^l = \bigoplus_{p+q=l}\mathcal{A}^{p,q}$:

$$G^r\mathcal{B}^l = \bigoplus_{p+q=l} L^{r-q}\mathcal{A}^{p,q}.$$

This filtration is compatible with the two differentials. Moreover, we have two natural inclusions of filtered complexes:

$$(\phi_* \mathcal{A}^q, d, L) \hookrightarrow (\mathcal{B}^q, D, G) \tag{4.1}$$

and

$$(\mathcal{B}^q, D, H) \hookrightarrow (\mathcal{B}^q, D, G), \tag{4.2}$$

and by lemma 4.5, it suffices to show that on the level of the complexes of global sections, which we denote respectively by (A^l, d, L), (B^l, D, H) and (B^l, D, G), each of these inclusions induces an isomorphism on the terms E_2 of these spectral sequences.

Let us first check that the term $E_2^{p,q}$ of the spectral sequence of the complex (A, d, L) has the desired form: the graded sheaves $\mathrm{Gr}_L^p(\phi_* \mathcal{A}^{p+q}, d)$ are equal to $\mathcal{A}_Y^p \otimes \phi_* \mathcal{A}_{X/Y}^q$, and the differential induced by d (and thus equal to the differential d_0) is equal to the relative differential $d_{X/Y}$. We thus find that for the spectral sequence associated to L on $A^q(X)$, we have

$$E_1^{p,q} = A_Y^p(R^q \phi_* \mathbb{R}) := \Gamma\big(Y, \mathcal{A}_Y^p \otimes R^q \phi_* \mathbb{R}\big).$$

Furthermore, we shall see (see proposition 5.5) that the differential d_1 is exactly the differential induced at the level of the global sections by the Gauss–Manin connection

$$\nabla : R^q \phi_* \mathbb{R} \otimes \mathcal{A}_Y^p \to R^q \phi_* \mathbb{R} \otimes \mathcal{A}_Y^{p+1}.$$

To see this d_1 is induced by d in the following way. A form in $A_Y^p(R^q \phi_* \mathbb{R})$ is represented by a form α of degree $p+q$ on X in L^p, and killed by $d_{X/Y}$. Then we consider $d\alpha$, which lies in L^{p+1}, and we set

$$d_1 \alpha = d\alpha \bmod L^{p+2}. \tag{4.3}$$

The equality of (4.3) with $\nabla\alpha$ is one way of formulating the Cartan–Lie formula of vI.9.2.2.

As $d_1 = \nabla$ and the de Rham complex of the flat bundle $R^q \phi_* \mathbb{R}$ is an acyclic resolution of the locally constant sheaf $R^{q-p} \phi_* \mathbb{R}$ (see proposition 5.3), we conclude that for the filtered complex (A^l, d, L), we have

$$E_2^{p,q} = H^p(Y, R^q \phi_* \mathbb{R}).$$

Now consider the complex (B^l, D, G). We have

$$E_0^{p,l-p} = \mathrm{Gr}_G^p B^l = \bigoplus\nolimits_{r+q=l} L^{p-q} A^{r,q} / L^{p-q+1} A^{r,q}$$

$$= \bigoplus\nolimits_{r+q=l} C^q\big(Y, \mathcal{A}_Y^{p-q} \otimes \phi_* \mathcal{A}_{X/Y}^{r-p+q}\big),$$

and the differential d_0 is induced by $D = d + \delta$. Clearly, we actually have $\delta = 0$ and $d = d_{X/Y}$ in the quotient, since $\delta L^p \subset L^p$. Thus,

$$E_1^{p,l-p} = \bigoplus_{r+q=l} C^q\big(Y, \mathcal{A}_Y^{p-q} \otimes R^{r-p+q}\phi_*\mathbb{R}\big),$$

and the differential d_1 is induced by $D = \nabla + \delta$. Therefore, it is in fact the double complex obtained by taking the Čech resolution of the complex $\mathcal{A}_Y^p \otimes R^{l-p}\phi_*\mathbb{R}$ equipped with the Gauss–Manin differential. As this is a complex of fine sheaves, the inclusion

$$\Gamma\big(\mathcal{A}_Y^p \otimes R^{l-p}\phi_*\mathbb{R}, \nabla\big) \hookrightarrow \Big(\bigoplus_q C^q\big(Y, \mathcal{A}_Y^{p-q} \otimes R^{l-p}\phi_*\mathbb{R}\big), \nabla + \delta\Big)$$

is a quasi-isomorphism, so the inclusion (4.1) does induce an isomorphism of spectral sequences starting from E_2.

Finally, consider $(\mathrm{Gr}_H^p B^l, d_0)$. By definition, it is equal to

$$(A^{l-p,p}, d) = C^p\big(\phi_*\mathcal{A}_X^{l-p}, d\big),$$

and its cohomology is given by

$$E_1^{p,q} = C^p(R^q\phi_*\mathbb{R}),$$

where the differential d_1 is the Čech differential δ. But then it is clear that the inclusion (4.2) induces the inclusion

$$C^p(R^q\phi_*\mathbb{R}, \delta) \subset \Big(\bigoplus_{l \leq p} C^{p-l}\big(\mathcal{A}_Y^l(R^q\phi_*\mathbb{R})\big), \nabla + \delta\Big)$$

on the level of the terms $E_1^{p,q}$. But this is a quasi-isomorphism, since the de Rham complex associated to ∇ is a fine resolution of $R^q\phi_*\mathbb{R}$. □

Example: sphere bundles. Let $\phi : X \to Y$ be an oriented n-sphere bundle, S^n. By definition, there exists an oriented vector bundle E on Y of rank $n+1$, and a metric on E such that $X \subset E$ can be identified with the set of vectors of norm 1. The Leray spectral sequence of such a sphere bundle is necessarily very simple, since there is very little cohomology in the fibres. Indeed, we have

$$R^l\phi_*\mathbb{Z} = 0 \text{ for } l \neq 0, n, \quad R^l\phi_*\mathbb{Z} = \mathbb{Z} \text{ for } l = 0 \text{ or } n,$$

where the last equality is given by the choice of an orientation on the fibres. Thus, for the corresponding Leray spectral sequence, we have

$$E_2^{p,q} = 0 \text{ for } q \neq 0,\ n.$$

A spectral sequence satisfying this condition is called spherical. As d_r sends $E_r^{p,q}$ to $E_r^{p+r,q-r+1}$, we see that for $r \geq 2$, we have $d_r = 0$, $r \neq n+1$. Thus,

$E_2^{p,q} = \cdots = E_{n+1}^{p,q}$ and

$$E_\infty^{p,q} = H^p\big(E_{n+1}^{\bullet,q}, d_{n+1}\big) = \begin{cases} \operatorname{Ker} d_{n+1} & \text{if } q = n, \\ \operatorname{Coker} d_{n+1} & \text{if } q = 0. \end{cases}$$

The exact sequence

$$0 \to E_\infty^{k,0} \to H^k(X, \mathbb{Z}) \to E_\infty^{k-n,n} \to 0$$

combined with the preceding equalities thus gives the long exact sequence

$$\cdots H^k(X, \mathbb{Z}) \to H^{k-n}(Y, R^n\phi_*\mathbb{Z}) \stackrel{d_{n+1}}{\to} H^{k+1}(Y, R^0\phi_*\mathbb{Z}) \to H^{k+1}(X, \mathbb{Z}) \cdots.$$

Obviously, the last arrow is the arrow ϕ^*. If we furthermore identify $R^n\phi_*\mathbb{Z}$ and $R^0\phi_*\mathbb{Z}$ with $\mathbb{Z}$, setting $k = n$, we find a element $d_{n+1}(1) \in H^{n+1}(Y, \mathbb{Z})$. Using the description of the Leray spectral sequence via de Rham cohomology given above, we easily show that this element is equal to the Euler class of the vector bundle E (see Bott 1957).

Moreover, $d_{n+1}(1)$ in fact determines the entire long exact sequence above, thanks to the equality $d_{n+1}(\alpha) = \alpha \cup d_{n+1}(1)$ which we will prove in the following section (see lemma 4.13).

4.2 Deligne's theorem

4.2.1 The cup-product and spectral sequences

Consider as above a continuous map $\phi : X \to Y$ between two topological spaces X and Y, and let $\alpha \in H^k(X, \mathbb{Z})$ be a cohomology class. The cup-product by α gives a map

$$\alpha\cup : H^*(X, \mathbb{Z}) \to H^{*+k}(X, \mathbb{Z}). \tag{4.4}$$

Moreover, using the cup-product on the open sets $\phi^{-1}(U)$ for open subsets $U \subset Y$, the restrictions $\alpha_{|\phi^{-1}(U)}$ provide a morphism of sheaves which we denote by

$$\alpha\cup : R^l\phi_*\mathbb{Z} \to R^{l+k}\phi_*\mathbb{Z}, \tag{4.5}$$

since $R^l\phi_*\mathbb{Z}$ is the sheaf associated to the presheaf $U \mapsto H^l(f^{-1}(U), \mathbb{Z})$.

Lemma 4.13 *The cup-product $\alpha\cup$ induces an endomorphism of degree k of the Leray spectral sequence of ϕ, i.e. for each r it induces a morphism of* *complexes*

$$\alpha_r\cup : \left(E_r^{p,q}, d_r\right) \to \left(E_r^{p,q+k}, d_r\right)$$

which is the morphism induced in cohomology by (4.5) at E_2. The morphism $\alpha\cup$ of (4.4) is compatible with the Leray filtration, and on the graded group $E_\infty^{p,q}$ for the Leray filtration on $H^{p+q}(X, \mathbb{Z})$, the induced morphism is equal to $\alpha_\infty\cup$.

Proof To explicitly construct the cup-product, we can show that there exists a flasque or fine resolution $\mathbb{Z} \hookrightarrow \mathcal{A}^\bullet$ of the constant sheaf $\mathbb{Z}$ on X, equipped with a morphism of complexes

$$\mu : \mathcal{A}^\bullet \otimes \mathcal{A}^\bullet \to \mathcal{A}^\bullet, \tag{4.6}$$

which makes the diagram

$$\begin{array}{ccc} \mathbb{Z} \otimes \mathbb{Z} & \longrightarrow & \mathbb{Z} \\ \downarrow & & \downarrow \\ \mu : \mathcal{A}^\bullet \otimes \mathcal{A}^\bullet & \longrightarrow & \mathcal{A}^\bullet \end{array}$$

commute, where the top arrow is the multiplication, and is such that the induced arrow

$$\begin{array}{ccccc} \mathbb{H}^k(\mathcal{A}^\bullet) \otimes \mathbb{H}^l(\mathcal{A}^\bullet) & \longrightarrow & \mathbb{H}^{k+l}(\mathcal{A}^\bullet \otimes \mathcal{A}^\bullet) & \xrightarrow{\mu} & \mathbb{H}^{k+l}(\mathcal{A}^\bullet) \\ \| & & & & \| \\ H^k(X, \mathbb{Z}) \otimes H^l(X, \mathbb{Z}) & & \longrightarrow & & H^{k+l}(X, \mathbb{Z}) \end{array} \tag{4.7}$$

is equal to the cup-product on $H^*(X, \mathbb{Z})$.

In Godement (1958), such a map μ is constructed on the Čech complex. In the differentiable case and for the cohomology with real coefficients, we saw in vI.5.3.2 that the cup-product could be computed using the exterior product on the de Rham complex.

The class α is thus represented by a d_A-closed global section $\tilde{\alpha}$ of $\mathcal{A}^k$. As μ is a morphism of complexes, we have

$$d_A(\mu(\tilde{\alpha} \otimes \beta)) = \mu(d_A\tilde{\alpha} \otimes \beta) \pm \mu(\tilde{\alpha} \otimes d_A\beta),$$

and as $d_A\tilde{\alpha} = 0$, the morphism

$$\mu(\tilde{\alpha} \otimes \cdot) : \mathcal{A}^* \to \mathcal{A}^{*+k}$$

is (up to sign) a morphism of complexes, which induces the morphism $\alpha\cup$ in cohomology by (4.7).

Applying ϕ_* to the complex $\mathcal{A}^*$, we obtain a complex of Γ-acyclic sheaves $\phi_*\mathcal{A}^*$ on Y, whose hypercohomology is equal to $H^*(X, \mathbb{Z})$. The Leray spectral sequence of ϕ is then equal to the spectral sequence of this complex relative to the functor $\Gamma(Y, \cdot)$. The morphism $\mu(\tilde{\alpha} \otimes \cdot)$ gives a morphism of complexes of acyclic sheaves

$$\mu_{\tilde{\alpha}} : \phi_*\mathcal{A}^* \to \phi_*\mathcal{A}^{*+k},$$

which (by the preceding argument restricted to the open sets $\phi^{-1}(U)$ of X) induces the morphism (4.5) on the cohomology $R^*\phi_*\mathbb{Z}$ of $\phi_*\mathcal{A}^*$. In particular, passing to the cohomology, $\mu_{\tilde{\alpha}}$ induces

$$\alpha\cup : H^p(Y, R^q\phi_*\mathbb{Z}) \to H^p(Y, R^{q+k}\phi_*\mathbb{Z}). \tag{4.8}$$

But the functoriality (see proposition 4.8) of the spectral sequence of a complex of sheaves relative to a left exact functor shows that $\mu_{\tilde{\alpha}}$ induces a morphism of the Leray spectral sequences which is equal to (4.8) at $E_2^{p,q}$ and which is compatible in the limit with the morphism induced in cohomology by $\mu_{\tilde{\alpha}}$, which can be identified with $\alpha\cup$:

$$\begin{array}{lccc} \mu_{\tilde{\alpha}} : & \mathbb{H}^l(Y, \phi_*\mathcal{A}^*) & \longrightarrow & \mathbb{H}^{k+l}(Y, \phi_*\mathcal{A}^*) \\ & \| & & \| \\ \alpha\cup : & H^l(X, \mathbb{Z}) & \longrightarrow & H^{k+l}(X, \mathbb{Z}). \end{array}$$

□

4.2.2 The relative Lefschetz decomposition

Let $\phi : X \to Y$ be a submersive and proper morphism of complex varieties. Recall the following.

Definition 4.14 *ϕ is said to be projective if there exists a holomorphic embedding*

$$i : X \hookrightarrow Y \times \mathbb{P}^n$$

such that $\phi = \mathrm{pr}_1 \circ i$.

Giving such a embedding provides a class $\omega \in H^2(X, \mathbb{Z})$ defined by

$$\omega = (\mathrm{pr}_2 \circ i)^* c_1(\mathcal{O}_{\mathbb{P}^n}(1)).$$

As $\mathrm{pr}_2 \circ i_{|X_t}$ is a holomorphic immersion on each fibre X_t of ϕ, the restriction

$$\omega_t := \omega_{|X_t} \in H^2(X_t, \mathbb{Z})$$

is a Kähler class, and the morphism

$$\omega_t \cup : H^k(X_t, \mathbb{Q}) \to H^{k+2}(X_t, \mathbb{Q})$$

is a Lefschetz operator on $H^*(X_t, \mathbb{Q})$. Moreover, as ω is closed, it induces as above a morphism of local systems

$$L := \omega \cup : R^* \phi_* \mathbb{Q} \to R^{*+2} \phi_* \mathbb{Q},$$

which is equal to $L_t = \omega_t \cup$ on the stalk at the point t. The operator L is called the relative Lefschetz operator. If $n = \dim X_t$ is the relative dimension of ϕ, we know that L_t satisfies the hard Lefschetz theorem, i.e.

$$L_t^{n-k} : H^k(X_t, \mathbb{Q}) \to H^{2n-k}(X_t, \mathbb{Q}) \tag{4.9}$$

is an isomorphism for $k \leq n$.

We deduce from this the Lefschetz decomposition

$$H^k(X_t, \mathbb{Q}) = \bigoplus\nolimits_{2r \leq k} L^r H^{k-2r}(X_t, \mathbb{Q})_{\text{prim}} \quad \text{for } k \leq n, \tag{4.10}$$

where

$$H^{k-2r}(X_t, \mathbb{Q})_{\text{prim}} := \operatorname{Ker}\big(L^{n-k+2r+1} : H^{k-2r}(X_t, \mathbb{Q})_{\text{prim}} \\ \to H^{2n-k+2r+2}(X_t, \mathbb{Q})\big).$$

The corresponding decomposition for $k \geq n$ is obtained thanks to the isomorphism (4.9).

The relative Lefschetz operator thus gives relative Lefschetz isomorphisms

$$L^{n-k} : R^k \phi_* \mathbb{Q} \cong R^{2n-k} \phi_* \mathbb{Q}$$

and a relative Lefschetz decomposition

$$R^k \phi_* \mathbb{Q} = \bigoplus\nolimits_{2r \leq k} L^r R^{k-2r} \phi_* \mathbb{Q}_{\text{prim}}, \ k \leq n.$$

By lemma 4.13, the relative Lefschetz operator L and its powers L^k induce endomorphisms (of degree $2k$) of the Leray spectral sequence of ϕ, i.e. morphisms L_r^k of the complexes $(E_r^{p,q}, d_r)$ (of degree $2k$ on the second index), and the absolute Lefschetz operator L^k of cup-product with $[\omega]^k$ is compatible with the Leray filtration and with each L_∞^k on $H^*(X, \mathbb{Q})$. Finally, the Lefschetz isomorphism (4.9) shows that

$$L_2^{n-k} : H^l(Y, R^k \phi_* \mathbb{Q}) \to H^l(Y, R^{2n-k} \phi_* \mathbb{Q}) \tag{4.11}$$

is an isomorphism for $k \leq n$.

4.2.3 Degeneration of the spectral sequence

We will use the Lefschetz operators L_r^k introduced above to prove the following theorem, due to Deligne (1968).

Theorem 4.15 *If $\phi : X \to Y$ is a submersive projective morphism, the Leray spectral sequence of ϕ with rational coefficients degenerates at E_2.*

Proof Let us first show that $d_2 = 0$. For this, note that if $q \geq n$, we have the following commutative diagram:

$$\begin{array}{ccc} E_2^{p,2n-q} = H^p(Y, R^{2n-q}\phi_*\mathbb{Q}) & \xrightarrow{L_2^{q-n}} & E_2^{p,q} = H^p(Y, R^q\phi_*\mathbb{Q}) \\ \downarrow d_2 & & \downarrow d_2 \\ E_2^{p+2,2n-q-1} = H^{p+2}(Y, R^{2n-q-1}\phi_*\mathbb{Q}) & \xrightarrow{L_2^{q-n}} & E_2^{p+2,q-1} = H^{p+2}(Y, R^{q-1}\phi_*\mathbb{Q}), \end{array}$$

where the upper horizontal arrow is an isomorphism. It thus suffices to show that $d_2 = 0$ on $E_2^{p,q}$ with $q \leq n$. Next we have the decomposition

$$E_2^{p,q} = \bigoplus_{2r \leq q} L_2^r H^p(Y, R^{q-2r}\phi_*\mathbb{Q}_{\text{prim}})$$

induced by the relative Lefschetz decomposition, and it suffices to show that $d_2 = 0$ on $L_2^r H^p(Y, R^{q-2r}\phi_*\mathbb{Q}_{\text{prim}})$. As L_2^r commutes with d_2, it suffices to show that $d_2 = 0$ on $H^p(Y, R^{q-2r}\phi_*\mathbb{Q}_{\text{prim}}) \subset E_2^{p,q-2r}$.

Setting $k = q - 2r$, we have the following commutative diagram:

$$\begin{array}{cccc} L_2^{n-k+1}: & H^p(Y, R^k\phi_*\mathbb{Q}_{\text{prim}}) & \longrightarrow & H^p(Y, R^{2n-k+2}\phi_*\mathbb{Q}) \\ & \downarrow d_2 & & \downarrow d_2 \\ L_2^{n-k+1}: & H^{p+2}(Y, R^{k-1}\phi_*\mathbb{Q}_{\text{prim}}) & \longrightarrow & H^{p+2}(Y, R^{2n-k+1}\phi_*\mathbb{Q}). \end{array}$$

The upper arrow is 0 by definition of the primitive cohomology, while the lower arrow is the isomorphism (4.11). Thus, the first arrow d_2 is zero.

To show that the arrows d_r, $r \geq 2$ are also zero, we proceed in exactly the same way, using the morphisms of spectral sequences L_r^k and noting that if $d_s = 0$ for $2 \leq s < r$, then $E_r^{p,q} = E_2^{p,q}$, so that we can use the Lefschetz decomposition as above on $E_r^{p,q}$. □

Remark 4.16 *In this entire section, we could replace the hypothesis that ϕ is projective by the hypothesis that ϕ is Kähler in the following weak sense: there exists a cohomology class $\alpha \in H^2(X, \mathbb{R})$, whose restriction to a fibre X_t (assuming that Y is connected) is a Kähler class. Indeed, this suffices to obtain* *the relative Lefschetz decomposition. Of course, one has then to work throughout with the cohomology with real coefficients.*

4.3 The invariant cycles theorem

4.3.1 Application of the degeneracy of the Leray spectral sequence

If $\phi : X \to Y$ is a continuous map between two topological spaces, then for every sheaf $\mathcal{F}$ on X, we have a natural map

$$H^k(X, \mathcal{F}) \to \Gamma(Y, R^k\phi_*\mathcal{F})$$

given by the restrictions $H^k(X, \mathcal{F}) \to H^k(X_U, \mathcal{F}_{|X_U})$, U open in Y. This map can be interpreted using the Leray spectral sequence of $(\mathcal{F}, \phi)$ as follows: the right-hand term is the term $E_2^{0,k}$ of the spectral sequence. By definition of the $E_r^{p,q}$ and for reasons of degree, we now have

$$E_\infty^{0,k} \subset E_2^{0,k},$$

since no arrow d_r for $r \geq 2$ can lead to $E_r^{0,k}$. Finally, by the definition of a spectral sequence, the term $E_\infty^{0,k}$ is the first quotient of $H^k(X, \mathcal{F})$ for the Leray filtration. Thus, we have a composition arrow

$$H^k(X, \mathcal{F}) \to E_\infty^{0,k} \subset E_2^{0,k} = \Gamma(Y, R^k\phi_*\mathcal{F}).$$

It is not difficult to show that this is indeed the restriction arrow above.

Assume now that $\phi : X \to Y$ is a proper submersion between differentiable varieties. Then $R^k\phi_*\mathbb{Q}$ is a local system on Y.

Lemma 4.17 *If $\mathcal{L}$ is a local system on an arcwise connected topological space Y, and $y \in Y$, then $\Gamma(Y, \mathcal{L})$ can be identified with the space of invariants*

$$\mathcal{L}_y^{\text{inv}} = \{\alpha \in \mathcal{L}_y \mid \rho(\gamma)(\alpha) = \alpha, \quad \forall \gamma \in \pi_1(Y, y)\}$$

under the monodromy action $\rho : \pi_1(Y, y) \to \operatorname{Aut} \mathcal{L}_y$.

Proof This is a consequence of corollary 3.10, or rather of its functorial version. Indeed, a non-zero section of $\mathcal{L}$ can be identified with a non-zero morphism of local systems $\mathbb{Z} \to \mathcal{L}$. By restriction to y, this is equivalent to giving a morphism of stalks $\mathbb{Z}_y \to \mathcal{L}_y$ which is equivariant for the monodromy actions. As $\mathbb{Z}$ is a trivial system with trivial monodromy, such an arrow is equivalent to giving an element of $\mathcal{L}_y$ which is invariant under ρ. □

If we now combine the above with Deligne's degeneracy theorem, we obtain the following result.

Theorem 4.18 *Let $\phi : X \to Y$ be a projective and submersive morphism between complex varieties, and let $y \in Y$. Then for every k, the restriction map*

$$H^k(X, \mathbb{Q}) \to H^k(X_y, \mathbb{Q})^{\mathrm{inv}}$$

is surjective, where $H^k(X_y, \mathbb{Q})^{\mathrm{inv}}$ is the space of elements of the cohomology of the fibre $H^k(X_y, \mathbb{Q})$ which are invariants under the monodromy action.

Proof We know by theorem 4.15 that the Leray spectral sequence satisfies $E_\infty^{0,k} = E_2^{0,k}$, and thus by the description given above of the restriction arrow, we find that

$$H^k(X, \mathbb{Q}) \to \Gamma(Y, R^k\phi_*\mathbb{Q})$$

is surjective. We may then conclude by applying lemma 4.17. □

4.3.2 Some background on mixed Hodge theory

Let us recall the definition of a mixed Hodge structure.

Definition 4.19 *A rational (real) mixed Hodge structure of weight n is given by a $\mathbb{Q}$-vector space ($\mathbb{R}$-vector space) H equipped with an increasing filtration $W_i H$ called the weight filtration, and a decreasing filtration on $H_\mathbb{C} := H \otimes \mathbb{C}$, called the Hodge filtration $F^k H_\mathbb{C}$. The induced Hodge filtration on each $\mathrm{Gr}_i^W H$ is required to equip $\mathrm{Gr}_i^W H$ with a Hodge structure of weight $n + i$.*

Of course, these filtrations are also required to satisfy $F^i H = 0$ for sufficiently large i, $F^i H = H$ for sufficiently small i, and similarly $W_i H = 0$ for sufficiently small i, $W_i H = H$ for sufficiently large i.

Equivalently, for each i, k, we must have

$$\mathrm{Gr}_i^W H_\mathbb{C} = F^k \mathrm{Gr}_i^W H_\mathbb{C} \oplus \overline{F^{n+i-k+1} \mathrm{Gr}_i^W H_\mathbb{C}},$$

where

$$F^k \mathrm{Gr}_i^W H_\mathbb{C} = \mathrm{Im}\big(F^k H_\mathbb{C} \cap W_i H_\mathbb{C} \to \mathrm{Gr}_i^W H_\mathbb{C}\big).$$

We have the obvious notion of a morphism of mixed Hodge structures. Next recall that a morphism α of filtered vector spaces (U, F) and (V, G) is said to be strict if

$$\operatorname{Im}\alpha \cap G^p V = \alpha(F^p U).$$

It has been shown (vI.7.3.1) that the morphisms of rational Hodge structures are strict for the Hodge filtration.

Theorem 4.20 (Deligne 1971) *The morphisms*

$$\alpha : (H, W, F) \to (H', W', F')$$

of (rational or real) mixed Hodge structures are strict for the filtrations W and F.

This is a consequence of the following fact.

Lemma 4.21 *Let (H, W, F) be a mixed Hodge structure. There exists a decomposition as a direct sum*

$$H_{\mathbb{C}} = \bigoplus_{p,q} H^{p,q}, \tag{4.12}$$

with $H^{p,q} \subset F^p H_{\mathbb{C}} \cap W_{p+q-n} H_{\mathbb{C}}$, such that under the projection $W_{p+q-n} H_{\mathbb{C}} \to \operatorname{Gr}^W_{p+q-n} H_{\mathbb{C}}$, $H^{p,q}$ can be identified with

$$H^{p,q}\big(\operatorname{Gr}^W_{p+q-n} H_{\mathbb{C}}\big) := F^p \operatorname{Gr}^W_{p+q-n} H_{\mathbb{C}} \cap \overline{F^q \operatorname{Gr}^W_{p+q-n} H_{\mathbb{C}}}.$$

More generally, we have

$$W_i H_{\mathbb{C}} = \bigoplus_{p+q \le n+i} H^{p,q}, \tag{4.13}$$

$$F^i H_{\mathbb{C}} = \bigoplus_{p \ge i} H^{p,q}. \tag{4.14}$$

This decomposition is preserved by the morphisms of mixed Hodge structures.

Remark 4.22 *This is a generalised Hodge decomposition corresponding to a mixed Hodge structure. However, careful attention must be paid to the fact that unlike the pure case (where only one of the $\operatorname{Gr}^W_i H$ is non-zero), the decomposition above is not in general invariant under conjugation, i.e.*

$$H^{p,q} \neq \overline{H^{q,p}},$$

although this does become true after projection to $\operatorname{Gr}^W_{p+q} H_{\mathbb{C}}$.

Proof of theorem 4.20 If $l' \in \alpha(H_{\mathbb{C}}) \cap W_i H'$, let us write $l' = \alpha(l)$ and decompose $l = \sum l^{p,q}$ as in (4.12). Then $\alpha(l)$ admits the decomposition $\alpha(l) = \sum \alpha(l^{p,q})$ with $\alpha(l^{p,q}) \in H'^{p,q}$. But as $l' \in W_i' H_{\mathbb{C}}'$, we have $\alpha(l^{p,q}) = 0$ for $p+q > n+i$. Thus,

$$l' = \alpha\left(\sum_{p+q \leq n+i} l^{p,q}\right) \in \alpha(W_i H_{\mathbb{C}}).$$

Therefore,

$$\operatorname{Im} \alpha \cap W_i' H_{\mathbb{C}}' = \alpha(W_i H_{\mathbb{C}}).$$

It is then easy to see that this still holds when $\mathbb{C}$ is replaced by $\mathbb{R}$ or $\mathbb{Q}$.

The same argument shows that α is also strict for the filtration F. □

Proof of lemma 4.21 Using the lighter notation $H = H_{\mathbb{C}}$, set

$$H^{p,q} = F^p H \cap W_l H \cap \left(W_l H \cap \overline{F^q H} + \sum_{i \geq 2} W_{l-i} H \cap \overline{F^{q-i+1} H}\right)$$

for $l + n = p + q$. Clearly, under the projection

$$W_l H \to \operatorname{Gr}_l^W H,$$

$H^{p,q}$ is sent to $H^{p,q}(\operatorname{Gr}_l^W H)$. Moreover, by definition, we have $H^{p,q} \subset F^p H_{\mathbb{C}}$ and $H^{p,q} \subset W_{p+q-n} H_{\mathbb{C}}$. It thus remains to prove the equalities (4.13) and (4.14). Now, clearly it suffices to show that the map

$$\sigma : H^{p,q} \to H^{p,q}(\operatorname{Gr}_l^W H)$$

given by the projection above is an isomorphism for every p, q. Indeed, by the Hodge decomposition, this will show that for each $\operatorname{Gr}_l^W H$, the right and left-hand spaces in (4.13) and (4.14) have the same rank, and by induction on $p+q$, this implies the injectivity of the arrows

$$\bigoplus_{p+q \leq n+i} H^{p,q} \to W_i H_{\mathbb{C}}, \tag{4.15}$$

$$\bigoplus_{p \geq i} H^{p,q} \to F^i H_{\mathbb{C}}, \tag{4.16}$$

and thus also their surjectivity.

Injectivity In order to simplify the notation, we will write W_i for $W_i H_{\mathbb{C}}$ in what follows. Let $a \in \operatorname{Ker} \sigma = H^{p,q} \cap W_{l-1}$; we can write $a = b + \sum_{i \geq 2} a_i$ with $b \in W_{l-1} \cap \overline{F^q H}$, $a_i \in W_{l-i} \cap \overline{F^{q-i-1} H}$ and $a \in F^p H \cap W_{l-1}$. In particular, modulo W_{l-2}, we have $a = b$. Thus, the projections of a and b to

$\mathrm{Gr}_{l-1}^W H = W_{l-1}H/W_{l-2}H$ lie in $F^p\mathrm{Gr}_{l-1}^W H \cap \overline{F^q\mathrm{Gr}_{l-1}^W H}$. Now, the Hodge structure induced by F on W_{l-1}/W_{l-2} is of weight $n+l-1 < p+q$. Thus, these projections are zero, and in fact, we have $a \in W_{l-2}$, $b \in W_{l-2}$. Now, projecting the preceding equality modulo W_{l-3}, we find that $a = b + a_2$ mod W_{l-3}, with $a \in F^pH \cap W_{l-2}$, $b \in \overline{F^qH} \cap W_{l-2}$, $a_2 \in \overline{F^{q-1}H} \cap W_{l-2}$. But then, in $\mathrm{Gr}_{l-2}^W H$ the projection of a lies in

$$F^p\mathrm{Gr}_{l-2}^W H \cap \overline{F^{q-1}\mathrm{Gr}_{l-2}^W H},$$

and this is also zero, since the induced Hodge structure on $\mathrm{Gr}_{l-2}^W H$ is of weight $n+l-2 < p+q-1$. Continuing similarly, we end up with $a = 0$.

Surjectivity We use induction on l. Assume that the surjectivity is proved for $l' < l$. Then it follows that the arrows (4.15) are isomorphisms for $i < l$.

Now let $\alpha \in H^{p,q}\mathrm{Gr}_l^W H = F^p\mathrm{Gr}_l^W H \cap \overline{F^q\mathrm{Gr}_l^W H}$. There exists $a \in F^pH \cap W_l$ which projects onto α. Similarly, there exists $b \in \overline{F^qH} \cap W_l$ which projects onto α. Then $c = a - b \in W_{l-1}$. Let us write

$$c = \sum_{r+s\leq l-1+n} c^{r,s},$$

with $c_{r,s} \in H^{r,s}$. Set

$$c' = \sum_{r+s\leq l-1+n, r\geq p} c^{r,s}, \quad c'' = \sum_{r+s=l-1+n, r<p} c^{r,s}, \quad c''' = \sum_{r+s<l-1+n, r<p} c^{r,s}.$$

Then $c' \in W_{l-1} \cap F^pH$, $c''' \in W_{l-2}$, and furthermore, for $r+s = l-1+n$, we have $r < p \Rightarrow s \geq q$ since $l = p+q-n$. Thus, by definition of the $H^{r,s}$, we have

$$\begin{aligned} c'' &\in W_{l-1} \cap \overline{F^qH} + \sum_{i\geq 2} W_{l-1-i} \cap \overline{F^{q-i+1}H} \\ &\subset W_l \cap \overline{F^qH} \quad + \sum_{i\geq 2} W_{l-i} \cap \overline{F^{q-i+1}H}. \end{aligned}$$

Setting $a' = a - c'$, $b' = b + c''$, we thus have $a' - b' = c''' \in W_{l-2}$ and $a' = \alpha \bmod W_{l-1}$, $b' = \alpha \bmod W_{l-1}$. Moreover, $a' \in F^pH \cap W_l$ and $b' \in W_l \cap \overline{F^qH} + \sum_{i\geq 2} W_{l-i} \cap \overline{F^{q-i+1}H}$. Using the same reasoning with $c''' \in W_{l-2}$, and using the decomposition (4.13) for $i < l$, we can construct for each $i \geq 1$ an element

$$a_i \in F^pH \cap W_l, \quad b_i \in W_l \cap \overline{F^qH} + \sum_{i\geq 2} W_{l-i} \cap \overline{F^{q-i+1}H}$$

such that a_i and b_i project onto α and $a_i - b_i \in W_{l-i}$. Finally, for sufficiently large i, we conclude that

$$A := a_i = b_i \in F^pH \cap W_l \cap \left(W_l \cap \overline{F^qH} + \sum_{i \geq 2} W_{l-i} \cap \overline{F^{q-i+1}H} \right) = H^{p,q}.$$

Then $A \in H^{p,q}$ projects onto $\alpha \in H^{p,q}Gr_l^W H$. □

4.3.3 The global invariant cycles theorem

Recall (vI.8.4) that if X is a projective or Kähler variety and $j : U \to X$ is the inclusion of a Zariski open set, i.e. the complement of an algebraic (or analytic in the Kähler case) subset into X, then the cohomology groups $H^k(U, \mathbb{Q})$ are equipped with a mixed Hodge structure of weight k satisfying the following properties:

$$W_iH^k(U, \mathbb{Q}) = 0, \quad i < 0,$$

and

$$W_0H^k(U, \mathbb{Q}) = \operatorname{Im} j^* : H^k(X, \mathbb{Q}) \to H^k(U, \mathbb{Q}).$$

Now assume that $Y \subset X$ is a closed complex subvariety contained in U. Consider the restriction map

$$H^k(U, \mathbb{Q}) \to H^k(Y, \mathbb{Q}).$$

It is a morphism of mixed Hodge structures, as $H^k(Y, \mathbb{Q})$ is equipped with its Hodge structure, considered as a mixed Hodge structure with trivial filtration W: $W_0H^k(Y, \mathbb{Q}) = H^k(Y, \mathbb{Q})$ and $W_iH^k(Y, \mathbb{Q}) = 0$ for $i < 0$. Theorem 4.20 thus applies and gives the following result.

Proposition 4.23 *Under the preceding hypotheses, the restriction maps* $\alpha : H^k(U, \mathbb{Q}) \to H^k(Y, \mathbb{Q})$ *and* $\alpha \circ j^* : H^k(X, \mathbb{Q}) \to H^k(Y, \mathbb{Q})$ *have the same image.*

Proof We have

$$\begin{aligned} \operatorname{Im} \alpha &= \operatorname{Im} \alpha \cap W_0H^k(Y, \mathbb{Q}) \\ &= \alpha(W_0H^k(U, \mathbb{Q})) = \alpha(j^*H^k(X, \mathbb{Q})). \end{aligned}$$

□

This result combines with theorem 4.18 to give the global invariant cycles theorem, due to Deligne (1971).

Theorem 4.24 *Let $\phi : X \to Y$ be a proper dominant morphism between smooth projective varieties. Let $U \subset Y$ be a Zariski open set over which ϕ is a submersion. Then, writing X_t, $t \in U$ for the fibre $\phi^{-1}(t)$, the image of the restriction map*

$$H^k(X, \mathbb{Q}) \to H^k(X_t, \mathbb{Q})$$

is the invariant cohomology $H^k(X_t, \mathbb{Q})^{\mathrm{inv}}$ under the monodromy action $\pi_1(U, t) \to \mathrm{Aut}\, H^k(X_t, \mathbb{Q})$.

Proof By theorem 4.18, letting $X_U = \phi^{-1}(U) \subset X$, we know that the image of the restriction map $H^k(X_U, \mathbb{Q}) \to H^k(X_t, \mathbb{Q})$ is the invariant cohomology $H^k(X_t, \mathbb{Q})^{\mathrm{inv}}$. In order to conclude, it suffices then to apply proposition 4.23 to the inclusions $X_t \subset X_U \subset X$. □

Corollary 4.25 *Let $\phi : X \to Y$ be a submersive projective morphism, with X and Y smooth and quasi-projective. Then for $y \in Y$, the space of invariants $H^k(X_y, \mathbb{Q})^{\mathrm{inv}}$ under the monodromy action is a rational Hodge substructure of $H^k(X_y, \mathbb{Q})$.*

Proof We can find projective smooth compactifications $\overline{X}$ and $\overline{Y}$ of X and Y equipped with a morphism $\overline{\phi} : \overline{X} \to \overline{Y}$ extending ϕ. The corollary then follows from theorem 4.24 together with the fact that the image of a morphism of Hodge structures is a Hodge substructure (see vI.7.3.1). □

Exercises

1. *The Leray spectral sequence and coherent sheaves.* Let $f : X \to Y$ be a holomorphic, proper and submersive map between complex varieties. Let $\mathcal{F}$ be a holomorphic vector bundle on X.
   (a) Show that if $H^i(X_y, \mathcal{F}_{|X_y}) = 0$ for a point y of Y and an integer i, then for every y' in a neighbourhood of y, we also have $H^i(X_{y'}, \mathcal{F}_{|X_{y'}}) = 0$ (use the semicontinuity theorem (see vI.9.3.1)).
   (b) Show that if for every $y \in Y$, we have $H^i(X_y, \mathcal{F}_{|X_y}) = 0$, then $R^i f_* \mathcal{F} = 0$. (This statement is local, so one should work in open sets U satisfying the property that every $\overline{\partial}$-closed $(0, q)$-form of degree

$q > 0$ on U is $\overline{\partial}$-exact. Then filter the complex

$$\left(A_U^{0,\bullet}(\mathcal{F}), \overline{\partial}\right)$$

of $(0, q)$-forms on U with values in $\mathcal{F}$ by the Leray filtration, or rather, its antiholomorphic analogue. Study the corresponding spectral sequence.)

(c) Let $E \to Y$ be a holomorphic vector bundle, and let $\mathbb{P}(E) \to Y$ be the associated projective bundle. Show that for $q \geq 0$ and for every i, we have

$$H^i(\mathbb{P}(E), \mathcal{O}_{\mathbb{P}(E)}(q)) = H^i(Y, S^q E^*).$$

(Use (b) and the Leray spectral sequence.)

NB. The results of (a) and (b) still hold under much weaker hypotheses (flatness of $\mathcal{F}$).

2. *Leray spectral sequence and Euler–Poincaré characteristic.* Let $X \to Y$ be a topological fibration between compact topological varieties, where Y is also assumed to be connected. (This means that locally over Y, we have homeomorphisms

$$X \cong X_y \times Y, \quad X_y := f^{-1}(y).)$$

Let F denote the fibre, which is defined up to homeomorphism.

(a) Show that if the Leray spectral sequence of f degenerates at E_2, and Y is simply connected, then the restriction map

$$H^q(X, \mathbb{Q}) \to H^q(X_y, \mathbb{Q})$$

is surjective for every q. Deduce from the Leray–Hirsch theorem (see vI.7.3.3, vI.8.1.3), that we then have a (non-canonical) decomposition

$$H^q(X, \mathbb{Q}) = \bigoplus_{0 \leq r \leq q} H^r(F, \mathbb{Q}) \otimes H^{q-r}(Y, \mathbb{Q}).$$

(b) Deduce that under the same hypothesis, the Betti numbers of X, Y and F satisfy

$$b_q(X) = \sum_{0 \leq r \leq q} b_r(F) b_{q-r}(Y).$$

Show that the Euler–Poincaré characteristics

$$\chi_{\mathrm{top}} = \sum_i (-1)^i b_i$$

of X, Y and F then satisfy

$$\chi_{\text{top}}(X) = \chi_{\text{top}}(F)\chi_{\text{top}}(Y).$$

We propose to show that this formula still holds without the hypotheses of degeneracy of the Leray spectral sequence, and of simple connectedness of Y.

(c) Let $(A^\bullet, d)$ be a left and right bounded complex of $\mathbb{Q}$-vector spaces of finite rank. Show that

$$\sum_i (-1)^i \dim A^i = \sum_i (-1)^i \dim H^i(A^\bullet, d).$$

(d) Let $\mathbb{H}$ be a local system of $\mathbb{Q}$-vector spaces of finite rank on Y. Using the equalities

$$H^q(Y, \mathbb{H}) = \check{H}^q(\mathcal{U}, \mathbb{H})$$

and

$$H^q(Y, \mathbb{Q}) = \check{H}^q(\mathcal{U}, \mathbb{Q})$$

for a suitable (finite) covering $\mathcal{U}$ of Y (see vI.4.3.1), and applying (c) to the Čech complexes of $\mathbb{H}$ and $\mathbb{Q}$ relative to $\mathcal{U}$, show that

$$\chi(Y, \mathbb{H}) := \sum_i (-1)^i \dim H^i(Y, \mathbb{H}) = \chi_{\text{top}}(Y) \text{rank}\, \mathbb{H}.$$

(e) Deduce that if $f : X \to Y$ is a topological fibration as above, we have

$$\chi(Y, R^q f_* \mathbb{Q}) = b_q(F)\chi_{\text{top}}(Y).$$

(f) Deduce from part (e), together with the Leray spectral sequence of f and part (c), that

$$\chi_{\text{top}}(X) = \chi_{\text{top}}(F)\chi_{\text{top}}(Y).$$

Part II
Variations of Hodge Structure

5
Transversality and Applications

In this chapter, we introduce the main objects in the theory of variations of Hodge structure; the three following chapters will be devoted to studying them.

We have already introduced the local systems $R^k\phi_*\mathbb{Z}$ or $R^k\phi_*\mathbb{R}$ associated to a locally trivial fibration $\phi : X \to Y$. In vI.9.2, we considered the associated vector bundle

$$\mathcal{H}^k := R^k\phi_*\mathbb{Z} \otimes \underline{\mathcal{C}}^\infty$$

on Y, and its Gauss–Manin connection ∇. In the complex framework, we also consider the holomorphic vector bundle

$$\mathcal{H}^k := R^k\phi_*\mathbb{Z} \otimes \mathcal{O}_Y.$$

The local system $H^k_{\mathbb{C}} := R^k\phi_*\mathbb{C}$ is the sheaf of ∇-flat sections of $\mathcal{H}^k$.

More generally, we will consider the holomorphic de Rham complex $\mathrm{DR}(H^k)$ of the local system $H^k_{\mathbb{C}}$, which thanks to the flatness of ∇ gives a holomorphic resolution of $H^k_{\mathbb{C}}$. This complex admits a filtration $F^i\mathrm{DR}(H^k)$ induced by the Hodge filtration $F^i\mathcal{H}^k$, using Griffiths transversality

$$\nabla F^i\mathcal{H}^k \subset F^{i-1}\mathcal{H}^k \otimes \Omega_Y.$$

The complexes $\mathcal{K}_{p,q} := \mathrm{Gr}^p_F\mathrm{DR}(\mathcal{H}^k),\ p+q=k$ can be written

$$0 \to \mathcal{H}^{p,q} \xrightarrow{\overline{\nabla}} \mathcal{H}^{p-1,q+1} \otimes \Omega_Y \to \cdots \to \mathcal{H}^{p-r,q+r} \otimes \Omega^r_Y \to \cdots,$$

and the geometric interpretation of the cohomology of this complex is the basis of the results introduced in this part.

On the one hand, the differentials $\overline{\nabla}$ describe the infinitesimal variations of Hodge structure (IVHS) on the cohomology of the fibres X_y; on the other hand we can interpret the complexes $\mathcal{K}_{p,q}$ as the E_1 term of the spectral sequence of the, sheaves Ω_X^s, relative to the derived functors $R^l\phi_*$ and to the Leray filtration

$$L^i\Omega_X^s = \phi^*\Omega_Y^i \wedge \Omega_X^{s-i}.$$

The precise statement is as follows.

Proposition 5.1 *The complex $(E_1^{p,q}, d_1)$ (graded by p), which is the first term of the spectral sequence of the sheaf Ω_X^k relative to the derived functors $R^l\phi_*$ and the Leray filtration, can be identified with the complex $\mathcal{K}_{k,q}$ (graded by the integer r above).*

This enables us to define the first infinitesimal invariant of a Dolbeault cohomology class $\eta \in H^p(X, \Omega_X^q)$. In chapter 7, we will give a concrete interpretation of this invariant in the case where η is the class of a cycle $Z \subset X$ homologous to 0 on the fibres of ϕ. We can then identify the first infinitesimal invariant with the infinitesimal invariant of the normal function associated to the cycle Z.

In the first section of this chapter, we summarise the main results of chapter 10 of volume I, namely Griffiths transversality and the computation of the differential of the period map. We also define the complexes $\mathcal{K}_{p,q}$, which play an essential role later on.

In the second section, we introduce the holomorphic Leray spectral sequence and give the proof of proposition 5.1. We also introduce the notion of an infinitesimal invariant of a Dolbeault cohomology class on the total space.

The last section is devoted to the study of Hodge loci, which illustrate the notion of a variation of Hodge structure, and whose general properties are obtained as an application of transversality. Given a variation of Hodge structure parametrised by a base B, these Hodge loci are the (closed analytic) subsets of B on which a locally constant class remains of a given Hodge level.

5.1 Complexes associated to IVHS

5.1.1 The de Rham complex of a flat bundle

Let Y be a differentiable variety, or a complex variety, and let H be a local system of $\mathbb{R}$ or $\mathbb{C}$-vector spaces on Y. To H, we associate the sheaf

$$\mathcal{H} = H \otimes \underline{\mathcal{C}}^\infty,$$

(resp. $\mathcal{H} = H \otimes \mathcal{O}_Y$ in the complex case). These are locally free sheaves of $\underline{\mathcal{C}}^\infty$-modules (resp. $\mathcal{O}_Y$-modules), i.e. sheaves of differentiable (resp. holomorphic) sections of a vector bundle on Y. This vector bundle is equipped with a flat connection

$$\nabla : \mathcal{H} \to \mathcal{H} \otimes \mathcal{A}^1_Y$$

(see vI.9.2), called the Gauss–Manin connection (resp.

$$\nabla : \mathcal{H} \to \mathcal{H} \otimes \Omega_Y$$

in the holomorphic case). The connection ∇ is defined as being equal to the usual differentiation in every local trivialisation of the vector bundle $\mathcal{H}$ induced by a local trivialisation of the local system H.

The connection ∇ extends to the sections of $\mathcal{A}^k_Y(H) := \mathcal{H} \otimes \mathcal{A}^k_Y$ by Leibniz' rule

$$\nabla(\sigma \otimes \alpha) = \sigma \otimes d\alpha + \nabla\sigma \wedge \alpha \in \mathcal{H} \otimes \mathcal{A}^{k+1}_Y$$

for $\alpha \in \mathcal{A}^k_Y$, $\sigma \in \mathcal{H}$.

In the holomorphic case, we have a similar extension

$$\nabla : \mathcal{H} \otimes \Omega^k_Y \to \mathcal{H} \otimes \Omega^{k+1}_Y.$$

The flatness of ∇ translates to the equality $\nabla \circ \nabla = 0$, which is clearly satisfied since ∇ can be identified locally with the exterior differential in suitable trivialisations of $\mathcal{H}$.

This allows us to construct the de Rham complex (and the holomorphic de Rham complex) of H.

Definition 5.2 *The de Rham complex (resp. the holomorphic de Rham complex)* $\mathrm{DR}(H)$ *of* H *is the complex*

$$0 \to \mathcal{H} \stackrel{\nabla}{\to} \mathcal{H} \otimes \mathcal{A}^1_Y \stackrel{\nabla}{\to} \mathcal{H} \otimes \mathcal{A}^2_Y \to \cdots$$

(resp.

$$0 \to \mathcal{H} \stackrel{\nabla}{\to} \mathcal{H} \otimes \Omega^1_Y \stackrel{\nabla}{\to} \mathcal{H} \otimes \Omega^2_Y \to \cdots.)$$

In a local trivialisation

$$H \cong \mathbb{R}^N \text{ or } H \cong \mathbb{C}^N$$

of H, these complexes become

$$0 \to (\underline{\mathcal{C}}^\infty)^N \stackrel{\nabla}{\to} (\mathcal{A}^1_Y)^N \stackrel{\nabla}{\to} (\mathcal{A}^2_Y)^N \to \cdots,$$

and respectively, in the holomorphic case,

$$0 \to \mathcal{O}_Y^N \stackrel{\nabla}{\to} \left(\Omega_Y^1\right)^N \stackrel{\nabla}{\to} \left(\Omega_Y^2\right)^N \to \cdots,$$

i.e. the direct sum of N copies of the de Rham complex (resp. of the holomorphic de Rham complex) of Y. As we know (see vI.4.1.3) that the de Rham complex is a resolution of the constant sheaf, we obtain the following result.

Proposition 5.3 *The de Rham complex* DR(H) *is a resolution of H by sheaves of free $\underline{\mathcal{C}}^\infty$-modules (or of free $\mathcal{O}_X$-modules in the holomorphic case).*

In particular, as sheaves of $\underline{\mathcal{C}}^\infty$-modules are acyclic for the functor Γ, we obtain the following corollary.

Corollary 5.4 *Letting $A_Y^p(H)$ denote the space of global sections of $\mathcal{A}_Y^k(H)$, we have a natural isomorphism*

$$H^q(Y,H) = \frac{\operatorname{Ker}(\nabla : A_Y^q(H) \to A_Y^{q+1}(H))}{\operatorname{Im}(\nabla : A_Y^{q-1}(H) \to A_Y^q(H))}.$$

Now consider the case of a local system $H = R^p\phi_*\mathbb{R}$ or $H = R^p\phi_*\mathbb{C}$, where $\phi : X \to Y$ is a proper submersion of differentiable varieties. The following description of ∇ can be found in vI.9.2.2. Let σ be a $\mathcal{C}^\infty$ local section of H. By definition, σ is represented by a family of closed $\mathcal{C}^\infty$ forms $(\alpha_y)_{y\in Y}$ of degree p on the fibres X_y, in other words by a $d_{X/Y}$-closed section α of $\phi_*\mathcal{A}_{X/Y}^p$. Let $y \in Y$ and $u \in T_{Y,y}$. We have the Cartan–Lie formula

$$\nabla_u\sigma = \operatorname{class}((\operatorname{int}(\tilde{u})\, d\tilde{\alpha})_{|X_y}), \tag{5.1}$$

where $\tilde{\alpha}$ is a p-form on X, i.e. a section of $\mathcal{A}_X^p$, whose image in $\mathcal{A}_{X/Y}^p$ is equal to α, and $\tilde{u}$ is a vector field on X, defined along X_y and lifting u, i.e. satisfying the equality $\phi_*\tilde{u} = u$.

This can be formulated in another way using the Leray filtration

$$L^l\mathcal{A}_X^p = \phi^*\mathcal{A}_Y^l \wedge \mathcal{A}_X^{p-l}$$

introduced in subsection 4.1.3. Note that we have

$$\mathcal{H}^p = R^p\phi_*(\mathcal{A}_{X/Y}^\bullet) = R^p\phi_*(L^0/L^1\mathcal{A}_X^\bullet),$$

so it is also the term $E_1^{0,p}$ of the spectral sequence of the complex $(\mathcal{A}_X^\bullet, d)$ equipped with the Leray filtration, for the functor ϕ_*. Similarly, we have

$$\mathcal{A}_Y^1 \otimes \mathcal{H}^p = E_1^{1,p}$$

for this spectral sequence. Moreover, the differential

$$d_1 : E_1^{0,p} \to E_1^{1,p}$$

is described as follows (see vI.8.3.2). As above, the section σ of $\mathcal{H}^p$ admits a representative which is a $d_{X/Y}$-closed section α of $\phi_*(L^0\mathcal{A}_X^p/L^1\mathcal{A}_X^p)$. The lifting $\tilde{\alpha}$ of α to $\phi_*\mathcal{A}_X^p$ thus satisfies the property that the projection of $d\tilde{\alpha}$ to $\mathcal{A}_{X/Y}^{p+1} = L^0\mathcal{A}_X^{p+1}/L^1\mathcal{A}_X^{p+1}$ is zero. Thus, $d\tilde{\alpha} \in L^1\mathcal{A}_X^{p+1}$ and its projection to

$$\phi_*\big(L^1\mathcal{A}_X^{p+1}/L^2\mathcal{A}_X^{p+1}\big) = \mathcal{A}_Y^1 \otimes \phi_*\mathcal{A}_{X/Y}^p$$

is $d_{X/Y}$-closed, since it is the projection of a d-closed form. This then gives the section $\beta = d_1\sigma$ of

$$\mathcal{A}_Y^1 \otimes R^p\phi_*\mathbb{R}.$$

Proposition 5.5 *The section β thus constructed is equal to $\nabla\sigma$. In other words, ∇ can be identified with the differential d_1 of the spectral sequence of the complex $(\mathcal{A}_X^\bullet, d)$ equipped with the Leray filtration, for the functor ϕ_*.*

Proof This follows immediately from the Cartan–Lie formula (5.1) and the fact that the identification

$$\phi_*\big(L^1\mathcal{A}_X^{p+1}/L^2\mathcal{A}_X^{p+1}\big) = \mathcal{A}_Y^1 \otimes \phi_*\mathcal{A}_{X/Y}^p$$

sends a form $\gamma \in \phi_*(L^1\mathcal{A}_X^{p+1}/L^2\mathcal{A}_X^{p+1})$ to the section γ' of $\mathrm{Hom}\,(T_Y, \phi_*\mathcal{A}_{X/Y}^p)$, which is given at the point y by

$$\gamma'(u) = (\mathrm{int}(\tilde{u})\,\tilde{\gamma})_{|X_y},$$

where $\tilde{u}$ is a lifting of $u \in \Gamma(X_u, \phi^*T_{Y,y})$ and $\tilde{\gamma}$ is a lifting of γ to $\phi_*L^1\mathcal{A}_X^{p+1}$. (The result is independent of the choice of $\tilde{u}$ because $\tilde{\gamma}_{|X_y} = 0$.) □

5.1.2 Transversality

We now consider a proper and holomorphic submersion $\phi : X \to Y$ between two complex varieties. Assume that the fibres of ϕ are Kähler. The Hodge decomposition

$$H^p(X_y, \mathbb{C}) = \bigoplus_{r+s=p} H^{r,s}(X_y)$$

on each fibre gives the Hodge filtration

$$F^l H^p(X_y) = \bigoplus_{r \ge l} H^{r,s}(X_y).$$

We know that the Hodge filtration varies holomorphically, i.e. that there exists a filtration on the holomorphic vector bundle (or sheaf of free $\mathcal{O}_X$-modules) $\mathcal{H}^p = R^p\phi_* H \otimes \mathcal{O}_Y$ by holomorphic vector subbundles $F^l\mathcal{H}^p$ such that for $y \in Y$, the inclusion

$$F^l\mathcal{H}^p_y \subset \mathcal{H}^p_y = H^p(X_y, \mathbb{C})$$

can be identified with the inclusion $F^l H^p(X_y) \subset H^p(X_y, \mathbb{C})$.

Let us write $\mathcal{H}^{l,p-l} = F^l\mathcal{H}^p / F^{l+1}\mathcal{H}^p$ for the quotient holomorphic vector bundle.

The theory of variations of Hodge structure is entirely based on the property of transversality, which was first investigated by Griffiths (1968). Let us summarise here the main points proved in vI.10.1 and vI.10.2.

Theorem 5.6 (Griffiths transversality; see vI.10.2.2) *The Hodge filtration satisfies the property*

$$\nabla F^l\mathcal{H}^p \subset F^{l-1}\mathcal{H}^p \otimes \Omega_Y.$$

The local period map $\mathcal{P}^p$ for the family $(X_y)_{y\in Y}$ is defined on the connected and simply connected open sets U of Y. On such an open set, the local system $R^p\phi_*\mathbb{C}$ is trivial, and we thus have a canonical identification

$$H^p(X_y, \mathbb{C}) \cong H, \quad y \in U,$$

where $H = H^p(X_0, \mathbb{C})$ is a fixed vector space. The map $\mathcal{P}^p$ then associates to each $y \in Y$ the Hodge filtration $F^{\cdot}H^p(X_y)$ on $H^p(X_y, \mathbb{C}) = H$. $\mathcal{P}^p$ can thus be considered as a map with values in a flag variety on the fixed vector space H. By the universal property of flag varieties (cf. [116], 10.1.3), the existence of the filtration $F^l\mathcal{H}^p$ by holomorphic subbundles is equivalent to the fact that $\mathcal{P}^p$ is holomorphic.

Finally, transversality translates to a statement on the differential $d\mathcal{P}^p$ as follows. If $W \subset H$ is a complex vector subspace, the tangent space at the point W of the Grassmannian G parametrising the complex vector subspaces of H can be canonically identified with $\mathrm{Hom}(W, H/W)$ (see vI.10.1.1). The flag variety on H parametrises the filtrations $F^{\cdot}H$ on H of given ranks, and via the map which to a filtration $F^{\cdot}$ associates the p-tuple of subspaces $F^i H$, can thus be viewed as a subvariety of the product of the corresponding Grassmannians G_i. The differential

$$d\mathcal{P}^p : T_{Y,y} \to \oplus_l T_{G_l, F^l H}$$

is thus a map

$$d\mathcal{P}^p : T_{Y,y} \to \bigoplus_l \mathrm{Hom}\,(F^l H^p(X_y), H^p(X_y, \mathbb{C})/F^l H^p(X_y)).$$

Griffiths transversality says that $d\mathcal{P}^p$ actually has values in

$$\bigoplus_l \mathrm{Hom}\,(F^l H^p(X_y), F^{l-1} H^p(X_y)/F^l H^p(X_y)).$$

Finally, since a tangent vector

$$(h_l)_{l\leq p} \in \bigoplus_l \mathrm{Hom}\,(F^l H^p(X_y), H^p(X_y, \mathbb{C})/F^l H^p(X_y)) = \bigoplus_l T_{G_l, F^l H}$$

is tangent to the flag variety if and only if

$$h_{l|F^{l+1}H^p(X_y)} = h_{l+1} \mod F^l H^p,$$

we find that $\mathrm{Im}\, d\mathcal{P}^p$ actually has values in the subspace

$$\bigoplus_l \mathrm{Hom}\,(F^l H^p(X_y)/F^{l+1} H^p(X_y), F^{l-1} H^p(X_y)/F^l H^p(X_y)),$$

which is called the horizontal tangent space of the period domain.

Every component $d\mathcal{P}^p_l$ of the map

$$d\mathcal{P}^p : T_{Y,y} \to \bigoplus_l \mathrm{Hom}\,(F^l H^p(X_y)/F^{l+1} H^p(X_y), F^{l-1} H^p(X_y)/F^l H^p(X_y))$$

thus obtained can be deduced simply from the map $\overline{\nabla}_l$ defined by the following diagram, whose columns are exact, and in which the two upper rows are given by the transversality property.

$$\begin{array}{lccc}
 & 0 & & 0 \\
 & \downarrow & & \downarrow \\
\nabla: & F^{l+1}\mathcal{H}^p & \longrightarrow & F^l\mathcal{H}^p \otimes \Omega_Y \\
 & \downarrow & & \downarrow \\
\nabla: & F^l\mathcal{H}^p & \longrightarrow & F^{l-1}\mathcal{H}^p \otimes \Omega_Y \\
 & \downarrow & & \downarrow \\
\overline{\nabla}_l: & F^l\mathcal{H}^p/F^{l+1}\mathcal{H}^p & \longrightarrow & F^{l-1}\mathcal{H}^p/F^l\mathcal{H}^p \otimes \Omega_Y \\
 & \downarrow & & \downarrow \\
 & 0 & & 0.
\end{array} \tag{5.2}$$

It follows immediately from Leibniz' formula that $\overline{\nabla}_l$ is $\mathcal{O}_Y$-linear, so by evalution at the point y, it gives

$$\overline{\nabla}_{l,y} : F^l H^p(X_y)/F^{l+1} H^p(X_y) \to F^{l-1} H^p(X_y)/F^l H^p(X_y) \otimes \Omega_{Y,y}.$$

For $u \in T_{Y,y}$, $\alpha \in F^l H^p(X_y)/F^{l+1} H^p(X_y)$, we then have

$$d\mathcal{P}_l^p(u)(\alpha) = \langle \overline{\nabla}_{l,y}(\alpha), u\rangle \in F^{l-1} H^p(X_y)/F^l H^p(X_y).$$

This formula follows immediately from the explicit description of the isomorphism $T_{G_l, F^l H} \cong \operatorname{Hom}(F^l H, H/F^l H)$. From now on, we will frequently omit the index l.

Finally, we can give a geometric description of $d\mathcal{P}_l^p$ or of $\overline{\nabla}_l$ as follows. We have the canonical identifications for $y \in Y$ given by

$$(F^l \mathcal{H}^p/F^{l+1}\mathcal{H}^p)_y = H^{l,p-l}(X_y) = H^{p-l}\left(X_y, \Omega^l_{X_y}\right),$$

$$(F^{l-1} \mathcal{H}^p/F^{l}\mathcal{H}^p)_y = H^{l-1,p-l+1}(X_y) = H^{p-l+1}\left(X_y, \Omega^{l-1}_{X_y}\right),$$

and the Kodaira–Spencer map

$$\rho : T_{Y,y} \to H^1(X_y, T_{X_y})$$

induced by the first-order deformation of the complex structure on X_y parametrised by the first infinitesimal neighbourhood of y in Y (see vI.9.1.2). Recall that ρ can be constructed as the connecting map associated to the exact sequence

$$0 \to T_{X_y} \to T_{X|X_y} \to \phi^* T_{Y,y} \to 0.$$

Theorem 5.7 (Griffiths) *The map*

$$\overline{\nabla}_y : (F^l \mathcal{H}^p/F^{l+1}\mathcal{H}^p)_y \to \operatorname{Hom}\left(T_{Y,y}, (F^{l-1} \mathcal{H}^p/F^{l}\mathcal{H}^p)_y\right)$$

can be identified via the identifications above with the composition of the map

$$H^{p-l}\left(X_y, \Omega^l_{X_y}\right) \to \operatorname{Hom}\left(H^1(X_y, T_{X_y}), H^{p-l+1}\left(X_y, \Omega^{l-1}_{X_y}\right)\right)$$

given by the cup-product and the map

$$\rho^* : \operatorname{Hom}\left(H^1(X_y, T_{X_y}), H^{p-l+1}\left(X_y, \Omega^{l-1}_{X_y}\right)\right)$$
$$\to \operatorname{Hom}\left(T_{Y,y}, H^{p-l+1}\left(X_y, \Omega^{l-1}_{X_y}\right)\right).$$

A proof of this theorem can be given by using the description of the Gauss–Manin connection of proposition 5.5. The identification of ∇ with the d_1 of the spectral sequence associated to the Leray filtration L on the complex $\mathcal{A}_X^\bullet$ tells us that

$$\nabla : \mathcal{H}^k \to \mathcal{H}^k \otimes \Omega_Y$$

is the connecting map associated to the exact sequence of holomorphic de Rham complexes

$$0 \to \Omega_{X/Y}^{\bullet-1} \otimes \Omega_Y \to \Omega_X^\bullet / L^2\Omega_X^\bullet \to \Omega_{X/Y}^\bullet \to 0,$$

which induces

$$\delta = d_1 = \nabla : R^k\phi_*(\Omega_{X/Y}^\bullet) = \mathcal{H}^k \to R^k\phi_*(\Omega_{X/Y}^\bullet) \otimes \Omega_Y = \mathcal{H}^k \otimes \Omega_Y.$$

Applying the naive filtration to this exact sequence of complexes, and recalling the definition of the Kodaira–Spencer map, we obtain the desired result. In vI.10.2.3, we gave the proof due to Griffiths, which uses the Cartan–Lie formula directly.

5.1.3 Construction of the complexes $\mathcal{K}_{l,r}$

In the same situation as above, consider the holomorphic de Rham complex $\mathrm{DR}(H^p)$ of $R^p\phi_*\mathbb{C} =: H^p$, given by

$$0 \to \mathcal{H}^p \xrightarrow{\nabla} \mathcal{H}^p \otimes \Omega_Y \to \cdots \xrightarrow{\nabla} \mathcal{H}^p \otimes \Omega_Y^N \to 0,$$

where $N = \dim Y$.

The transversality property allows us to filter this complex by setting

$$F^l\mathrm{DR}(H^p) = 0 \to F^l\mathcal{H}^p \xrightarrow{\nabla} F^{l-1}\mathcal{H}^p \otimes \Omega_Y \to \cdots \xrightarrow{\nabla} F^{l-N}\mathcal{H}^p \otimes \Omega_Y^N \to 0.$$

We write $\mathcal{K}_{l,r}$, $l + r = p$ for the complexes obtained by taking the graded complex of $DR(H^p)$ for this filtration:

$$\mathcal{K}_{l,r} = F^l\mathrm{DR}(H^p)/F^{l+1}\mathrm{DR}(H^p). \tag{5.3}$$

By the definition of $\overline{\nabla}$ (see (5.2)), the first two terms of the complex $\mathcal{K}_{l,r}$ are given by

$$0 \to \mathcal{H}^{l,r} \xrightarrow{\overline{\nabla}} \mathcal{H}^{l-1,r+1} \otimes \Omega_Y,$$

where

$$\mathcal{H}^{l,r} := F^l\mathcal{H}^p/F^{l+1}\mathcal{H}^p,\ p = l + r.$$

Note that r is not necessarily positive here. If $r < 0$, we set $\mathcal{H}^{l,r} = 0$. More generally, in every degree k, we have

$$\mathcal{K}^k_{l,r} = (F^{l-k}\mathcal{H}^p/F^{l+1-k}\mathcal{H}^p) \otimes \Omega^k_Y = \mathcal{H}^{l-k,p-l+k} \otimes \Omega^k_Y,$$

and also the following result.

Lemma 5.8 *The differentials*

$$\overline{\nabla}^k_{l-k} : \quad \begin{array}{ccc} \mathcal{K}^k_{l,r} & \longrightarrow & \mathcal{K}^{k+1}_{l,r} \\ \| & & \| \\ \mathcal{H}^{l-k,p-l+k} \otimes \Omega^k_Y & \longrightarrow & \mathcal{H}^{l-k-1,p-l+k+1} \otimes \Omega^{k+1}_Y \end{array}$$

are $\mathcal{O}_Y$-linear, and they are given by

$$\overline{\nabla}^k_{l-k}(\sigma \otimes \alpha) = \overline{\nabla}_{l-k}(\sigma) \wedge \alpha, \tag{5.4}$$

where σ is a section of $\mathcal{H}^{l-k,p-l+k}$ and α is a section of Ω^k_Y.

Proof This follows immediately from Leibniz' formula and the definition of $\overline{\nabla}_{l-k}$. Indeed, let $\tilde{\sigma}$ be a lifting of σ to $F^{l-k}\mathcal{H}^p$. Then

$$\overline{\nabla}^k_{l-k}(\sigma \otimes \alpha) = \nabla(\tilde{\sigma} \otimes \alpha) \text{ mod } F^{l-k}\mathcal{H}^p \otimes \Omega^{k+1}_Y.$$

Now, we have

$$\nabla(\tilde{\sigma} \otimes \alpha) = \tilde{\sigma} \otimes d\alpha + \nabla\tilde{\sigma} \wedge \alpha,$$

and the first term is in $F^{l-k}\mathcal{H}^p \otimes \Omega^{k+1}_Y$. Moreover, again by definition of $\overline{\nabla}$, we have

$$\nabla\tilde{\sigma} \text{ mod } F^{l-k}\mathcal{H}^p \otimes \Omega_Y = \overline{\nabla}_{l-k}\sigma.$$

Thus, formula (5.4) is proved, and in particular it implies that the differential $\overline{\nabla}^k_{l-k}$ is $\mathcal{O}_Y$-linear. □

5.2 The holomorphic Leray spectral sequence

5.2.1 The Leray filtration on Ω^p_X and the complexes $\mathcal{K}_{p,q}$

Let $\phi : X \to Y$ be a holomorphic submersive map between smooth complex varieties. We have an inclusion of holomorphic vector bundles

$$\phi^*\Omega_Y \hookrightarrow \Omega_X,$$

whose quotient $\Omega_{X/Y}$ is the bundle of relative holomorphic differential forms. As in subsection 4.1.3, we thus have a Leray filtration $L^{\cdot}\Omega_X^k$ on each bundle Ω_X^k, given by

$$L^l\Omega_X^k = \phi^*\Omega_Y^l \wedge \Omega_X^{k-l}.$$

The graded sheaf associated to this filtration is given by

$$\mathrm{Gr}_L^l\Omega_X^k = \phi^*\Omega_Y^l \otimes \Omega_{X/Y}^{k-l}.$$

Now assume that ϕ is projective or has compact Kähler fibres, and consider the coherent sheaf $R^p\phi_*\Omega_X^k$ on Y. It is equipped with the filtration associated to the filtration L, given by

$$L^lR^p\phi_*\Omega_X^k = \mathrm{Im}\left(R^p\phi_*L^l\Omega_X^k \to R^p\phi_*\Omega_X^k\right).$$

Moreover, we have a spectral sequence

$$E_r^{l,q} \Rightarrow R^{l+q}\phi_*\Omega_X^k,$$

in which each complex $(E_r^{l,q}, d_r)$ is a complex of coherent sheaves on Y, and such that

$$E_\infty^{l,q} = \mathrm{Gr}_L^l R^{l+q}\phi_*\Omega_X^k.$$

Proposition 5.9 *For fixed q, the complex $(E_1^{l,q}, d_1)$ can be identified with the complex $(\mathcal{K}_{k,q}^l, \overline{\nabla})$ of (5.3).*

Proof Let us introduce the Dolbeault resolution

$$\left(\mathcal{A}^{0,s}(\Omega_X^k), \overline{\partial}\right)$$

of Ω_X^k. It induces the Dolbeault resolution $(\mathcal{A}^{0,s}(L^l\Omega_X^k), \overline{\partial})$ of each sheaf $L^l\Omega_X^k$, and as these resolutions are acyclic for the functor ϕ_*, we have

$$R^{l+q}\phi_*\Omega_X^k = \mathcal{H}^{l+q}\left(\phi_*\mathcal{A}^{0,\bullet}(\Omega_X^k), \overline{\partial}\right).$$

By definition (see vI.8.3), the spectral sequence $E_r^{l,q} \Rightarrow R^{l+q}\phi_*\Omega_X^k$ is that of the filtered complex $(\phi_*\mathcal{A}^{0,\bullet}(\Omega_X^k), \overline{\partial})$, where the filtration is given by

$$L^l\phi_*\mathcal{A}^{0,\bullet}(\Omega_X^k) = \phi_*\mathcal{A}^{0,\bullet}(L^l\Omega_X^k).$$

We thus obtain

$$\left(E_0^{l,q}, d_0\right) = \left(\phi_*\mathcal{A}^{0,l+q}(\mathrm{Gr}_L^l\Omega_X^k), \overline{\partial}\right),$$

hence

$$E_1^{l,q} = R^{l+q}\phi_*\big(\mathrm{Gr}_L^l \Omega_X^k\big) = \Omega_Y^l \otimes R^{l+q}\phi_*\big(\Omega_{X/Y}^{k-l}\big). \tag{5.5}$$

Now, by the definition of $\mathcal{K}_{k,q}$, we have

$$R^{l+q}\phi_*\big(\Omega_{X/Y}^{k-l}\big) = \mathcal{H}^{k-l,l+q}$$

and

$$\Omega_Y^l \otimes \mathcal{H}^{k-l,l+q} = \mathcal{K}_{k,q}^l. \tag{5.6}$$

Comparing (5.5) and (5.6) it thus remains only to show that the differentials d_1 and $\overline{\nabla}$ coincide. But we know (see vI.8.3.2) that the differential

$$d_1 : R^{l+q}\phi_*\big(\mathrm{Gr}_L^l \Omega_X^k\big) \to R^{l+q+1}\phi_*\big(\mathrm{Gr}_L^{l+1}\Omega_X^k\big)$$

is the connecting morphism induced by the exact sequence

$$0 \to \mathrm{Gr}_L^{l+1}\Omega_X^k \to L^l\Omega_X^k / L^{l+2}\Omega_X^k \to \mathrm{Gr}_L^l \Omega_X^k \to 0,$$

i.e.

$$0 \to \phi^*\Omega_Y^{l+1} \otimes \Omega_{X/Y}^{k-l-1} \to L^l\Omega_X^k / L^{l+2}\Omega_X^k \to \phi^*\Omega_Y^l \otimes \Omega_{X/Y}^{k-l} \to 0. \tag{5.7}$$

If $\alpha \in \Omega_Y^l$ is a local section, the cup-product with $\phi^*\alpha$ gives a commutative diagram

$$\begin{array}{ccccccccc}
0 & \longrightarrow & \phi^*\Omega_Y^1 \otimes \Omega_{X/Y}^{k-l-1} & \longrightarrow & L^l\Omega_X^{k-l} / L^{l+2}\Omega_X^{k-l} & \longrightarrow & \Omega_{X/Y}^{k-l} & \longrightarrow & 0 \\
 & & \Big\downarrow{\scriptstyle \phi^*(\wedge\alpha)} & & \Big\downarrow{\scriptstyle \wedge\phi^*\alpha} & & \Big\downarrow{\scriptstyle \phi^*\alpha\otimes} & & \\
0 & \longrightarrow & \phi^*\Omega_Y^{l+1} \otimes \Omega_{X/Y}^{k-l-1} & \longrightarrow & L^l\Omega_X^k / L^{l+2}\Omega_X^k & \longrightarrow & \phi^*\Omega_Y^l \otimes \Omega_{X/Y}^{k-l} & \longrightarrow & 0.
\end{array}$$

It follows immediately that

$$d_1(\sigma \otimes \alpha) = d_1\sigma \wedge \alpha \tag{5.8}$$

for a section σ of $\mathcal{H}^{k-l,l+q}$. Moreover, we also have

$$\overline{\nabla}(\sigma \otimes \alpha) = \overline{\nabla}\sigma \wedge \alpha \tag{5.9}$$

by lemma 5.8. Comparing (5.8) and (5.9), we see that it suffices to show the equality $d_1 = \overline{\nabla}$ for every k on the terms $E_1^{0,q} = \mathcal{H}^{k,q}$. But this follows from proposition 5.5, from the description given above of d_1, with $l = 0$, and the definition of $\overline{\nabla}$ as induced by ∇. So the two differentials coincide. □

5.2.2 Infinitesimal invariants

The preceding result now enables us to construct the first infinitesimal invariant $\delta\alpha$ of a class

$$\alpha \in \mathrm{Ker}\left(H^p(X, \Omega_X^k) \to H^0(Y, R^p\phi_*\Omega_{X/Y}^k)\right), \tag{5.10}$$

i.e. of a class which vanishes on the fibres of ϕ.

We begin by considering the image of α in $H^0(Y, R^p\phi_*\Omega_X^k)$, which we also denote by α. Note that for the filtration $L^\cdot$ on $R^p\phi_*\Omega_X^k$, we have

$$L^1R^p\phi_*\Omega_X^k = \mathrm{Ker}\left(R^p\phi_*\Omega_X^k \to R^p\phi_*\Omega_{X/Y}^k\right).$$

Indeed, the exact sequence of sheaves on X

$$0 \to L^1\Omega_X^k \to \Omega_X^k \to \Omega_{X/Y}^k \to 0$$

gives the long exact sequence

$$\cdots \to R^p\phi_*L^1\Omega_X^k \to R^p\phi_*\Omega_X^k \to R^p\phi_*\Omega_{X/Y}^k \to \cdots,$$

and by definition of the filtration $L^\cdot$ on $R^p\phi_*\Omega_X^k$, we have

$$L^1R^p\phi_*\Omega_X^k = \mathrm{Im}\left(R^p\phi_*L^1\Omega_X^k \to R^p\phi_*\Omega_X^k\right).$$

The class α as in (5.10) thus belongs to $H^0(Y, L^1R^p\phi_*\Omega_X^k)$. Therefore, it has an image in $H^0(Y, L^1R^p\phi_*\Omega_X^k/L^2R^p\phi_*\Omega_X^k)$, with

$$L^1R^p\phi_*\Omega_X^k/L^2R^p\phi_*\Omega_X^k = E_\infty^{1,p-1}$$

for the spectral sequence introduced above. But for degree reasons, no non-zero arrow d_r, $r \geq 2$ can lead to $E_r^{1,p-1}$, so we have $E_\infty^{1,p-1} \subset E_2^{1,p-1}$. As

$$E_2^{1,p-1} = \frac{\mathrm{Ker}\left(d_1 : E_1^{1,p-1} \to E_1^{2,p-1}\right)}{\mathrm{Im}\left(d_1 : E_1^{0,p-1} \to E_1^{1,p-1}\right)},$$

we can apply proposition 5.9 to conclude that we have a natural inclusion

$$g : L^1R^p\phi_*\Omega_X^k/L^2R^p\phi_*\Omega_X^k \hookrightarrow \frac{\mathrm{Ker}\left(\overline{\nabla} : \mathcal{H}^{k-1,p} \otimes \Omega_Y \to \mathcal{H}^{k-2,p+1} \otimes \Omega_Y^2\right)}{\mathrm{Im}\,(\overline{\nabla} : \mathcal{H}^{k,p-1} \to \mathcal{H}^{k-1,p} \otimes \Omega_Y)}. \tag{5.11}$$

Thus, we can make the following definition.

Definition 5.10 *The first infinitesimal invariant $\delta\alpha$ of the class*

$$\alpha \in \mathrm{Ker}\left(H^p(X, \Omega_X^k) \to H^0(Y, R^p\phi_*\Omega_{X/Y}^k)\right)$$

is the image of α *under the map* g*, in the space of sections of the cohomology sheaf*

$$\frac{\operatorname{Ker}\left(\overline{\nabla}:\mathcal{H}^{k-1,p}\otimes\Omega_Y\to\mathcal{H}^{k-2,p+1}\otimes\Omega_Y^2\right)}{\operatorname{Im}(\overline{\nabla}:\mathcal{H}^{k,p-1}\to\mathcal{H}^{k-1,p}\otimes\Omega_Y)}=H^1(\mathcal{K}^\bullet_{k,p-1}).$$

For $y\in Y$*, the infinitesimal invariant* $\delta\alpha_y$ *of* α *at the point* y *is the value of* $\delta\alpha$ *at the point* y*, i.e. the image of* $\delta\alpha$ *in the cohomology* $H^1(\mathcal{K}^\bullet_{k,p-1,y})$ *of the fibre at the point* y *of the complex* $\mathcal{K}^\bullet_{k,p-1}$.

The terminology ‘infinitesimal invariant’ comes from the fact that this invariant generalises the infinitesimal invariant of normal functions defined by Griffiths (1983), which can be computed by differentiation of normal functions (see chapter 7).

It is possible to refine this invariant (see Voisin 1994), by studying more precisely the spectral sequence associated to the Leray filtration on Ω_X^k and the functor ϕ_*. Indeed, assume that for this spectral sequence, we have

$$E_1^{l,q}=0\ \text{ for }\ l<l_0-1,\ \ l+q=p-1,$$

$$E_1^{l,q}=0\ \text{ for }\ 0<l<l_0,\ \ l+q=p.$$

Then for every $r\geq 1$, we have

$$E_r^{l,q}=0\ \text{ for }\ l<l_0-1,\ \ l+q=p-1,$$

$$E_r^{l,q}=0\ \text{ for }\ 0<l<l_0,\ \ l+q=p.$$

Thus, we have $L^1R^p\phi_*\Omega_X^k=L^{l_0}R^p\phi_*\Omega_X^k$, and a map

$$L^1R^p\phi_*\Omega_X^k=L^{l_0}R^p\phi_*\Omega_X^k\to E_\infty^{l_0,q},\quad l_0+q=p.$$

The first hypothesis then shows that $E_\infty^{l_0,q}\subset E_2^{l_0,q}$, and by proposition 5.9, we can represent this last sheaf as the cohomology in degree l_0 of the complex $\mathcal{K}_{k,q}$, $q=p-l_0$. This enables us to associate to a class

$$\alpha\in L^1H^p(X,\Omega_X^k)=\operatorname{Ker}\left(H^p(X,\Omega_X^k)\to H^0(Y,R^p\phi_*\Omega_X^k)\right)$$

an infinitesimal invariant $\delta\alpha$ which is a section of the sheaf of cohomology at the middle of the sequence

$$\mathcal{H}^{k-l_0+1,p-1}\otimes\Omega_Y^{l_0-1}\stackrel{\overline{\nabla}}{\to}\mathcal{H}^{k-l_0,p}\otimes\Omega_Y^{l_0}\stackrel{\overline{\nabla}}{\to}\mathcal{H}^{k-l_0-1,p+1}\otimes\Omega_Y^{l_0+1}.$$

Example 5.11 *Consider the smooth hypersurfaces X_y of $\mathbb{P}^{k+1}$ of given degree. By the Lefschetz theorem 1.23, such a hypersurface satisfies*

$$H^q(X_y, \Omega^r_{X_y}) = 0 \text{ for } q \neq r, \ \ q + r \neq k. \tag{5.12}$$

Now let $\phi : X \to Y$ be a family of such hypersurfaces. For the spectral sequence of the sheaf Ω^k_X relative to the derived functors of ϕ_* and the Leray filtration, we have $E_1^{l,q} = R^{l+q}\phi_*\Omega_X^{k-l} \otimes \Omega^l_Y$. This is zero by (5.12) if $k + q \neq k$ and $l + q \neq k - l$, i.e. if $q \neq 0$ and $2l \neq k - q$. Thus, the terms $E_1^{l,q}$ are zero for $l + q = k$, $l < k$, and for $l + q = k - 1$, $l < k - 1$. Thus we can associate to a class $\alpha \in L^1H^k(X, \Omega^k_X)$ an infinitesimal invariant $\delta\alpha$ which is a section of

$$\text{Coker}\left(\mathcal{H}^{1,k-1} \otimes \Omega_Y^{k-1} \overset{\overline{\nabla}}{\to} \mathcal{H}^{0,k} \otimes \Omega^k_Y\right).$$

In Voisin (1994), this is used to show that certain cycle classes $[Z]$, for $Z \subset X$ a cycle of codimension 2, are non-zero, where $X \to Y$ is a locally complete family of surfaces in $\mathbb{P}^3$. This result is used to study rational equivalence (see subsection 9.1.1) on the 0-cycles on a general surface of $\mathbb{P}^3$ of degree at least 7. It is shown in that paper that for such a general surface, no two distinct points are rationally equivalent.

5.3 Local study of Hodge loci

5.3.1 General properties

Let B be a complex variety. A variation of integral Hodge structure of weight k on B is given by a local system H of free $\mathbb{Z}$-modules and a decreasing filtration $F^l\mathcal{H}$, $0 \leq l \leq k$ on the holomorphic vector bundle $\mathcal{H} = H \otimes_{\mathbb{Z}} \mathcal{O}_B$, satisfying the transversality property

$$\nabla(F^p\mathcal{H}) \subset F^{p-1}\mathcal{H} \otimes \Omega_B.$$

We can also speak of rational, real or complex variations of Hodge structure, and everything that follows remains valid. However, integral variations of Hodge structure are privileged insofar as the variations of Hodge structure of geometric origin, i.e. those coming from Kähler or projective fibrations are integral. Furthermore, from the point of view of the Hodge conjecture, studying *integral* classes of given Hodge level makes more sense. Nevertheless, most of what follows also holds in the context of complex variations of Hodge structure.

Let $U \subset B$ be a connected open set and λ a section of H on U. For every $u \in U$, let $\lambda_u \in H_u$ denote the value of λ at u. Note that by the inclusion $H \subset \mathcal{H}$, λ can also be considered as a holomorphic section of $\mathcal{H}$. Then λ_u is also the value at u of the section λ of $\mathcal{H}$.

Definition 5.12 *The Hodge locus U_λ^p defined by λ is the set*

$$U_\lambda^p := \{u \in U \mid \lambda_u \in F^p\mathcal{H}_u\}.$$

In the geometric framework, we are most interested in the case where $k = 2p$ and $\lambda \in H$ is integral. Then the Hodge locus U_λ^p is the set of points u where the class λ is a Hodge class (see vI.11.3). If the variation of Hodge structure is the variation of Hodge structure of a family of projective varieties, this locus is then also conjecturally the set of points u for which a multiple of the class λ is the class of an algebraic cycle. The study of Hodge loci is thus a first step in the study of the Hodge conjecture.

The first important fact concerning Hodge loci is the following result.

Lemma 5.13 *Every $U_\lambda^p \subset U$ is a complex analytic subset of U, equipped with a natural scheme structure.*

Proof Indeed, U_λ^p is described as the locus of zeros of the section $\overline{\lambda}$ of $\mathcal{H}/F^p\mathcal{H}$ obtained as the projection of $\lambda \in \mathcal{H}$. As λ is holomorphic, its projection is also holomorphic, so that U_λ^p is the locus of zeros of a holomorphic section of a holomorphic bundle on U, which endows it with a natural structure of an analytic subscheme. □

Remark *It is a difficult theorem due to Cattani, Deligne and Kaplan (see Cattani, Deligne & Kaplan 1995), that in the geometric context described above, these loci are in fact algebraic. This is a remarkable piece of evidence for the Hodge conjecture.*

A priori, the description of U_λ^p given above shows that this analytic set can be defined locally by $N = \operatorname{rank} \mathcal{H}/F^p\mathcal{H}$ equations (obtained by locally trivialising $\mathcal{H}/F^p\mathcal{H}$). The property of transversality in fact implies the following result.

Proposition 5.14 *The analytic set U_λ^p can be defined locally by*

$$h^{p-1} := \dim (F^{p-1}\mathcal{H}/F^p\mathcal{H})$$

equations.

By this, we mean that its sheaf of ideals as described above can be locally generated by h^{p-1} equations.

Proof Let $u \in U_\lambda^p$. In the neighbourhood of u, choose a decomposition

$$\mathcal{H}/F^p\mathcal{H} = F^{p-1}\mathcal{H}/F^p\mathcal{H} \oplus \mathcal{F}$$

as a direct sum of two holomorphic vector bundles. The section $\overline{\lambda}$ decomposes correspondingly as

$$\overline{\lambda} = \overline{\lambda}_{p-1} + \overline{\lambda}_F,$$

where $\overline{\lambda}_{p-1}$ is a holomorphic section of $F^{p-1}/F^p\mathcal{H}$. Then U_λ^p is contained in the analytic subset $V_\lambda^p \subset U$ defined by the vanishing of $\overline{\lambda}_{p-1}$. The proposition is then a consequence of the following lemma.

Lemma 5.15 *U_λ^p and V_λ^p coincide as schemes in the neighbourhood of u.*

Proof Assume for simplicity that V_λ^p is smooth. Replacing U by V_λ^p, we thus know that the section $\overline{\lambda}_{p-1}$ is zero on U, and we want to deduce that $U_\lambda^p = U$. By Nakayama's lemma, it suffices to show that U_λ^p coincides formally with U at the point u, i.e. that for every l, we have

$$U_\lambda^p \cap U_l = U_l,$$

where

$$U_l = \operatorname{Spec} \mathcal{O}_{U,u}/\mathcal{M}_u^{l+1}$$

is the lth infinitesimal neighbourhood of u in U. We prove this by induction on l. Assume that $U_\lambda^p \cap U_l = U_l$. This means that the section λ of $\mathcal{H}$ lies in $F^p\mathcal{H}$ up to order l. In other words, there exists a section μ of $F^p\mathcal{H}$ on U in the neighbourhood of u such that

$$\lambda = \mu + \sum_i \alpha_i \sigma_i, \tag{5.13}$$

where $\alpha_i \in \mathcal{M}_u^{l+1}$ and $\sigma_i \in \mathcal{H}$ form a local basis of flat sections of $\mathcal{H}$. We may assume that the basis σ_i is such that $\sigma_i(u) \in F^{p-1}\mathcal{H}_u$, $i \leq N$, and $\sigma_i(u) \in \mathcal{F}$, $i > N$.

As $\overline{\lambda}_{p-1} = 0$ on U, we see that $\lambda - \mu$, which lies in $\mathcal{M}_u^{l+1}\mathcal{H}$, projects to zero in

$$\frac{\mathcal{M}_u^{l+1}}{\mathcal{M}_u^{l+2}} F^{p-1}\mathcal{H}/F^p\mathcal{H}.$$

It follows that up to modifying μ, we may assume that $\alpha_i = 0 \bmod \mathcal{M}_u^{l+2}$ for $i \leq N$.

Now if we apply the Gauss–Manin connection to (5.13), as λ is flat, we obtain

$$-\nabla\mu = \sum_i d\alpha_i \otimes \sigma_i \text{ in } \mathcal{H} \otimes \Omega_U,$$

and by transversality, this is a section of $F^{p-1}\mathcal{H} \otimes \Omega_U$. This implies that the image of $\sum_i d\alpha_i \otimes \sigma_i$ in

$$\mathcal{H}_u \otimes \frac{\mathcal{M}_u^l}{\mathcal{M}_u^{l+1}} \otimes \Omega_{U,u}$$

lies in

$$F^{p-1}\mathcal{H}_u \otimes \frac{\mathcal{M}_u^l}{\mathcal{M}_u^{l+1}} \otimes \Omega_{U,u}.$$

Now, as $F^{p-1}\mathcal{H}_u$ is generated by the σ_i, $i \leq N$, this implies that the image of $d\alpha_i$ in $\frac{\mathcal{M}_u^l}{\mathcal{M}_u^{l+1}} \otimes \Omega_{U,u}$ is zero for $i > N$. Thus, we also have $\alpha_i \in \mathcal{M}_u^{l+2}$ for $i > N$, and therefore

$$\lambda \in F^p\mathcal{H} \operatorname{mod} \mathcal{M}_u^{l+2}\mathcal{H},$$

i.e. $U_\lambda^p \cap U_{l+1} = U_{l+1}$. □

5.3.2 Infinitesimal study

Let $x \in U_\lambda^p$ as before. The analytic subspace U_λ^p can be singular, but it has a Zariski tangent space at x defined by

$$T_{U_\lambda^p,x} = \left\{ v \in T_{U,x} \mid df(v) = 0, \ \ \forall f \in \mathcal{I}_{U_\lambda^p} \right\}.$$

We will now describe this tangent space. Let $\overline{\lambda}_x$ be the projection of $\lambda_x \in F^p\mathcal{H}_x$ in $F^p\mathcal{H}_x / F^{p+1}\mathcal{H}_x$. As in section 5.1.2, we have maps

$$\overline{\nabla}_x : F^p\mathcal{H}_x / F^{p+1}\mathcal{H}_x \to F^{p-1}\mathcal{H}_x / F^p\mathcal{H}_x \otimes \Omega_{U,x}$$

defined by

$$\overline{\nabla}_x(\alpha) = \nabla_x(\tilde{\alpha}) \text{ mod } F^p\mathcal{H}_x \otimes \Omega_{U,x},$$

where $\tilde{\alpha}$ is a section of $F^p\mathcal{H}$ such that $\alpha = \tilde{\alpha}_x$ mod $F^{p+1}\mathcal{H}_x$.

Lemma 5.16 *We have*

$$T_{U_\lambda^p,x} = \operatorname{Ker} \overline{\nabla}_x(\overline{\lambda}_x) \subset T_{U,x},$$

where

$$\overline{\nabla}_x(\overline{\lambda}_x) \in F^{p-1}\mathcal{H}_x / F^p\mathcal{H}_x \otimes \Omega_{U,x} \cong \operatorname{Hom}(T_{U,x}, F^{p-1}\mathcal{H}_x / F^p\mathcal{H}_x).$$

Proof A vector v is tangent to U_λ^p if and only if the flat section λ remains in $F^p\mathcal{H}$ to first order along v. This is equivalent to saying that there exists a section μ of $F^p\mathcal{H}$ which coincides with λ to first order along v, or that there exists a section μ of $F^p\mathcal{H}$ such that $\mu_x = \lambda_x$ and $\nabla_v\mu = 0$. Let μ' be an arbitrary section of $F^p\mathcal{H}$ such that $\mu'_x = \lambda_x$. Then every other section μ satisfying the same condition can be written $\mu = \mu' + \eta$, where η vanishes at the point x and lies in $F^p\mathcal{H}$. By Leibniz' rule, such an η satisfies the condition

$$\nabla_v \eta \in F^p\mathcal{H}_x,$$

and conversely, every element of $F^p\mathcal{H}_x$ can be written in the form $\nabla_v\eta$, where η is a section of $F^p\mathcal{H}$ which is zero at u. The existence of $\mu = \mu' + \eta$ satisfying the condition $\nabla_v\mu = 0$ is thus equivalent to the condition $\nabla_v\mu' \in F^p\mathcal{H}_x$. But this is equivalent to $\overline{\nabla}_x(\overline{\lambda}_x)(v) = 0$ by the definition of $\overline{\nabla}_x$. □

Corollary 5.17 *Assume that U is connected and the variation of Hodge structure satisfies the following condition: there exists $x \in U$ such that*

$$\overline{\nabla}_x : F^p\mathcal{H}_x/F^{p+1}\mathcal{H}_x \to F^{p-1}\mathcal{H}_x/F^p\mathcal{H}_x \otimes \Omega_{U,x}$$

is injective. Then for every $\lambda \in H$ such that $U^p_\lambda = U$, we also have $U^{p+1}_\lambda = U$.

Proof If $U^p_\lambda = U$, the Zariski tangent space $T_{U^p_\lambda,x}$ is equal to $T_{U,x}$. By the lemma, this then implies that $\overline{\nabla}_p(\overline{\lambda}_x) = 0$, and by the hypothesis on x, this implies that $\overline{\lambda}_x = 0$, i.e. that $\lambda_x \in F^{p+1}\mathcal{H}_x$. But the hypothesis on x is satisfied on a Zariski open set, and is thus satisfied in an dense open subset of U. Thus, for the same reason as above, we also have $\lambda_{x'} \in F^{p+1}\mathcal{H}_{x'}$ for x' in the same dense open subset of U, so $\lambda \in F^{p+1}\mathcal{H}$, i.e. $U = U^{p+1}_\lambda$. □

We also have the following important corollary.

Corollary 5.18 *If $x \in U^p_\lambda$ and $\overline{\lambda}_x \in F^p\mathcal{H}_x/F^{p+1}\mathcal{H}_x$ is such that*

$$\overline{\nabla}_x(\overline{\lambda}_x) : T_{U,x} \to F^{p-1}/F^p\mathcal{H}_x$$

is surjective, then U^p_λ is smooth of codimension equal to h^{p-1} in U.

Proof We know that U^p_λ can be defined locally by h^{p-1} holomorphic equations. Thus, we have codim $U^p_\lambda \leq h^{p-1}$. Moreover, we have

$$\operatorname{codim}(U^p_\lambda \subset U) \geq \operatorname{codim}(T_{U^p_\lambda,x} \subset T_{U,x}),$$

with equality if and only if U^p_λ is smooth at u. Now, by lemma 5.16, the hypothesis on (x, λ) says that

$$\operatorname{codim}\,(T_{U^p_\lambda,x} \subset T_{U,x}) = h^{p-1}.$$

Thus, under the hypotheses of the corollary, we have the chain of inequalities

$$h^{p-1} \geq \operatorname{codim} U^p_\lambda \geq \operatorname{codim} T_{U^p_\lambda,x} = h^{p-1},$$

so that we have equality throughout. □

5.3.3 The Noether–Lefschetz locus

In the preceding situation, assume that H is a local system of free $\mathbb{Z}$-modules, and that $k = 2$, $p = 1$. We thus consider an integral variation of Hodge structure of weight 2. For $U \subset B$ a connected and simply connected open set and λ a section of H on U, the Hodge loci U_λ^p are known as local components of the Noether–Lefschetz locus. The terminology can be explained by the fact that if $H = R^2\phi_*\mathbb{Z}$, where $\phi : X \to Y$ is a projective morphism, equipped with its variation of Hodge structure, each set U_λ^p can be identified via the Lefschetz theorem on $(1, 1)$-classes with the set of points $u \in U$ where the class λ is the first Chern class of a holomorphic line bundle on the fibre X_u, so that the Hodge locus is then a local component of the Noether–Lefschetz locus studied in section 3.3.

Assume that ϕ is of relative dimension 2. Then the fibre at the point u of the bundle $\mathcal{H}^{0,2} := \mathcal{H}^2/F^1\mathcal{H}^2$, which is canonically isomorphic to $H^2(X_u, \mathcal{O}_{X_u})$, is a vector space, which is dual by Serre duality to the space $H^0(X_u, K_{X_u})$. We have the following result, which in certain cases enables us to give a geometric interpretation of codimension defect of the Noether–Lefschetz locus.

Proposition 5.19 *Let $C \subset X_u$ be a curve, and let*

$$\lambda_u = [C] \in H^2(X_u, \mathbb{Z}) \cap F^1H^2(X_u)$$

be its cohomology class. Then the image of

$$\overline{\nabla}_u(\overline{\lambda}_u) : T_{U,u} \to H^2(X_u, \mathcal{O}_{X_u})$$

is contained in the orthogonal complement of

$$H^0(X_u, K_{X_u}(-C)) \subset H^0(X_u, K_{X_u}).$$

Here $H^0(X_u, K_{X_u}(-C))$ is the set of sections of K_{X_u} which vanish along C.

Proof Note first that if $\eta \in H^0(X_u, K_{X_u})$ and $v \in T_{U,u}$, then

$$\langle \overline{\nabla}_u(\overline{\lambda}_u)(v), \eta \rangle = -\langle \overline{\nabla}_u(\eta)(v), \overline{\lambda}_u \rangle, \tag{5.14}$$

where the first $\langle\, , \rangle$ is the duality between $H^2(X_u, \mathcal{O}_{X_u})$ and $H^0(X_u, K_{X_u})$, and the second $\langle\, , \rangle$ is the intersection form on $H^1(X_u, \Omega_{X_u})$.

Indeed, we have the intersection form $\langle\, , \rangle$ on the local system $R^2\phi_*\mathbb{C}$, and the induced intersection form on $\mathcal{H}^2$ is obviously flat for ∇, i.e. it satisfies

$$d\langle \alpha, \beta \rangle = \langle \alpha, \nabla\beta \rangle + \langle \nabla\alpha, \beta \rangle$$

for two local sections α, β of $\mathcal{H}^2$. Let us apply this to sections

$$\tilde{\eta} \in F^2\mathcal{H}^2, \quad \tilde{\lambda} \in F^1\mathcal{H}^2.$$

As $F^2\mathcal{H}^2$ and $F^1\mathcal{H}^2$ are orthogonal for $\langle\, , \rangle$, the left-hand term in the equation above is equal to 0, and we obtain

$$0 = \langle \tilde{\eta}, \nabla\tilde{\lambda}\rangle + \langle \nabla\tilde{\eta}, \tilde{\lambda}\rangle.$$

Applying the definition of $\overline{\nabla}$, we see immediately that this equation taken at the point u and applied to a tangent vector v gives the relation

$$\langle \eta, \overline{\nabla}_u(\overline{\lambda}_u)(v)\rangle + \langle \overline{\nabla}_u(\eta)(v), \overline{\lambda}_u\rangle = 0,$$

where $\eta = \tilde{\eta}_u$ and $\overline{\lambda}_u$ is the projection of $\tilde{\lambda}_u$ in $H^1(X_u, \Omega_{X_u}) = F^1\mathcal{H}^2_u/F^2\mathcal{H}^2_u$.

Now, let $\eta \in H^0(X_u, K_{X_u}(-C))$ and $v \in T_{U,u}$. By theorem 5.7, we have

$$\overline{\nabla}_u(\eta)(v) = \mathrm{int}(\rho(v))(\eta),$$

where $\rho(v) \in H^1(X_u, T_{X_u})$. As $\eta \in H^0(X_u, K_{X_u}(-C))$, $\mathrm{int}(\rho(v))(\eta)$ in fact lies in the image of the map

$$H^1(X_u, \Omega_{X_u}(-C)) \to H^1(X_u, \Omega_{X_u})$$

induced by the inclusion $\Omega_{X_u}(-C) \subset \Omega_{X_u}$, where by definition $\Omega_{X_u}(-C) = \Omega_{X_u} \otimes \mathcal{I}_C$. Therefore

$$\mathrm{int}(\rho(v))(\eta)_{|C} = 0 \ \text{ in } \ H^1(C, \Omega_{X_u|C}).$$

A fortiori, the restriction $\mathrm{int}(\rho(v))(\eta)_{|C}$ vanishes in $H^1(C, \Omega_C)$. Now, the composition

$$H^1(X_u, \Omega_{X_u}) \to H^1(C, \Omega_C) \xrightarrow{\int_C} \mathbb{C},$$

where the last arrow is given by integration, is equal, by the definition of the class of C, to the linear form given by the intersection with $\overline{\lambda}_u$ (which is the class of C viewed as an element of $H^1(X_u, \Omega_{X_u})$). Thus, we have

$$\int_C (\mathrm{int}(\rho(v))(\eta)) = 0 = \langle \mathrm{int}(\rho(v))(\eta), \overline{\lambda}_u\rangle = \langle \overline{\nabla}_u(\eta)(v), \overline{\lambda}_u\rangle,$$

so by (5.14), we also have

$$\langle \overline{\nabla}_u(\overline{\lambda}_u)(v), \eta\rangle = 0.$$

□

This proposition shows that if λ_u is the class of a curve $C \subset X_u$, then the infinitesimal codimension of the Noether–Lefschetz locus U_λ at the point u is at most equal to

$$\begin{aligned}\operatorname{rank} \mathcal{H}^{0,2} - h^0(X_u, K_{X_u}(-C)) &= h^0(X_u, K_{X_u}) - h^0(X_u, K_{X_u}(-C))\\ &= \dim H^0(X_u, K_{X_u})_{|C},\end{aligned}$$

while the expected codimension is equal to $\operatorname{rank} \mathcal{H}^{0,2} = \dim H^0(X_u, K_{X_u})$ by lemma 5.16.

Example. In certain geometric cases, proposition 5.19 explains why certain components of the Noether–Lefschetz locus are not of maximal codimension $h^{2,0} = \operatorname{rank} \mathcal{H}^{0,2}$. Consider the family of smooth surfaces of degree d in $\mathbb{P}^3$. Let $\phi : \mathcal{X}_U \to U$ be the universal hypersurface, where $U \subset \mathbb{P}H^0(\mathbb{P}^3, \mathcal{O}_{\mathbb{P}^3}(d))$ is the open set parametrising smooth surfaces. We easily see by dimension-counting that the algebraic subset $U_\Delta \subset U$ consisting of the surfaces containing a line Δ is of codimension equal to $d-3$, and that it is a global component of the Noether–Lefschetz locus in U. Now, for the variation of Hodge structure of this family of surfaces, we have $\operatorname{rank} \mathcal{H}^{0,2} = \dim H^0(X_u, K_{X_u})$. By the adjunction formula, we have $K_{X_u} = \mathcal{O}_{X_u}(d-4)$, and thus

$$\dim H^0(X_u, K_{X_u}) > d-3$$

whenever $d > 4$. The codimension of this family is thus not the 'expected' codimension, and this is in fact explained by proposition 5.19. For this family, the proposition says that the infinitesimal codimension of $U_\Delta \subset U$ is less than or equal to

$$d-3 = h^0(\mathcal{O}_\Delta(d-4)) = \dim H^0(X_u, K_{X_u})_{|\Delta}, \ \Delta \subset X_u.$$

In this example, we can show that the inclusion of proposition 5.19 is in fact an equality.

In Voisin (1988) and Green (1989a), it is actually shown that this component U_Δ is the only component of the Noether–Lefschetz locus for surfaces of degree $d \geq 5$ which is of codimension $\leq d-3$. This result is refined in Voisin (1989), where it is shown that the only components of the Noether–Lefschetz locus of codimension $\leq 2d-7$ are the family of surfaces containing a conic and the family of surfaces containing a line.

5.3.4 A density criterion

The following result concerns mostly variations of Hodge structure of weight 2, since the hypotheses are practically never satisfied (because of the transversality condition) in weight $k = 2p > 2$. Consider a rational variation of Hodge structure $(H^2, F^{\cdot}\mathcal{H}^2)$ of weight 2 on a connected complex variety B. Let us impose the condition, which is satisfied in the geometric case, that

$$F^2\mathcal{H}^2_u \oplus \overline{F^1\mathcal{H}^2_u} = \mathcal{H}^2_u, \quad \forall u \in U.$$

(This means that the Hodge filtration induces a rational Hodge structure on each stalk H^2_u.)

Let $\mathcal{H}^2_{\mathbb{R}}$ denote the real vector bundle whose sheaf of differentiable sections is equal to $H^2 \otimes_{\mathbb{Q}} \mathcal{C}^\infty_{\mathbb{R}}$. The fibre of this vector bundle at the point u is equal to $H^2_{u,\mathbb{R}} := H^2_u \otimes_{\mathbb{Q}} \mathbb{R}$. This vector bundle contains a vector subbundle, which we denote by $\mathcal{H}^{1,1}_{\mathbb{R}}$, whose sheaf of differentiable sections is equal to

$$F^1\mathcal{H}^2 \otimes \mathcal{C}^\infty_{\mathbb{C}} \cap H^2 \otimes_{\mathbb{Q}} \mathcal{C}^\infty_{\mathbb{R}}.$$

The fibre of this vector bundle at the point u is equal to

$$\mathcal{H}^{1,1}_{u,\mathbb{R}} = F^1\mathcal{H}^2_u \cap H^2_{u,\mathbb{R}} = F^1\mathcal{H}^2_u \cap \overline{F^1\mathcal{H}^2_u} \cap H^2_{u,\mathbb{R}}.$$

Letting $\mathcal{H}^{1,1}_{u,\mathbb{C}}$ denote the intersection $F^1\mathcal{H}^2_u \cap \overline{F^1\mathcal{H}^2_u}$, we can also identify $\mathcal{H}^{1,1}_{u,\mathbb{R}}$ with the real part of $\mathcal{H}^{1,1}_{u,\mathbb{C}}$, i.e. with $\mathcal{H}^{1,1}_{u,\mathbb{C}} \cap H^2_{u,\mathbb{R}}$. In the case of a variation of Hodge structure coming from geometry, this fibre is the set of real classes of type $(1, 1)$. Note that via the projection, $\mathcal{H}^{1,1}_{u,\mathbb{C}}$ can be identified with $F^1\mathcal{H}^2_u/F^2\mathcal{H}^2_u$.

If $u \in B$ and $\lambda \in H^2_u \cap F^1\mathcal{H}^2_u$, then λ gives an elements of the fibre of $\mathcal{H}^{1,1}_{\mathbb{R}}$ at u, since λ is rational and in particular real. The following result due to Mark Green gives an infinitesimal criterion for the density of the Hodge classes, that is elements of $\mathcal{H}^2_{\mathbb{R}}$ which are rational and of type $(1, 1)$, in the vector bundle $\mathcal{H}^{1,1}_{\mathbb{R}}$.

Proposition 5.20 *Assume that there exist $u \in B$ and an element*

$$\overline{\lambda} \in F^1\mathcal{H}^2_u/F^2\mathcal{H}^2_u$$

such that

$$\overline{\nabla}_u(\overline{\lambda}) : T_{U,u} \to \mathcal{H}^2_u/F^1\mathcal{H}^2_u = \mathcal{H}^{0,2}_u$$

is surjective. Then the set of rational Hodge classes is dense in $\mathcal{H}^{1,1}_{\mathbb{R}}$. In particular, for every simply connected open set U of B, the union of the components U^p_λ for $\lambda \neq 0$ of the Noether–Lefschetz locus is dense in U.

Proof We have a natural identification

$$\mathcal{H}_u^{1,1} \cong \mathcal{H}_{u,\mathbb{C}}^{1,1} = \mathcal{H}_{u,\mathbb{R}}^{1,1} \otimes \mathbb{C},$$

given by the projection $p : \mathcal{H}_{u,\mathbb{R}}^{1,1} \to (F^1\mathcal{H}^2/F^2\mathcal{H}^2)_u = \mathcal{H}_u^{1,1}$.

Note also that the property of the class $\overline{\lambda}$, stated in the hypothesis, is Zariski open on the class $\overline{\lambda}$, which is an element of the holomorphic vector bundle $\mathcal{H}^{1,1} = F^1\mathcal{H}^2/F^2\mathcal{H}^2$. It follows that it is satisfied in a Zariski open set of $\mathcal{H}^{1,1}$.

Thus, if the hypothesis is satisfied by a complex class $\overline{\lambda}$, it is also satisfied in a dense open subset of $p(\mathcal{H}_{u,\mathbb{R}}^{1,1})$, so we can restrict ourselves to proving the density in the neighbourhood of an element $\lambda \in \mathcal{H}_{u,\mathbb{R}}^{1,1}$ such that $p(\lambda) =: \overline{\lambda}$ satisfies the hypothesis of the proposition.

Let U be a simply connected neighbourhood of u. The local system H^2 is trivial on U, so we have a trivialisation of the vector bundle $\mathcal{H}_{\mathbb{R}}^2$ restricted to U:

$$t : \mathcal{H}_{\mathbb{R}}^2 \cong H_{u,\mathbb{R}}^2 \times U.$$

Let $\phi : \mathcal{H}_{\mathbb{R}}^{1,1} \to H_{u,\mathbb{R}}^2$ be the composed map

$$\phi : \mathcal{H}_{\mathbb{R}}^{1,1} \hookrightarrow \mathcal{H}_{\mathbb{R}}^2 \xrightarrow{t} H_{u,\mathbb{R}}^2 \times U \to H_{u,\mathbb{R}}^2,$$

where the last arrow is the projection. Proposition 5.20 then follows from the following lemma.

Lemma 5.21 *If the image $\overline{\lambda} := p(\lambda)$ of λ in $\mathcal{H}_u^{1,1}$ satisfies the condition of the proposition, then the map ϕ is a submersion at the point λ.*

Temporarily admitting this lemma, we conclude as follows. ϕ is submersive and thus open in the neighbourhood of λ. As the rational points $\alpha \in H_u^2$ are dense in $H_{u,\mathbb{R}}^2$, the inverse images of these points under ϕ are dense in $\mathcal{H}_{\mathbb{R}}^{1,1}$. Now, clearly these are exactly the Hodge classes, since for $\lambda' \in \mathcal{H}_{u',\mathbb{R}}^{1,1}$, $\phi(\lambda')$ is simply the class $\lambda' \in H_{u',\mathbb{R}}^2$, viewed in $H_{u,\mathbb{R}}^2$ thanks to the trivialisation of the local system. But this trivialisation preserves the rational classes. Thus, $\phi(\lambda')$ is rational if and only if λ' is. □

Proof of lemma 5.21 As $\mathcal{H}_{\mathbb{R}}^{1,1}$ can be identified with the intersection of $\mathcal{H}_{\mathbb{R}}^2$ with the subbundle $F^1\mathcal{H}^2 \subset \mathcal{H}_{\mathbb{C}}^2$, it suffices to show that the composed map

$$\psi : F^1\mathcal{H}^2 \hookrightarrow \mathcal{H}_{\mathbb{C}}^2 \xrightarrow{t} H_{u,\mathbb{C}}^2 \times U \xrightarrow{\pi} H_{u,\mathbb{C}}^2,$$

where π is the projection, is submersive at the point λ. Indeed,

$$\mathcal{H}_{\mathbb{R}}^2 = (\pi \circ t)^{-1}\big(H_{u,\mathbb{R}}^2\big),$$

so $\mathcal{H}^{1,1}_{\mathbb{R}} = \psi^{-1}(H^2_{u,\mathbb{R}})$, and by definition, ϕ is the restriction of ψ to $\mathcal{H}^{1,1}_{\mathbb{R}}$. Thus, ϕ is submersive if ψ is.

Now, by the definition of the Gauss–Manin connection ∇, the map $\pi \circ t$ is such that for every differentiable section $\alpha : U \to \mathcal{H}^2_{\mathbb{C}}$, we have

$$\nabla\alpha = d(\pi \circ t)(\alpha) : T_{U,u} \to H^2_{u,\mathbb{C}}.$$

Applying this to the sections α of $F^1\mathcal{H}^2$ such that $\alpha(u) = \lambda$, we find by the definition of $\overline{\nabla}(\overline{\lambda})$ that the following diagram commutes:

$$\begin{array}{llcll} d\psi_\lambda : & T_{F^1\mathcal{H}^2,\lambda} & \longrightarrow & T_{H^2_{u,\mathbb{C}},\psi(\lambda)} & = \ H^2_{u,\mathbb{C}} \\ & \downarrow & & \downarrow & \\ \overline{\nabla}(\overline{\lambda}) : & T_{U,u} & \longrightarrow & H^2_{u,\mathbb{C}}/F^1\mathcal{H}^2_u. & \end{array}$$

As it is obvious that $\operatorname{Im} d\psi_\lambda$ contains $F^1\mathcal{H}^2_u$, we conclude that $d\psi$ is surjective if and only if $\overline{\nabla}(\overline{\lambda})$ is surjective. □

Exercises

1. *Complex tori and algebraic tori.* Let X be an n-dimensional complex torus and $W := H^0(X, \Omega_X)$.
   (a) Show that $H^1(X, \mathcal{O}_X) \cong \overline{W} \cong H^1(X, \mathbb{C})/W$ and

   $$H^1(X, T_X) \cong W^* \otimes \overline{W}. \qquad (5.15)$$

   (Use the fact that the tangent bundle of X is trivial.)
   (b) Let $\Gamma = H^1(X, \mathbb{Z})$, $\Gamma_{\mathbb{C}} = H^1(X, \mathbb{C})$. Construct a family of complex tori $\mathcal{X} \to U$ parametrised by an open set U of the Grassmannian $\operatorname{Grass}(n, \Gamma_{\mathbb{C}})$ containing the point 0 corresponding to $W \subset \Gamma_{\mathbb{C}}$, such that $\mathcal{X}_0 \cong X$. (Set $X_K = K^*/\Gamma^*_{\mathbb{Z}}$, $K \in \operatorname{Grass}(n, \Gamma_{\mathbb{C}})$.)
   (c) Show that the Kodaira–Spencer map

   $$\rho : T_{U,0} \cong \operatorname{Hom}(W, \Gamma_{\mathbb{C}}/W) \to H^1(X, T_X)$$

   is the inverse of the isomorphism (5.15). Deduce that in the neighbourhood of 0, $\mathcal{X} \to U$ is a universal local deformation of X.
   (d) Show that the map

   $$\overline{\nabla} : H^{1,1}(X) \to \operatorname{Hom}(H^1(X, T_X), H^{0,2}(X))$$

can be identified, via the natural isomorphisms

$$H^{1,1}(X) \cong W \otimes \overline{W}, \quad H^{0,2}(X) \cong \bigwedge^2 \overline{W},$$

with the map

$$W \otimes \overline{W} \to \mathrm{Hom}\left(W^* \otimes \overline{W}, \bigwedge^2 \overline{W}\right)$$

which is the composition of the contraction between W and W^* and the exterior product $\overline{W} \otimes \overline{W} \to \bigwedge^2 \overline{W}$.

(e) Deduce from (d), together with proposition 5.20 and Kodaira's embedding theorem (see vI.7.1.3), that the algebraic tori (or abelian varieties) are dense in the set of complex tori, and more precisely that every pair (X, η) where η is a Kähler class on X can be approximated by pairs (X_n, η_n) such that an integer multiple of η_n is the Chern class of a very ample invertible bundle on X_n.

2. *A non-reduced component of the Noether–Lefschetz locus.* Consider the open set $U \subset H^0(\mathbb{P}^3, \mathcal{O}_{\mathbb{P}^3}(5))$ consisting of the polynomials f such that the surface S_f with equation f is smooth. By the adjunction formula, these surfaces satisfy $K_{S_f} = \mathcal{O}_{S_f}(1)$, and thus $h^{2,0}(S_f) = 4$. Let $f \in U$ be such that the surface S_f contains two distinct lines Δ_1, Δ_2 which meet in a point. For $\alpha, \beta \in \mathbb{Z}$, set

$$\delta = \alpha[\Delta_1] + \beta[\Delta_2] \in H^2(S_f, \mathbb{Z}) \cap H^{1,1}(S).$$

Let $V \subset U$ be a connected and simply connected open set containing f.

(a) Using proposition 5.19, show that the Zariski tangent space at f of the component $V_\delta \subset V$ of the Noether–Lefschetz locus defined by δ is of codimension at most 3 in $T_{V,f}$. (Note that the 'expected' codimension is 4.)

Let us admit that codim $T_{V_\delta,f} \subset T_{V,f} = 3$ if $\alpha\beta \neq 0$.

(b) Show that the family U_{Δ_1,Δ_2} of polynomials f such that the surface X_f contains two lines which meet in a point is of codimension 4 in U and smooth. Show that

$$U_{\Delta_1,\Delta_2} \cap V \subset V_\delta.$$

(c) Admitting that the only component $V_{\delta'}$ of the Noether–Lefschetz locus for quintic surfaces which is smooth of codimension 3 is the family of the surfaces containing a conic C, where the class δ' is proportional to the class $[C]$ modulo the class $h = c_1(\mathcal{O}_S(1))$, show that V_δ is not smooth along U_{Δ_1,Δ_2} if $\alpha \neq \beta$. (Note that in the case where $\alpha = \beta$, δ is the class of the singular conic $\Delta_1 \cup \Delta_2$, and thus

V_δ is the set of surfaces containing a conic which is a deformation of $\Delta_1 \cup \Delta_2$, so it is smooth of codimension 3.) Deduce that if

$$\alpha\beta \neq 0 \text{ and } \alpha \neq \beta,$$

then $U_{\Delta_1 \Delta_2} \cap V$ is an irreducible component of V_δ, but V_δ is non-reduced along this component.

6

Hodge Filtration of Hypersurfaces

This chapter, which is devoted to the variations of Hodge structure of complete families of sufficiently ample hypersurfaces, reveals the algebraic nature of the theory of infinitesimal variations of Hodge structure. We give an explicit description of the complexes $\mathcal{K}_{p,q}$ introduced above, at least in the case of hypersurfaces of projective space. (One can find a similar description in Green (1984c) for the sufficiently ample hypersurfaces of any smooth variety.)

The first result, and maybe the most important one conceptually speaking, is the realisation of the vanishing cohomology of a hypersurface $Y \subset X$ as above, defined by a homogeneous equation $f = 0$, using residues $\mathrm{Res}_Y \eta/f^p$ for $\eta \in H^0(X, K_X(pY))$ of meromorphic forms with poles along Y, together with the fact that this realisation relates the Hodge level and the order p of the pole, as in the following theorem.

Theorem 6.1 (Griffiths) *The residues* $\mathrm{Res}_Y \eta/f^p$ *for* $\eta \in H^0(X, K_X(pY))$ *generate* $F^{n-p}H^{n-1}(Y)_{\mathrm{van}}$, *where* $n = \dim X$.

The second result, more computational in nature, allows us to deduce the following description of the Dolbeault cohomology of a hypersurface $Y \subset \mathbb{P}^n$ of degree d from the preceding theorem. Letting S denote the ring of homogeneous polynomials on $\mathbb{P}^n$ and Ω a generator of $H^0(\mathbb{P}^n, K_{\mathbb{P}^n}(n+1))$, we have the following result.

Proposition 6.2 *The kernel of the map* $\overline{\alpha}_p : S^{pd-n-1} \to \mathrm{Gr}_F^{n-p} H^{n-1}(Y)_{\mathrm{van}}$, *which to* P *associates*

$$\overline{\alpha}_p(P) = \mathrm{Res}_Y P\Omega/f^p \mod F^{n-p+1}H^{n-1}(Y)_{\mathrm{van}},$$

is equal to the degree $pd - n - 1$ component J_f^{pd-n-1} of the Jacobian ideal of f, and thus induces an isomorphism $R_f^{pd-n-1} \cong H^{n-p,p-1}(Y)_{\text{van}}$, where $R_f = S/J_f$ is the Jacobian ring of f.

The last important result we introduce here concerns the computation of the map

$$\overline{\nabla} : H^{n-p,p-1}(Y)_{\text{van}} \to \operatorname{Hom}(S^d, H^{n-p-1,p}(Y)_{\text{van}}),$$

which describes the infinitesimal variation of Hodge structure on the vanishing (or primitive) cohomology of the hypersurfaces Y_t at the point f. Here, S^d is viewed as the tangent space of the universal family of hypersurfaces at the point f. One can show that via the preceding identification, this arrow can be identified up to a coefficient with the map given by multiplication

$$R_f^{pd-n-1} \to \operatorname{Hom}\left(S^d, R_f^{(p+1)d-n-1}\right).$$

The remainder of the chapter is devoted to the study of the algebraic properties of the Jacobian rings of hypersurfaces of projective space. On the one hand, we prove Macaulay's theorem, which in modern language states that these rings are Gorenstein rings, and on the other hand, we prove the symmetriser lemma. From the point of view of infinitesimal variations of Hodge structure, these statements imply the exactness of the complexes $\mathcal{K}_{p,q}$, $p + q = n - 1$, in degree 0 for $q < n - 1$ and in degree 1 for $q < n - 2$, whenever d is sufficiently large. We can deduce a generalisation of the Noether–Lefschetz theorem in the following form. Let $U \subset S^d$ be an open set parametrising smooth hypersurfaces, and let λ be a non-zero section of the local system $H^{n-1}_{\mathbb{C},\text{ev}}$ on U.

Theorem 6.3 *If the integer d is sufficiently large, then the Hodge locus*

$$U_\lambda^1 = \{t \in U \mid \lambda_t \in F^1 H^{n-1}(Y_t)\}$$

is a proper analytic subset of U.

In chapter 7, we will give an application of the symmetriser lemma to the study of the Abel–Jacobi map.

6.1 Filtration by the order of the pole

6.1.1 Logarithmic complexes

Let us recall some results from vI.8.2 and vI.8.4. Let X be a projective variety, and $Y \overset{l}{\hookrightarrow} X$ a smooth hypersurface. Set

$$U = X - Y \overset{j}{\hookrightarrow} X.$$

Recall (see vI.8.2.2) that the logarithmic de Rham complex $(\Omega_X^\bullet(\log Y), d)$ is the complex of free $\mathcal{O}_X$-modules defined as follows:

$$\Omega_X^k(\log Y) = \bigwedge^k \Omega_X(\log Y),$$

where $\Omega_X(\log Y)$ is the sheaf of free $\mathcal{O}_X$-modules locally generated by Ω_X and df/f, where f is a local holomorphic equation for Y. The differential is the exterior differential. This complex can be viewed as a subcomplex of the complex $j_*\mathcal{A}_U^\bullet$, and we have the following result, which was proved in vI.8.2.3.

Theorem 6.4 *The inclusion*

$$\Omega_X^\bullet(\log Y) \hookrightarrow j_*\mathcal{A}_U^\bullet$$

is a quasi-isomorphism.

As the de Rham complex $(\mathcal{A}_U^\bullet, d)$ is a resolution of the constant sheaf $\mathbb{C}$ on U, this implies that

$$H^k(U, \mathbb{C}) \cong \mathbb{H}^k(X, j_*\mathcal{A}_U^\bullet) \cong \mathbb{H}^k(X, \Omega_X^\bullet(\log Y)),$$

where the first equality follows from the fact that each $\mathcal{A}_U^k$ is fine and thus acyclic for the functor j_*. We then define the Hodge filtration $F^\cdot$ on each $H^k(U, \mathbb{C})$ as the filtration induced by the Hodge filtration $F^\cdot$ on the logarithmic de Rham complex

$$F^p\Omega_X^\bullet(\log Y) = 0 \to \Omega_X^p(\log Y) \to \Omega_X^{p+1}(\log Y) \to \cdots.$$

Furthermore, we have two morphisms of complexes: the inclusion

$$\Omega_X^\bullet \hookrightarrow \Omega_X^\bullet(\log Y),$$

which in cohomology induces the restriction

$$H^k(X, \mathbb{C}) \to H^k(U, \mathbb{C}), \tag{6.1}$$

and the residue

$$\text{Res} : \Omega_X^\bullet(\log Y) \to \Omega_Y^{\bullet-1},$$

which to $\alpha \wedge (df/f)$ associates $\text{Res}\,\alpha \wedge (df/f) = 2i\pi\alpha_{|Y}$. The map induced by Res in cohomology is the residue

$$H^k(U, \mathbb{C}) \to H^{k-1}(Y, \mathbb{C}), \tag{6.2}$$

which we can also define as the composition

$$H^k(U, \mathbb{C}) \to H^{k+1}(X, U, \mathbb{C}) \cong H^{k+1}(T, \partial T, \mathbb{C}) \cong H^{k-1}(Y, \mathbb{C}),$$

where T is a tubular neighbourhood of Y in X, the first arrow is the connecting morphism in the long exact sequence of relative cohomology, the second is the excision isomorphism, and the last is the Thom isomorphism.

In particular, we see that the morphisms (6.1) and (6.2) are defined over $\mathbb{Z}$ and compatible with the Hodge filtrations. (More precisely, the morphism Res sends $F^pH^k(U, \mathbb{C})$ to $F^{p-1}H^{k-1}(Y, \mathbb{C})$, so that it is compatible with the Hodge filtrations after shifting the bidegrees of the Hodge structure on $H^{k-1}(Y, \mathbb{C})$ by $(1, 1)$: we call this operation a Tate twist. It also makes it possible to ingeniously get rid of the coefficient $2\pi i$ in the definition of the residue.)

Note that by the description given above, the residue is part of the long exact sequence of relative cohomology of the pair (X, U), which thanks to the Thom isomorphism $H^{k+1}(X, U, \mathbb{Z}) \cong H^{k-1}(Y, \mathbb{Z})$ given above, can be written

$$\cdots H^k(X, \mathbb{Z}) \to H^k(U, \mathbb{Z}) \stackrel{\text{Res}}{\to} H^{k-1}(Y, \mathbb{Z}) \to H^{k+1}(X, \mathbb{Z}) \cdots.$$

One can show (see vI.8.4.2) that the last arrow is the Gysin morphism

$$l_* : H^{k-1}(Y, \mathbb{Z}) \to H^{k+1}(X, \mathbb{Z}).$$

Now, assume that $\dim X = n = k$ and that Y is an ample hypersurface of X. Then, passing to the rational cohomology, by the Lefschetz decomposition associated to the polarisation given by $\mathcal{O}_X(Y)$, we have

$$H^n(X, \mathbb{Q})/l_*H^{k-2}(Y, \mathbb{Q}) = H^n(X, \mathbb{Q})_{\text{prim}},$$
$$\text{Ker}\,(l_* : H^{n-1}(Y, \mathbb{Q}) \to H^{n+1}(X, \mathbb{Q})) =: H^{n-1}(Y, \mathbb{Q})_{\text{van}}$$

and a short exact sequence compatible with the Hodge filtrations:

$$0 \to H^n(X, \mathbb{Q})_{\text{prim}} \stackrel{j^*}{\to} H^n(U, \mathbb{Q}) \stackrel{\text{Res}}{\to} H^{n-1}(Y, \mathbb{Q})_{\text{van}} \to 0. \tag{6.3}$$

The morphisms j^* and Res are in fact strictly compatible with the Hodge filtrations by theorem 4.20, since the weight filtration W

$$W_0 H^n(U, \mathbb{Q}) = j^* H^n(X, \mathbb{Q})_{\text{prim}}, \quad W_1 H^n(U, \mathbb{Q}) = H^n(U, \mathbb{Q})$$

and the Hodge filtration equip $H^n(U, \mathbb{Q})$ with a mixed Hodge structure (after shifting the bidegree of the Hodge structure on $H^{n-1}(Y, \mathbb{Q})_{\text{van}}$ as above).

6.1.2 Hodge filtration and filtration by the order of the pole

Let us now make the following hypotheses:

(*) *For every $k > 0$, $i > 0$, $j \geq 0$, we have*

$$H^i\big(X, \Omega_X^j(kY)\big) = 0,$$

where $\Omega_X^j(kY) = \Omega_X^j \otimes \mathcal{O}_X(Y)^{\otimes k}$.

These hypotheses are satisfied by Serre's vanishing theorem (see volume I, chapter 7, exercise 2) if Y is sufficiently ample in X, i.e. if $\mathcal{O}_X(Y)$ is a sufficiently large multiple of an ample line bundle on X. They are also satisfied by Bott's vanishing theorem if X is the projective space $\mathbb{P}^n$.

Under these hypotheses (which can actually be weakened, as we easily see from the proof below), we have the following result.

Theorem 6.5 (Griffiths 1969) *For every integer p between 1 and n, the image of the natural map*

$$H^0(X, K_X(pY)) \to H^n(U, \mathbb{C}) \tag{6.4}$$

which to a section α (viewed as a meromorphic form on X of degree n, and therefore closed, holomorphic on U and having a pole of order p along Y) associates its de Rham cohomology class, is equal to $F^{n-p+1}H^n(U)$.

We begin by proving the following lemma.

Lemma 6.6 *Let α be a meromorphic closed differential form of degree k defined on an open set V of X, holomorphic outside $V \cap Y$ and having a pole of order l along $V \cap Y$. Then if $l \geq 2$, locally in V we can write*

$$\alpha = d\beta + \gamma,$$

where β and γ are meromorphic with a pole of order at most $l-1$ along $V \cap Y$, and holomorphic outside Y.

If $l = 1$, then α is a logarithmic form.

Proof The last assertion is a consequence of the following characterisation of forms with logarithmic singularities along Y: they are those which have a pole of order at most 1 and whose exterior differential has a pole of order at most 1 along Y (see vI.8.2.2).

Assume then that $l \geq 2$, and let us write

$$\alpha = \frac{dz_1 \wedge \alpha'}{z_1^l} + \frac{\alpha''}{z_1^l},$$

locally, where we have chosen holomorphic coordinates $z_1, \ldots, z_n$ such that Y is defined by $z_1 = 0$, and the forms α' and α'' are holomorphic and do not contain dz_1.

Considering the order of the pole, the fact that $d\alpha = 0$ then immediately implies that the form α'' vanishes along $z_1 = 0$, so that $\frac{\alpha''}{z_1^l}$ has a pole of order at most $l-1$ along Y.

Moreover, let $\beta = -\frac{\alpha'}{(l-1)z_1^{l-1}}$. Then clearly

$$\frac{dz_1 \wedge \alpha'}{z_1^l} = d\beta$$

modulo a form having a pole of order at most $l-1$ along Y. □

Corollary 6.7 *If moreover the degree k of α is at least 2, we can write $\alpha = d\beta$ locally, where β is a meromorphic form of degree $k-1$, having a pole of order at most $l-1$ along Y.*

Proof Reasoning by induction on the order l of the pole, we deduce from the lemma that α can be written $d\beta + \alpha'$, where β is a meromorphic form of degree $k-1$, with a pole of order at most $l-1$ along Y, and α' has logarithmic singularities along Y. But we showed in vI.8.2.3 that the holomorphic logarithmic de Rham complex is locally exact in degree ≥ 2 if the hypersurface Y is smooth. Thus, if $k \geq 2$, then $\alpha' = d\gamma$, where γ has logarithmic singularities along Y. □

Proof of theorem 6.5 Let $\Omega_X^{k,c}(lY)$ denote the sheaf of closed meromorphic differential forms of degree k, holomorphic on U and having a pole of order at most l along Y. Then corollary 6.7 yields the following exact sequences for $l \geq 2,\ k \geq 2$:

$$0 \to \Omega_X^{k-1,c}((l-1)Y) \to \Omega_X^{k-1}((l-1)Y) \xrightarrow{d} \Omega_X^{k,c}(lY) \to 0. \tag{6.5}$$

Finally, if $l = 1$, we have the equality

$$\Omega_X^{k,c}(\log Y) = \Omega_X^{k,c}(lY). \tag{6.6}$$

Starting from $K_X(pY) = \Omega_X^{n,c}(pY)$, and (if $p \geq 2$) applying (6.5) repeatedly, then because $n - p + 2 \geq 2$, we obtain

$$0 \to \Omega_X^{n-1,c}((p-1)Y) \to \Omega_X^{n-1}((p-1)Y) \xrightarrow{d} K_X(pY) \to 0,$$
$$0 \to \Omega_X^{n-2,c}((p-2)Y) \to \Omega_X^{n-2}((p-2)Y) \xrightarrow{d} \Omega_X^{n-1,c}((p-1)Y) \to 0,$$
$$\cdots$$
$$0 \to \Omega_X^{n-p+1,c}(\log Y) \to \Omega_X^{n-p+1}(Y) \xrightarrow{d} \Omega_X^{n-p+2,c}(2Y) \to 0.$$

Now consider the long exact sequences of cohomology associated to these short exact sequences. Applying the vanishing hypotheses (*), we obtain surjective maps

$$H^0(X, K_X(pY)) \twoheadrightarrow H^1\big(X, \Omega_X^{n-1,c}((p-1)Y)\big),$$
$$H^1\big(X, \Omega_X^{n-1,c}((p-1)Y)\big) \twoheadrightarrow H^2\big(X, \Omega_X^{n-2,c}((p-2)Y)\big),$$
$$\cdots$$
$$H^{p-2}\big(X, \Omega_X^{n-p+2,c}(2Y)\big) \twoheadrightarrow H^{p-1}\big(X, \Omega_X^{n-p+1,c}(\log Y)\big).$$

To conclude the proof of the theorem, note that by the definition of the Hodge filtration on $H^n(U, \mathbb{C})$, we have

$$H^{p-1}\big(X, \Omega_X^{n-p+1,c}(\log Y)\big) = F^{n-p+1}H^n(U, \mathbb{C}).$$

Indeed, the logarithmic de Rham complex is exact in degree ≥ 2, so for $k \geq 1$, the complex

$$F^k\Omega_X^\bullet(\log Y) = 0 \to \Omega_X^k(\log Y) \to \Omega_X^{k+1}(\log Y) \to \cdots$$

is a resolution of the sheaf $\Omega_X^{k,c}(\log Y)$. As $n - p + 1 \geq 1$, we obtain

$$H^{p-1}\big(X, \Omega_X^{n-p+1,c}(\log Y)\big) = \mathbb{H}^n(F^{n-p+1}\Omega_X^\bullet(\log Y)).$$

Now, by definition, we have

$$F^{n-p+1}H^n(U, \mathbb{C}) = \operatorname{Im}(\mathbb{H}^n(F^{n-p+1}\Omega_X^\bullet(\log Y)) \to \mathbb{H}^n(\Omega_X^\bullet(\log Y))),$$

and this last map is injective by the degeneracy at E_1 of the Frölicher spectral sequence (see vI.8.4.3).

We have thus obtained a surjective map

$$H^0(X, K_X(pY)) \twoheadrightarrow F^{n-p+1}H^n(U, \mathbb{C}),$$

and so all we have to do is check that this map is indeed the map which to a meromorphic form of degree n, holomorphic on U, associates its de Rham cohomology class. But this is easy. □

6.1.3 The case of hypersurfaces of $\mathbb{P}^n$

Now assume that X is the projective space $\mathbb{P}^n$, and that Y is a smooth hypersurface of degree d, with equation $f = 0$. We know that $K_{\mathbb{P}^n} = \mathcal{O}_{\mathbb{P}^n}(-n-1)$, a generator of $H^0(\mathbb{P}^n, K_{\mathbb{P}^n}(n+1))$ being given by

$$\begin{aligned}\Omega &= \sum_i (-1)^i X_i dX_0 \wedge \cdots \wedge d\hat{X}_i \wedge \cdots \wedge dX_n \\ &= X_0 \ldots X_n \sum_i (-1)^i \frac{dX_0}{X_0} \wedge \cdots \wedge \frac{d\hat{X}_i}{X_i} \wedge \cdots \wedge \frac{dX_n}{X_n},\end{aligned}$$

where the X_i are homogeneous coordinates on $\mathbb{P}^n$. As $\mathcal{O}_{\mathbb{P}^n}(Y) = \mathcal{O}_{\mathbb{P}^n}(d)$, theorem 6.5 shows that for every p, $1 \leq p \leq n$, we have a surjective map

$$\alpha_p : H^0(\mathbb{P}^n, \mathcal{O}_{\mathbb{P}^n}(pd - n - 1)) \to F^{n-p+1}H^n(U, \mathbb{C}) \cong F^{n-p}H^{n-1}(Y, \mathbb{C})_{\mathrm{van}},$$

which to a polynomial P associates the residue of the class of the meromorphic form $\frac{P\Omega}{f^p}$. Here, the last isomorphism follows from the exact sequence (6.3) and the fact that $H^n(\mathbb{P}^n, \mathbb{C})_{\mathrm{prim}} = 0$.

Remark 6.8 *For hypersurfaces in projective space, the vanishing middle degree cohomology is the same as the primitive cohomology. Hence we shall use the notation* $H^*(Y)_{\mathrm{prim}}$, *which is more common.*

Definition 6.9 *The Jacobian ideal* $J_f = \bigoplus J_f^l$ *of* f *is the homogeneous ideal of the ring of polynomials*

$$S = \bigoplus_l S^l, \quad S^l = \mathrm{Sym}^l S^1 = H^0(\mathbb{P}^n, \mathcal{O}_{\mathbb{P}^n}(l))$$

generated by the partial derivatives $\frac{\partial f}{\partial X_i}$, $i = 0, \ldots, n$.

Clearly, this ideal does not depend on the choice of coordinates, since every other coordinate system is obtained by applying a linear transformation to the coordinates X_i.

Theorem 6.10 (Griffiths 1969) *The kernel of the composed map*

$$\overline{\alpha}_p : H^0(\mathbb{P}^n, \mathcal{O}_{\mathbb{P}^n}(pd - n - 1)) \to F^{n-p}H^{n-1}(Y, \mathbb{C})$$
$$\to F^{n-p}H^{n-1}(Y, \mathbb{C})/F^{n-p+1}H^{n-1}(Y, \mathbb{C}) = H^{n-p,p-1}(Y) \tag{6.7}$$

is equal to J_f^{pd-n-1}.

Proof We know that the image of the map α_{p-1} is $F^{n-p+1}H^{n-1}(Y, \mathbb{C})_{\text{prim}}$, so we have

$$P \in \operatorname{Ker}\overline{\alpha}_p \Leftrightarrow \exists Q \in S^{(p-1)d-n-1}, \quad \alpha_{p-1}(Q) = \alpha_p(P).$$

As we clearly have

$$\alpha_{p-1}(Q) = \alpha_p(fQ),$$

this is equivalent to $P - fQ \in \operatorname{Ker}\alpha_p$.

Now, the kernel of α_p is in fact described in the proof of theorem 6.5. Indeed, as the vanishing hypotheses (*) are satisfied, we find that the maps

$$H^1\big(\mathbb{P}^n, \Omega_{\mathbb{P}^n}^{n-1,c}((p-1)Y)\big) \twoheadrightarrow H^2\big(\mathbb{P}^n, \Omega_{\mathbb{P}^n}^{n-2,c}((p-2)Y)\big),$$
$$\dots$$
$$H^{p-2}\big(\mathbb{P}^n, \Omega_{\mathbb{P}^n}^{n-p+2,c}(2Y)\big) \twoheadrightarrow H^{p-1}\big(\mathbb{P}^n, \Omega_{\mathbb{P}^n}^{n-p+1,c}(\log Y)\big)$$

are all isomorphisms. Moreover, as $H^n(\mathbb{P}^n, \mathbb{C})_{\text{prim}} = 0$, the arrow

$$\operatorname{Res} : H^{p-1}\big(\mathbb{P}^n, \Omega_{\mathbb{P}^n}^{n-p+1,c}(Y)\big) = F^{n-p+1}H^n(U, \mathbb{C}) \to F^{n-p}H^{n-1}(Y, \mathbb{C})_{\text{prim}}$$

is also an isomorphism. It follows that $\operatorname{Ker}\alpha_p$ is equal to the kernel of the map

$$H^0(\mathbb{P}^n, K_{\mathbb{P}^n}(pY)) \twoheadrightarrow H^1\big(\mathbb{P}^n, \Omega_{\mathbb{P}^n}^{n-1,c}((p-1)Y)\big),$$

or also, by the exact sequence

$$0 \to \Omega_{\mathbb{P}^n}^{n-1,c}((p-1)Y) \to \Omega_{\mathbb{P}^n}^{n-1}((p-1)Y) \xrightarrow{d} K_{\mathbb{P}^n}(pY) \to 0,$$

to $dH^0(\mathbb{P}^n, \Omega_{\mathbb{P}^n}^{n-1}((p-1)Y))$. The theorem is thus equivalent to the following statement.

Lemma 6.11 *Let* $P \in S^{pd-n-1}$. *Then there exists* $Q \in S^{(p-1)d-n-1}$ *such that*

$$\frac{(P - fQ)}{f^p}\Omega = d\gamma, \quad \gamma \in H^0\big(\mathbb{P}^n, \Omega_{\mathbb{P}^n}^{n-1}((p-1)d)\big)$$

if and only if $P \in J_f^{pd-n-1}$.

Proof of lemma 6.11 Let $\gamma \in H^0(\mathbb{P}^n, \Omega^{n-1}_{\mathbb{P}^n}((p-1)d))$. γ is considered as a meromorphic form of degree $n-1$ on $\mathbb{P}^n$, holomorphic on U and having a pole of order at most $p-1$ along Y. Moreover, by the interior product, we have a canonical isomorphism

$$\mathrm{int}()(\Omega) : T_{\mathbb{P}^n} \to \Omega^{n-1}_{\mathbb{P}^n}(n+1),$$

and thus an isomorphism

$$\mathrm{int}()(\Omega) : H^0(\mathbb{P}^n, T_{\mathbb{P}^n}((p-1)d-n-1)) \to H^0\big(\mathbb{P}^n, \Omega^{n-1}_{\mathbb{P}^n}((p-1)d)\big).$$

Finally, we have the Euler exact sequence

$$0 \to \mathcal{O}_{\mathbb{P}^n} \to V \otimes \mathcal{O}_{\mathbb{P}^n}(1) \to T_{\mathbb{P}^n} \to 0, \tag{6.8}$$

where $V = (S^1)^*$ is generated by the $\frac{\partial}{\partial X_i}$, $i = 0, \ldots, n$. Here, we consider $\frac{\partial}{\partial X_i}$ as a vector field on $\mathbb{C}^{n+1}$, which does not descend to a vector field on $\mathbb{P}^n$, whereas the vector fields $X_j \frac{\partial}{\partial X_i}$ on $\mathbb{C}^{n+1}$ do descend to vector fields on $\mathbb{P}^n$. The Euler vector field $E = \sum_i X_i \frac{\partial}{\partial X_i}$ generates the kernel of the exact sequence (6.8), as it is the vector field tangent to the fibres of

$$\pi : \mathbb{C}^{n+1} - \{0\} \to \mathbb{P}^n.$$

From the Euler exact sequence together with the fact that $H^1(\mathbb{P}^n, \mathcal{O}_{\mathbb{P}^n}(k)) = 0$ for $k \geq 0$, we deduce that $H^0(\mathbb{P}^n, T_{\mathbb{P}^n}((p-1)d-n-1))$ is generated by the vector fields $\sum_i P_i \frac{\partial}{\partial X_i}$ with

$$P_i \in H^0(\mathbb{P}^n, \mathcal{O}_{\mathbb{P}^n}((p-1)d-n)) = S^{(p-1)d-n}.$$

Thus, γ can be written

$$\frac{\sum_i P_i \,\mathrm{int}\big(\frac{\partial}{\partial X_i}\big)(\Omega)}{f^{p-1}},$$

with P_i arbitrary in $S^{(p-1)d-n}$. (More precisely, this expression should be understood as that of the pullback of γ on $\mathbb{C}^{n+1} - \{0\}$.)

We thus obtain

$$d\gamma = -(p-1)\frac{\sum_{i,j} P_i \frac{\partial f}{\partial X_j} dX_j \wedge \mathrm{int}\big(\frac{\partial}{\partial X_i}\big)(\Omega)}{f^p} + \frac{d\Big(\sum_i P_i \,\mathrm{int}\big(\frac{\partial}{\partial X_i}\big)(\Omega)\Big)}{f^{p-1}}.$$

Clearly, the second term in this expression is a meromorphic form with a pole of order $\leq p-1$ along Y. Recalling also that

$$\Omega = \sum_i (-1)^i X_i dX_0 \wedge \cdots \wedge d\hat{X}_i \wedge \cdots \wedge dX_n,$$

we find

$$dX_j \wedge \operatorname{int}\left(\frac{\partial}{\partial X_i}\right)(\Omega) = -\operatorname{int}\left(\frac{\partial}{\partial X_i}\right)(dX_j \wedge \Omega) + \delta_{ij}\Omega$$
$$= -\operatorname{int}\left(\frac{\partial}{\partial X_i}\right)(X_j dX_0 \wedge \cdots \wedge dX_n) + \delta_{ij}\Omega.$$

Finally, we obtain

$$d\gamma = -\frac{(p-1)}{f^p}\left(\sum_i P_i \frac{\partial f}{\partial X_i}\right)\Omega$$
$$+\frac{(p-1)}{f^p}\sum_{i,j} P_i X_j \frac{\partial f}{\partial X_j}\operatorname{int}\left(\frac{\partial}{\partial X_i}\right)(dX_0 \wedge \cdots \wedge dX_n)$$

modulo a meromorphic form having a pole of order at most $p-1$ along Y. Then, using the Euler relation

$$\sum_j X_j \frac{\partial f}{\partial X_j} = \deg(f) f,$$

we see that the last term admits a pole of order at most $p-1$ along Y, so that

$$d\gamma = -\frac{(p-1)}{f^p}\sum_i P_i \frac{\partial f}{\partial X_i}\Omega$$

modulo a meromorphic form having a pole of order at most $p-1$ along Y.

As the P_i are arbitrary, we deduce that a form $\frac{P\Omega}{f^p}$ can be written $d\gamma$ modulo a form having a pole of order at most $p-1$ along Y if and only if $P \in J_f^{pd-n-1}$. This proves the lemma, and thus concludes the proof of theorem 6.10. □

As the map $\overline{\alpha}_p : S^{pd-n-1} \to F^{n-p}H^{n-1}(Y,\mathbb{C})_{\text{prim}}/F^{n-p+1}H^{n-1}(Y,\mathbb{C})_{\text{prim}}$ is surjective by theorem 6.5, and moreover

$$F^{n-p}H^{n-1}(Y,\mathbb{C})_{\text{prim}}/F^{n-p+1}H^{n-1}(Y,\mathbb{C})_{\text{prim}} = H^{n-p,p-1}(Y)_{\text{prim}}$$
$$:= H^{n-p,p-1}(Y) \cap H^{n-1}(Y,\mathbb{C})_{\text{prim}},$$

we deduce from theorem 6.10 the following description of the primitive Dolbeault cohomology groups of Y.

Corollary 6.12 *The residue map induces a natural isomorphism*

$$R_f^{pd-n-1} \cong H^{n-p,p-1}(Y)_{\text{prim}},$$

where $R_f^l := S^l/J_f^l$ denotes the lth component of the Jacobian ring $R_f = S/J_f$.

6.2 IVHS of hypersurfaces

6.2.1 Computation of $\overline{\nabla}$

Let X be an n-dimensional projective variety, and let $Y \subset X$ be a hypersurface satisfying the conditions (*) of 6.1.2. Let $B \subset H^0(X, \mathcal{O}_X(Y))$ denote the Zariski open set consisting of the polynomials f such that the hypersurface Y with equation $f = 0$ is smooth. We have the universal smooth hypersurface

$$\pi : \mathcal{Y} \to B,$$

and we will describe the infinitesimal variation of Hodge structure on $H^{n-1}(Y, \mathbb{C})_{\text{van}}$ for $f \in B$, i.e. the maps

$$\overline{\nabla}_{l,f} : H^{l,n-1-l}(Y)_{\text{van}} \to \operatorname{Hom}(T_{B,f}, H^{l-1,n-l}(Y)_{\text{van}}).$$

By theorem 6.5, we have the surjective maps of (6.7)

$$\overline{\alpha}_p : H^0(X, K_X(pY)) \to H^{n-p,p-1}(Y)_{\text{van}}.$$

We have the following result due to Carlson & Griffiths (1980).

Theorem 6.13 *The map*

$$\overline{\nabla}_{n-p,f} : H^{n-p,p-1}(Y)_{\text{van}} \to \operatorname{Hom}(T_{B,f}, H^{n-p-1,p}(Y)_{\text{van}})$$

can be described as follows. For $P \in H^0(X, K_X(pY))$ and $H \in T_{B,f} = H^0(X, \mathcal{O}_X(Y))$, we have

$$\overline{\nabla}_{n-p,f}(\overline{\alpha}_p(P))(H) = -p\overline{\alpha}_{p+1}(PH).$$

In other words, the diagram

$$\begin{array}{cccc} & H^0(X, K_X(pY)) & \longrightarrow & \operatorname{Hom}(H^0(X, \mathcal{O}_X(Y)), H^0(X, K_X((p+1)Y))) \\ & \downarrow & & \downarrow \\ \overline{\nabla}: & H^{n-p,p-1}(Y)_{\text{van}} & \longrightarrow & \operatorname{Hom}(T_{B,f}, H^{n-p-1,p}(Y)_{\text{van}}), \end{array}$$

where the upper horizontal arrow is given by multiplication, and the vertical arrows are given by $\overline{\alpha}_p$, $\overline{\alpha}_{p+1}$ and the identification $T_{B,f} = H^0(X, \mathcal{O}_X(Y))$, is commutative up to a multiplicative coefficient.

Proof Note that B also parametrises the universal open set

$$\pi_U : \mathcal{U} \to B, \quad \mathcal{U} := X \times B - \mathcal{Y}.$$

We showed earlier that the residue induces a morphism of local systems

$$\mathrm{Res} : (R^n\pi_U)_*\mathbb{Z} \twoheadrightarrow R^{n-1}\pi_*\mathbb{Z}_{\mathrm{van}}.$$

Writing

$$\mathcal{H}^n_U, \ \text{resp.}\ \mathcal{H}^{n-1}_{\mathrm{van}}$$

for the corresponding flat vector bundles, the residue map then commutes with the Gauss–Manin connections, which we denote by ∇_U and ∇ respectively.

Recall that if $\lambda \in H^{n-p,p-1}(Y)_{\mathrm{van}}$ and $H \in T_{B,f}$, then

$$\overline{\nabla}_H(\lambda) \in H^{n-p-1,p}(Y)_{\mathrm{van}}$$

is obtained by choosing a section $\tilde{\lambda} \in F^{n-p}\mathcal{H}^{n-1}_{\mathrm{van}}$ such that the image of $\tilde{\lambda}(f)$ in

$$\left(F^{n-p}\mathcal{H}^{n-1}_{\mathrm{van}}/F^{n-p+1}\mathcal{H}^{n-1}_{\mathrm{van}}\right)_f = H^{n-p,p-1}(Y)_{\mathrm{van}}$$

is equal to λ. We then have

$$\overline{\nabla}_f(\lambda)(H) = \nabla_H\tilde{\lambda} \bmod \left(F^{n-p}\mathcal{H}^{n-1}_{\mathrm{van}}\right)_f.$$

Now let $P \in H^0(X, K_X(pY))$ be such that $\overline{\alpha}_p(P) = \lambda$. Here, we identify $K_X(pY)$ with $K_X \otimes \mathcal{L}^{\otimes p}$, $\mathcal{L} = \mathcal{O}_X(Y)$. Then $f \in H^0(X, \mathcal{L})$ and $\frac{P\Omega}{f^p}$ is a meromorphic form with a pole of order p along Y. By theorem 6.5, the section

$$g \mapsto \mathrm{Res}_{Y_g}\left[\frac{P\Omega}{g^p}\right], \quad g \in B$$

of $\mathcal{H}^{n-1}_{\mathrm{van}}$, where the bracket denotes the de Rham cohomology class of the closed form considered, is a section of $F^{n-p}\mathcal{H}^{n-1}_{\mathrm{van}}$ whose value at the point f projects to λ. It follows that

$$\begin{aligned}\overline{\nabla}_f(\lambda)(H) &= \nabla_H\left(\mathrm{Res}_{Y_g}\left[\frac{P\Omega}{g^p}\right]\right) \bmod F^{n-p}H^{n-1}(Y)_{\mathrm{van}}\\ &= \mathrm{Res}_Y\nabla_{U,H}\left(\left[\frac{P\Omega}{g^p}\right]\right) \bmod F^{n-p}H^{n-1}(Y)_{\mathrm{van}},\end{aligned}$$

where $g \mapsto [\frac{P\Omega}{g^p}]$ is a section of $\mathcal{H}^n_U$.

Lemma 6.14 *Let $\omega = (\omega_g)_{g\in B}$ be a family of closed singular differential forms on X, where ω_g is $\mathcal{C}^\infty$ on U_g and varies holomorphically with g. Then* *the corresponding section $\sigma \in \mathcal{H}_U^n$ defined by $\sigma_g = [\omega_g]$ is holomorphic, and we have*

$$\nabla_{U,H}\sigma = [d_H\omega],$$

where d_H is the derivative with respect to the tangent vector $H \in T_{B,f}$.

Temporarily admitting this lemma, we conclude that

$$\begin{aligned}\overline{\nabla}_f(\lambda)(H) &= \mathrm{Res}_Y \nabla_{U,H}\left(\left[\frac{P}{g^p}\right]\right) \mod F^{n-p}H^{n-1}(Y)_{\mathrm{van}} \\ &= \mathrm{Res}_Y\left[d_H\frac{P}{g^p}\right] = \mathrm{Res}_Y\left(\left[\frac{-pPH}{f^{p+1}}\right]\right) \mod F^{n-p}H^{n-1}(Y)_{\mathrm{van}},\end{aligned}$$

which proves theorem 6.13. □

Proof of lemma 6.14 The statement is local, so we can replace B by a small open set W containing f. We can then find a tubular neighbourhood T of Y which retracts by deformation to Y_g, for every $g \in W$. Then $X - T$ has the homotopy type of $U_g = X - Y_g$ for every g, and in the statement, we can replace the family of open sets

$$\pi_U : \pi_U^{-1}(W) \to W$$

by the constant open set $(X - T) \times W$. But then the lemma simply states that for a family of closed forms $(\omega_g)_{g\in W}$ on the constant variety $X - T$, and for a tangent vector $H \in T_{W,f}$, we have

$$d_H([\omega_g]) = [d_H\omega_g],$$

which is obvious. □

Now consider the case where $X = \mathbb{P}^n$ and B is the family of polynomials of degree d parametrising smooth hypersurfaces.

Lemma 6.15 *If $n \geq 2$, the kernel of the Kodaira–Spencer map*

$$\rho : T_{B,f} = S^d \to H^1(Y, T_Y)$$

is J_f^d. Moreover, this map is surjective if $n \geq 4$, or $n = 3$, $d \neq 4$.

Proof By vI.9.1.2, we know that ρ is the connecting morphism associated to the exact sequence

$$0 \to T_Y \to T_{\mathcal{Y}|Y} \to \pi^* T_{B,f} \to 0,$$

where $Y \subset \mathcal{Y}$ is identified with the fibre $\pi^{-1}(f)$. Now, as $\mathcal{Y} \subset \mathbb{P}^n \times B$, the map

$$\mathrm{pr}_{1*} : T_{\mathcal{Y}|Y} \to T_{\mathbb{P}^n|Y}$$

gives a commutative diagram

$$\begin{array}{ccccccccc} 0 & \longrightarrow & T_Y & \longrightarrow & T_{\mathcal{Y}|Y} & \longrightarrow & \pi^* T_{B,f} & \longrightarrow & 0 \\ & & \| & & \downarrow & & \downarrow & & \\ 0 & \longrightarrow & T_Y & \longrightarrow & T_{\mathbb{P}^n|Y} & \longrightarrow & \mathcal{O}_Y(d) & \longrightarrow & 0, \end{array} \tag{6.9}$$

in which the lower exact sequence is the normal exact sequence of Y in $\mathbb{P}^n$. Here, the last vertical arrow is simply the evaluation arrow

$$S^d \otimes \mathcal{O}_Y \to \mathcal{O}_Y(d).$$

The commutativity of this diagram shows that ρ is also the composed map

$$S^d \to H^0(\mathcal{O}_Y(d)) \stackrel{\overline{\rho}}{\to} H^1(Y, T_Y),$$

where the first arrow is surjective and $\overline{\rho}$ is the connecting arrow induced by the normal exact sequence. It follows immediately that

$$\mathrm{Ker}\,\overline{\rho} = \mathrm{Im}\, H^0(Y, T_{\mathbb{P}^n|Y}),$$

and as the restriction map

$$H^0(\mathbb{P}^n, T_{\mathbb{P}^n}) \to H^0(Y, T_{\mathbb{P}^n|Y})$$

is surjective for $n \geq 2$ by Bott's theorem (Bott 1957), and the composed map

$$H^0(\mathbb{P}^n, T_{\mathbb{P}^n}) \to H^0(Y, T_{\mathbb{P}^n|Y}) \to H^0(Y, \mathcal{O}_Y(d))$$

sends $X_i \frac{\partial}{\partial X_j}$ to $X_i \frac{\partial f}{\partial X_j}{}_{|Y}$ by definition of the normal exact sequence, we conclude that $\mathrm{Ker}\,\rho$ is generated by f and J_f^d. Finally, Euler's relation shows that $f \in J_f^d$, so $\mathrm{Ker}\,\rho$ is generated by J_f^d. The assertion concerning the surjectivity follows from the above argument, since by Bott's vanishing theorem, we have $H^1(Y, T_{\mathbb{P}^n|Y}) = 0$ for $n \geq 4$ or $n = 3$, $d \neq 4$. □

Remark 6.16 *Note that J_f^d is the tangent space at the point f of the orbit O_f of f under the action of the group* $\mathrm{Gl}(n+1)$. *Indeed, the tangent space to O_f is generated by the $df_t/dt_{|t=0}$, where*

$$f_t = g_t^*(f), \quad g_t = \mathrm{Id} + tA, \quad A \in M^{n+1}(\mathbb{C}).$$

Writing $g_t^* X_i = X_i + t \sum_j A_{i,j} X_j$, *we find*

$$g_t^* f = f(X_0 + tA_0, \ldots, X_n + tA_n), \quad A_i = \sum_j A_{i,j} X_j.$$

Thus, we have

$$\frac{d}{dt}(f_t)_{|t=0} = \sum_i A_i \frac{\partial f}{\partial X_i},$$

which shows that $T_{O_f,f} = J_f^d$.

If B' *is the quotient of* B^0 *by* $\mathrm{Gl}(n+1)$, *where* B^0 *is the open set of* B *parametrising the hypersurfaces without any non-trivial automorphisms, and* $\pi : \mathcal{Y}' \to B'$ *is the universal quotient hypersurface of* $\mathcal{Y}$ *by the action of* $\mathrm{Gl}(n+1)$ *on* $\mathbb{P}^n \times B$, *then lemma 6.15 shows that* $\mathcal{Y}'$ *is a universal deformation of each of the fibres* Y_f *of* π *for* $n \geq 4$ *or* $n = 3$, $d \neq 4$.

Theorem 6.13 can now be reformulated for hypersurfaces of $\mathbb{P}^n$ in the following way. The infinitesimal variation of Hodge structure for the family $\pi : \mathcal{Y}' \to B'$ is given by the maps

$$\overline{\nabla}_{n-p} : H^{n-p,p-1}(Y)_{\mathrm{prim}} \to \mathrm{Hom}\,(T_{B',f}, H^{n-p-1,p}(Y)_{\mathrm{prim}}),$$

and we have the following result.

Theorem 6.17 *Via the isomorphisms*

$$\overline{\alpha}_p : R_f^{pd-n-1} \cong H^{n-p,p-1}(Y)_{\mathrm{prim}}, \quad \overline{\alpha}_{p+1} : R_f^{(p+1)d-n-1} \cong H^{n-p-1,p}(Y)_{\mathrm{prim}}$$

of corollary 6.12, together with the isomorphism $R_f^d \cong T_{B',f}$, $\overline{\nabla}_{n-p}$ *can be identified up to a multiplicative coefficient with the map given by the product*

$$R_f^{pd-n-1} \to \mathrm{Hom}\left(R_f^d, R_f^{(p+1)d-n-1}\right).$$

6.2.2 Macaulay's theorem

The Jacobian ideal of a hypersurface $Y \subset \mathbb{P}^n$ is generated in the ring of polynomials S by the partial derivatives $G_i = \frac{\partial f}{\partial X_i}$, $i = 0, \ldots, n$ of the equation f of Y. The G_i have no common zeros in $\mathbb{P}^n$ when Y is smooth. Indeed, by Euler's relation, if $\frac{\partial f}{\partial X_i}(x) = 0$ for all i, we also have $f(x) = 0$, so $x \in Y$ is a singular point of Y.

Definition 6.18 *A sequence of homogeneous polynomials $G_i \in S^{d_i}$, $i = 0, \ldots, n$, with $d_i > 0$, is said to be regular if the G_i have no common zero.*

Given such a regular sequence, let $R_{G.}$ denote the quotient ring $S/J_{G.}$, where $J_{G.}$ is the ideal generated by the G_i. We know by Hilbert's Nullstellensatz that $J^k_{G.} = S^k$ for sufficiently large k, so that $R_{G.}$ is an Artinian ring.

Theorem 6.19 (Macaulay) *The ring $R_{G.}$ satisfies the following property: for $N = \sum_{i=0}^{n} d_i - n - 1$, we have* rank $R^N_{G.} = 1$, *and for every integer k, the pairing*

$$R^k_{G.} \times R^{N-k}_{G.} \to R^N_{G.} \tag{6.10}$$

is perfect.

Such a ring is called a graded Gorenstein ring. The 1-dimensional vector space $R^N_{G.}$ is called its socle. As a consequence of the duality (6.10), we have the following fact.

Corollary 6.20 *We have*

(i) $R^k_{G.} \neq 0 \Leftrightarrow 0 \leq k \leq N$.

(ii) For integers a, b such that $b \geq 0$ and $a + b \leq N$, the map given by the product

$$\mu : R^a_{G.} \to \mathrm{Hom}\left(R^b_{G.}, R^{b+a}_{G.}\right)$$

is injective.

Proof If $k > N$ we have $N - k < 0$, so $R^{N-k}_{G.} = 0$, i.e. $R^k_{G.} = 0$ by duality. Conversely, if $R^k_{G.} = 0$ for $k \geq 0$, then $R^l_{G.} = 0$ for all $l \geq k$, so $N < k$, which proves (i).

Let us now prove (ii). Let $A \in \mathrm{Ker}\,\mu \subset R^a_{G.}$. For every $B \in R^b_{G.}$, we have $AB = 0$ in $R^{b+a}_{G.}$. Thus, for every $\phi \in R^{N-a-b}_{G.}$, we have $AB\phi = 0 \in R^N_{G.}$. Now, as $b \geq 0$ and $N - a - b \geq 0$, the product

$$R^{N-a-b}_{G.} \otimes R^b_{G.} \to R^{N-a}_{G.},$$
$$\phi \otimes B \mapsto \phi B$$

is surjective. Thus, $AC = 0$ in $R^N_{G.}$ for every $C \in R^{N-a}_{G.}$, and by Macaulay's theorem, this implies that $A = 0$. □

Proof of theorem 6.19 Let $\mathcal{E}$ denote the holomorphic vector bundle $\bigoplus_i \mathcal{O}_{\mathbb{P}^n}(-d_i)$ on $\mathbb{P}^n$. Every G_i can be viewed as a morphism $G_i : \mathcal{O}_{\mathbb{P}^n}(-d_i) \to \mathcal{O}_{\mathbb{P}^n}$, and we will write $G_\cdot : \mathcal{E} \to \mathcal{O}_{\mathbb{P}^n}$ for the sum of these morphisms. The fact that the G_i have no common zero is equivalent to the surjectivity of the morphism $G_\cdot$. Moreover, by definition, the ideal $J_{G_\cdot}$ satisfies

$$J^k_{G_\cdot} = \operatorname{Im} G_\cdot : H^0(\mathbb{P}^n, \mathcal{E}(k)) \to H^0(\mathbb{P}^n, \mathcal{O}_{\mathbb{P}^n}(k)).$$

As the morphism $G_\cdot$ is a section of the dual of the bundle $\mathcal{E}$, by contraction it also induces a morphism

$$G^l_\cdot : \bigwedge^l \mathcal{E} \to \bigwedge^{l-1} \mathcal{E}$$

for each l. For $l = 1$, we have $G^l_\cdot = G_\cdot$; furthermore it is clear that $G^{l-1}_\cdot \circ G^l_\cdot = 0$, and the complex $\mathcal{E}^\cdot$ given by

$$0 \to \bigwedge^{n+1} \mathcal{E} \stackrel{G^{n+1}_\cdot}{\to} \bigwedge^n \mathcal{E} \to \cdots \to \mathcal{E} \stackrel{G_\cdot}{\to} \mathcal{O}_{\mathbb{P}^n} \to 0 \tag{6.11}$$

is exact. Indeed, these assertions can be proved pointwise. At a given point $x \in \mathbb{P}^n$, the fibre $\mathcal{E}_x$ is a vector space V, and the morphism $G_\cdot$ is a linear form $\sigma \in V^*$. We then use the fact that the interior product $\sigma_l : \bigwedge^l V \to \bigwedge^{l-1} V$ by σ satisfies $\sigma_l \circ \sigma_{l-1} = 0$, and the sequence

$$\bigwedge^{l+1} V \stackrel{\sigma_{l+1}}{\to} \bigwedge^l V \stackrel{\sigma_l}{\to} \bigwedge^{l-1} V$$

is exact at the middle for every l, which is an easy result of linear algebra. The complex (6.11) is called the Koszul resolution associated to $(\mathcal{E}, G_\cdot)$.

Let us assign the degree 0 to the first term $\bigwedge^{n+1} \mathcal{E}$. The twisted complex $\mathcal{E}^\cdot(k)$ is exact, so in particular we have $\mathbb{H}^{n+1}(\mathbb{P}^n, \mathcal{E}^\cdot(k)) = 0$ for all k. But in the hypercohomology spectral sequence of this complex associated to the naive filtration

$$F^p\mathcal{E}^\cdot(k) = 0 \to \mathcal{E}^p(k) \to \cdots \to \mathcal{E}^{n+1}(k) \to 0,$$

the term $E_1^{p,q}$ is given by $E_1^{p,q} = H^q(\mathbb{P}^n, \mathcal{E}^p(k))$, so by Bott's vanishing theorem, since the $\mathcal{E}^p$ are direct sums of line bundles, we have $E_1^{p,q} = 0$ for $q \neq 0, n$. Thus we are actually considering a spherical spectral sequence (see subsection 4.1.3), and we have $E_2^{p,q} = \cdots = E_{n+1}^{p,q}$ and

$$\begin{aligned} E_\infty^{p,0} &= \operatorname{Coker}\left(d_{n+1} : E_2^{p-n-1,n} \to E_2^{p,0}\right), \\ E_\infty^{p-n-1,n} &= \operatorname{Ker}\left(d_{n+1} : E_2^{p-n-1,n} \to E_2^{p,0}\right), \\ E_\infty^{p,q} &= 0, \ q \neq 0, n. \end{aligned}$$

As we also know that $\mathbb{H}^l(\mathbb{P}^n, \mathcal{E}^{\cdot}(k)) = 0$ for all l, we have in particular $E_\infty^{p-n-1,n} = E_\infty^{p,0} = 0$, so that the arrows d_{n+1} above are isomorphisms.

But since d_1 is induced by the differential of $\mathcal{E}^{\cdot}(k)$, we also have

$$\begin{aligned} E_2^{n+1,0} &= H^0(\mathbb{P}^n, \mathcal{E}^{n+1}(k))/G.H^0(\mathbb{P}^n, \mathcal{E}^n(k)) \\ &= H^0(\mathbb{P}^n, \mathcal{O}_{\mathbb{P}^n}(k))/G.H^0(\mathbb{P}^n, \mathcal{E}(k)) = R_{G.}^k. \end{aligned}$$

Similarly, we have

$$\begin{aligned} E_2^{0,n} &= \mathrm{Ker}\,(G. : H^n(\mathbb{P}^n, \mathcal{E}^0(k)) \to H^n(\mathbb{P}^n, \mathcal{E}^1(k))) \\ &= \mathrm{Ker}\left(G. : H^n\left(\mathbb{P}^n, \bigwedge^{n+1}\mathcal{E}(k)\right) \to H^n\left(\mathbb{P}^n, \bigwedge^{n}\mathcal{E}(k)\right)\right). \end{aligned}$$

We have thus constructed a canonical isomorphism

$$d_{n+1} : \mathrm{Ker}\left(G. : H^n\left(\mathbb{P}^n, \bigwedge^{n+1}\mathcal{E}(k)\right) \to H^n\left(\mathbb{P}^n, \bigwedge^{n}\mathcal{E}(k)\right)\right) \to R_{G.}^k.$$

Now, $\bigwedge^{n+1}\mathcal{E}(k) = \mathcal{O}_{\mathbb{P}^n}(-\sum_i d_i + k)$ and

$$\bigwedge^{n}\mathcal{E}(k) = \mathrm{Hom}\left(\mathcal{E}, \bigwedge^{n+1}\mathcal{E}(k)\right) = \mathcal{E}^*\left(-\sum_i d_i + k\right).$$

One easily checks that via Serre duality, the map

$$G. : H^n\left(\mathbb{P}^n, \bigwedge^{n+1}\mathcal{E}(k)\right) \to H^n\left(\mathbb{P}^n, \bigwedge^{n}\mathcal{E}(k)\right)$$

is the map dual to

$$\begin{aligned} G. : H^0\left(\mathbb{P}^n, \mathcal{E}\left(\sum_i d_i - k - n - 1\right)\right) &\to \\ H^0\left(\mathbb{P}^n, \mathcal{O}_{\mathbb{P}^n}\left(\sum_i d_i - k - n - 1\right)\right)&, \end{aligned}$$

so that we have constructed an isomorphism

$$d_{n+1} : \left(R_{G.}^{\sum_i d_i - k - n - 1}\right)^* \cong R_{G.}^k. \tag{6.12}$$

In particular, setting $k = 0$, we find that $R_{G.}^N$ is of rank 1 for $N = \sum_i d_i - n - 1$. Finally, to conclude the proof of theorem 6.19, it suffices to note that the isomorphisms (6.12) satisfy the following compatibility, where the vertical arrows are given by multiplication by $P \in S^l$ and its dual:

$$\begin{array}{cccc} d_{n+1}: & \left(R_{G.}^{N-k}\right)^* & \xrightarrow{\cong} & R_{G.}^k \\ & \downarrow P^* & & \downarrow P \\ d_{n+1}: & \left(R_{G.}^{N-k-l}\right)^* & \xrightarrow{\cong} & R_{G.}^{k+l}. \end{array} \tag{6.13}$$

Indeed, the commutativity of this diagram says that the pairing (6.10) satisfies the property

$$\langle A, PB \rangle = \langle PA, B \rangle$$

for $A \in R_{G.}^{N-k-l}$, $P \in R_{G.}^l$, $B \in R_{G.}^k$, in other words that the isomorphism (6.12) is given by the pairing of (6.10). □

6.2.3 The symmetriser lemma

As above, consider the ring $R_{G.} = S/J_{G.}$ associated to a regular sequence $G_0, \dots, G_n$ on $\mathbb{P}^n$. Let $d_i = deg\, G_i$ and $N = \sum_i d_i - n - 1$.

As the ring $R_{G.}$ is commutative, if $b \geq a$ are integers and $P \in R_{G.}^{b-a}$, the multiplication by P

$$\mu_P : R_{G.}^a \to R_{G.}^b$$

satisfies the following property:

$$\forall A,\ B \in R_{G.}^a, \quad A\mu_P(B) = B\mu_P(A) \in R_{G.}^{b+a}.$$

Furthermore, Macaulay's theorem 6.19 shows that the map

$$\mu : P \mapsto \mu_P, \quad R_{G.}^{b-a} \to \operatorname{Hom}\left(R_{G.}^a, R_{G.}^b\right)$$

is injective for $b \leq N, a \geq 0$. The symmetriser lemma (see Donagi & Green 1984) is the following statement.

Proposition 6.21 *Let*

$$T^{a,b} := \left\{\phi \in \operatorname{Hom}\left(R_{G.}^a, R_{G.}^b\right) \mid A\phi(B) = B\phi(A) \in R_{G.}^{b+a} \ \forall A,\ B \in R_{G.}^a\right\}.$$

If $a + b < N$ *and* $\sup_i(d_i + b) \leq N$, *then we have the equality*

$$\mu\left(R_{G.}^{b-a}\right) = T^{a,b} \subset \operatorname{Hom}\left(R_{G.}^a, R_{G.}^b\right).$$

Proof By definition, $T^{a,b}$ is the kernel of the map

$$\sigma : \mathrm{Hom}\,(R^a_{G.}, R^b_{G.}) \to \mathrm{Hom}\left(\bigwedge^2 R^a_{G.}, R^{a+b}_{G.}\right)$$

given by $\sigma(\phi)(A \wedge B) = A\phi(B) - B\phi(A)$. We want to show that the following sequence is exact at the middle:

$$R^{b-a}_{G.} \xrightarrow{\mu} \mathrm{Hom}\left(R^a_{G.}, R^b_{G.}\right) \xrightarrow{\sigma} \mathrm{Hom}\left(\bigwedge^2 R^a_{G.}, R^{a+b}_{G.}\right). \tag{6.14}$$

Equivalently, consider the dual exact sequence

$$\bigwedge^2 R^a_{G.} \otimes \left(R^{a+b}_{G.}\right)^* \xrightarrow{\sigma^*} R^a_{G.} \otimes \left(R^b_{G.}\right)^* \xrightarrow{\mu^*} \left(R^{b-a}_{G.}\right)^*. \tag{6.15}$$

Applying Macaulay's theorem, we have natural isomorphisms

$$\left(R^b_{G.}\right)^* \cong R^{N-b}_{G.}, \quad \left(R^{a+b}_{G.}\right)^* \cong R^{N-a-b}_{G.}, \quad \left(R^{b-a}_{G.}\right)^* \cong R^{N+a-b}_{G.},$$

so that (6.15) can be written

$$\bigwedge^2 R^a_{G.} \otimes R^{N-a-b}_{G.} \xrightarrow{\sigma^*} R^a_{G.} \otimes R^{N-b}_{G.} \xrightarrow{\mu^*} R^{N-b+a}_{G.}. \tag{6.16}$$

Furthermore, the compatibility of the diagram (6.13), i.e. the fact that the maps given by multiplication by P are self-dual for the Macaulay duality, implies immediately that we have

$$\mu^*(A \otimes B) = A \cdot B, \quad \sigma^*(A \wedge B \otimes C) = A \otimes (B \cdot C) - B \otimes (A \cdot C).$$

The following result on the Koszul cohomology of projective space is due to Green (1984b). Let S be the ring of polynomials with $n+1$ variables, and for every pair of integers k and l, write

$$\begin{aligned} &\mu^* : S^l \otimes S^k \to S^{k+l}, \quad \mu^*(A \otimes B) = A \cdot B \\ &\sigma^* : \bigwedge^2 S^l \otimes S^k \to S^l \otimes S^{k+l}, \\ &\qquad \sigma^*(A \wedge B \otimes C) = A \otimes (B \cdot C) - B \otimes (A \cdot C). \end{aligned}$$

Theorem 6.22 (Green) *The sequence*

$$\bigwedge^2 S^l \otimes S^{k-l} \xrightarrow{\sigma^*} S^l \otimes S^k \xrightarrow{\mu^*} S^{k+l}$$

is exact whenever $k > l$.

Consider the case where $l = a$, $k = N - b$. By the hypothesis $N > a + b$, we can apply the theorem to conclude that the sequence

$$\bigwedge^2 S^a \otimes S^{N-a-b} \xrightarrow{\sigma^*} S^a \otimes S^{N-b} \xrightarrow{\mu^*} S^{N-b+a} \tag{6.17}$$

is exact at the middle. Also, we have the following commutative diagram, in which the vertical arrows are surjective:

$$\begin{array}{ccccc}\bigwedge^2 S^a \otimes S^{N-a-b} & \xrightarrow{\sigma^*} & S^a \otimes S^{N-b} & \xrightarrow{\mu^*} & S^{N-b+a} \\ \downarrow & & \downarrow & & \downarrow \\ \bigwedge^2 R^a_{G.} \otimes R^{N-a-b}_{G.} & \xrightarrow{\sigma^*} & R^a_{G.} \otimes R^{N-b}_{G.} & \xrightarrow{\mu^*} & R^{N-b+a}_{G.}.\end{array} \tag{6.18}$$

Now let $\gamma \in \mathrm{Ker}\,(\mu^* : R^a_{G.} \otimes R^{N-b}_{G.} \to R^{N-b+a}_{G.})$, and let $\tilde{\gamma}$ be a lifting of γ to $S^a \otimes S^{N-b}$. Then we have $\mu^*(\tilde{\gamma}) \in J^{N-b+a}_{G.}$. But by the hypothesis $N - b \geq \sup(d_i)$, the ideal $J_{G.}$ is generated by its component of degree $N - b$. Thus, we have the surjectivity

$$S^a \otimes J^{N-b}_{G.} \twoheadrightarrow J^{N-b+a}_{G.},$$

which shows that up to modifying the lifting $\tilde{\gamma}$, we may assume that $\mu^*(\tilde{\gamma}) = 0$. The exactness of the sequence (6.17) thus implies that $\tilde{\gamma} \in \mathrm{Im}\,\sigma^*$, and by the diagram (6.18), we conclude that $\gamma \in \mathrm{Im}\,\sigma^*$. Hence (6.16) is exact at the middle.

□

Remark 6.23 *The same proof also shows that under the same hypotheses on a and b, the sequence*

$$R^{b-a}_{G.} \xrightarrow{\mu} \mathrm{Hom}\left(S^a, R^b_{G.}\right) \xrightarrow{\sigma} \mathrm{Hom}\left(\bigwedge^2 S^a, R^{a+b}_{G.}\right)$$

is exact at the middle.

6.3 First applications

6.3.1 Hodge loci for families of hypersurfaces

Using Macaulay's theorem and the results of section 5.3.2, one can show that the Hodge loci for the family of smooth hypersurfaces of projective space are proper analytic subsets. Precisely, let $\pi : \mathcal{Y} \to B$ be the family of smooth hypersurfaces of degree d of $\mathbb{P}^n$, and let $U \subset B$ be an open set. For every

$$\lambda \in \Gamma(U, R^{n-1}\pi_*\mathbb{C}_{\mathrm{van}})$$

and for every $p \leq n-1$, recall that the corresponding Hodge locus U^p_λ consists of the points $u \in U$ such that $\lambda_u \in F^p H^{n-1}(X_u, \mathbb{C})_{\mathrm{prim}}$. We have the following result (see Carlson *et al.* 1983).

Theorem 6.24 *If p is such that $d(n-p+1)-n-1 \leq (d-2)(n+1)$, then the Hodge loci U_λ^p are proper analytic subsets of U for every $0 \neq \lambda \in \Gamma(U, R^{n-1}\pi_*\mathbb{C}_{\text{prim}})$.*

Macaulay's theorem and its corollary 6.20 say that for $N = (d-2)(n+1)$ and $f \in U$, the Jacobian ring R_f is of rank 1 in degree N, and satisfies $R_f^k \neq 0$ if and only if $0 \leq k \leq N$. By the isomorphism

$$R_f^{(n-p+1)d-n-1} \cong H^{p-1,n-p}(Y_f)_{\text{prim}}$$

of corollary 6.12, the hypothesis is thus equivalent to the condition $H^{p-1,n-p}(Y_f)_{\text{van}} \neq 0$. One shows easily that the condition is equivalent to

$$F^p H^{n-1}(Y_f)_{\text{prim}} \neq H^{n-1}(Y_f)_{\text{prim}}, \quad \forall f \in B,$$

unless $F^p H^{n-1}(Y_f)_{\text{prim}} = 0$. Indeed the description given above of the primitive Dolbeault cohomology of a hypersurface, together with Macaulay's theorem, shows that the set

$$\{p' \in \mathbb{N} \mid H^{p',q'}(Y_f)_{\text{prim}} \neq 0\}$$

is an interval.

The theorem can thus be reformulated as follows.

Theorem 6.25 *If d, n and p are such that $F^p\mathcal{H}^{n-1} \neq \mathcal{H}^{n-1}$, then the Hodge locus U_λ^p is a proper analytic subset of U for every $0 \neq \lambda \in \Gamma(U, R^{n-1}\pi_*\mathbb{C}_{\text{prim}})$.*

Remark 6.26 *Note that if $d \geq n+1$, we have $d(n-p+1)-n-1 \leq (d-2)(n+1)$ for $p \geq 1$, i.e. $F^p\mathcal{H}^{n-1} \neq \mathcal{H}^{n-1}$ for every $p \geq 1$. Theorem 6.25 then says that a class λ which is locally constant on U is generically of maximal Hodge level, i.e. that its component of type $(0, n-1)$ is non-zero.*

Proof of theorem 6.24 Let $f \in U$, and let $\overline{\lambda}_f \in H^{p,n-p-1}(Y_f)_{\text{prim}}$. By theorem 6.13, if $P \in R_f^{(n-p)d-n-1}$ corresponds to $\overline{\lambda}_f$ under the isomorphism

$$\overline{\alpha}_{n-p} : R_f^{(n-p)d-n-1} \cong H^{p,n-p-1}(Y_f)_{\text{prim}},$$

the map $\overline{\nabla}(\overline{\lambda}_f)$ can be identified with the multiplication by P

$$\mu_P : S^d \to R_f^{(n-p+1)d-n-1}.$$

As $(n-p+1)d-n-1 \leq N$, corollary 6.20 shows that $\mu_P = 0$ if and only if $P = 0$. Thus, the map

$$\overline{\nabla}(\overline{\lambda}_f) : T_{U,f} \to H^{p-1,n-p}(Y_f)_{\mathrm{prim}}$$

is non-zero for $0 \neq \overline{\lambda}_f \in H^{p,n-p-1}(Y_f)_{\mathrm{prim}}$. It follows by corollary 5.17 that if $U_\lambda^p = U$, we also have $U_\lambda^{p+1} = U$. Reasoning by induction on $n-p$, we find that if $U_\lambda^p = U$, then $\lambda = 0$. □

Remark 6.27 *Consider the case $n = 3$ and $p = 1$. Then theorem 6.24 says that for $d \geq 4$, the Hodge loci U_λ^1 are proper analytic subsets of U for $\lambda \neq 0$. In particular, this holds for the integral classes λ. Now, we saw that the Hodge loci for the integral classes of degree 2 are the local components of the Noether–Lefschetz locus parametrising surfaces having a holomorphic line bundle which is not a multiple of $\mathcal{O}(1)$. As there is only a countable number of such components, theorem 6.24 gives another, infinitesimal, proof of Noether–Lefschetz theorem 3.32. Theorem 6.24 can thus be viewed as a generalised Noether–Lefschetz theorem.*

Finally, theorem 6.25 can be generalised to the hypersurfaces of high degree of any smooth projective variety X. For this, one uses theorem 6.5 and the analysis of the kernels of the maps $\overline{\alpha}_p$ (see Green 1984c), to obtain the following result.

Theorem 6.28 *Let L be a sufficiently ample invertible bundle on an n-dimensional variety X. Then for every open set $U \subset H^0(X, L)$ and for every locally constant non-zero vanishing class*

$$(\lambda_b)_{b \in U}, \quad \lambda_b \in H^{n-1}(Y_b, \mathbb{C})_{\mathrm{van}},$$

the Hodge loci U_λ^p for $p \geq 1$ are proper analytic subsets of U.

6.3.2 The generic Torelli theorem

Donagi (1983) used the symmetriser lemma to prove a generic Torelli theorem for hypersurfaces of the projective space, with a certain number of exceptions. Some of these, such as cubic surfaces in $\mathbb{P}^3$, are actually counterexamples to the Torelli theorem, whereas others satisfy the statement even though Donagi's method does not apply. This happens in the cases of quartic surfaces in $\mathbb{P}^3$, for which the result is due to Piatetski-Shapiro & Shafarevich (1971), quintics in $\mathbb{P}^4$ (see Voisin 1999a and cubics in $\mathbb{P}^5$ Voisin 1986).

Let $\pi : \mathcal{Y}' \to B'$ be the universal family of smooth hypersurfaces of degree d in $\mathbb{P}^n$ having no non-trivial automorphism. Let B' denote the quotient by $\mathrm{Gl}(n+1)$ of the open set of $H^0(\mathbb{P}^n, \mathcal{O}_{\mathbb{P}^n}(d))$ parametrising smooth hypersurfaces with no non-trivial automorphism. The tangent space of B' at a point f can be naturally identified with R_f^d (see remark 6.16).

On B', we have the global period map

$$\mathcal{P} : B' \to \mathcal{D}/\Gamma,$$

where $\mathcal{D}$ is the period domain parametrising the Hodge structures whose underlying lattice L is isomorphic to $H^{n-1}(Y_f, \mathbb{Z})_{\text{prim}}$, and that have the same Hodge numbers as $H^{n-1}(Y_f, \mathbb{C})_{\text{prim}}$ for any $f \in B'$. Here, the group Γ is the automorphism group of L.

Remark *It is in fact more reasonable to work as in vI.7.1.2 with the polarised period map, where we include the data of the intersection form on the lattices. In this last case,* Γ *has to be the group of automorphism of* $(L, < \ , \ >)$.

By definition, the global period map associates to f the Hodge structure on $H^{n-1}(Y_f, \mathbb{Z})_{\text{prim}}$, which can be considered as a Hodge structure on L via the choice of an isomorphism

$$H^{n-1}(Y_f, \mathbb{Z})_{\text{prim}} \cong L.$$

We know that the period map is holomorphic. The generic Torelli problem is the question of knowing whether $\mathcal{P}$ is of degree 1 over its image. In other words, given two hypersurfaces Y and Y' of degree d in $\mathbb{P}^n$, with Y generic, such that there exists an isomorphism of (polarised) Hodge structures

$$i : H^{n-1}((Y, \mathbb{Z})_{\text{prim}}, F^{\cdot}) \cong H^{n-1}((Y', \mathbb{Z})_{\text{prim}}, F^{\cdot}),$$

one asks whether Y and Y' are isomorphic. In vI.10.3.1 we showed how to obtain this statement in the case of curves of genus ≥ 5, using arguments from the theory of infinitesimal variations of Hodge structure.

In the case of hypersurfaces, Donagi used this method to prove the following result.

Theorem 6.29 *The generic Torelli theorem holds for hypersurfaces of degree d in $\mathbb{P}^n$, up to the possible exception of the following cases:*

(i) d *divides* $n + 1$;
(ii) $d = 3,\ n = 3$;
(iii) $d = 4,\ n \equiv 1 \mod 4$;
(iv) $d = 6,\ n \equiv 2 \mod 6$.

Remark 6.30 *Case (ii) is a true exception, since all cubic surfaces have isomorphic Hodge structures (they satisfy $h^{2,0} = 0$), whereas the quotient B' in this case is 4-dimensional.*

Case (iv) was considered by Cox & Green (1990). It is probable that the infinite series of exceptions in case (i) contains only a finite number of counterexamples to the generic Torelli theorem, maybe none. The first cases which occur, namely cubic curves in $\mathbb{P}^2$, surfaces of degree 4 in $\mathbb{P}^3$, and quintics in $\mathbb{P}^4$, have all been solved already.

Note first that thanks to Macaulay's theorem, the period map is an immersion in all the cases considered. This is known as the infinitesimal Torelli theorem for hypersurfaces.

Indeed, it suffices to check that at each point $f \in B'$ the map $d\mathcal{P}$, which by the results of subsection 5.1.2 is induced by the $\overline{\nabla}_p$ via the adjunction relation

$$d\mathcal{P}_f(u) = \bigoplus \langle \overline{\nabla}_{p,u} \rangle \in \bigoplus_p \operatorname{Hom}\left(H^{p,n-1-p}(Y_f)_{\text{van}}, H^{p-1,n-p}(Y_f)_{\text{van}}\right),$$

is injective.

Now, by theorem 6.13 and remark 6.16, this map $d\mathcal{P}_f$ can be identified up to a coefficient with the map given by the product

$$R_f^d \to \bigoplus_p \operatorname{Hom}\left(R_f^{(n-p)d-n-1}, R_f^{(n-p+1)d-n-1}\right).$$

By corollary 6.20, this map is injective if there exists $p \geq 0$ such that

$$(n-p)d - n - 1 \geq 0, \quad (n-p+1)d - n - 1 \leq N = (d-2)(n+1),$$

which only excludes cubic surfaces and quadratic hypersurfaces of any dimension. The latter have in fact no moduli, so we are left with the cubic surfaces, which is case (ii) above.

As $\mathcal{P}$ is an immersion, it suffices to see that if U and U' are two open sets of B' and $j : U \to U'$ is an isomorphism such that there exists an isomorphism of variations of Hodge structure

$$J : (R^{n-1}\pi_*\mathbb{Z}_{\text{prim}}, F^{\cdot}\mathcal{H}^{n-1}) \cong j^*(R^{n-1}\pi_*\mathbb{Z}_{\text{prim}}, F^{\cdot}\mathcal{H}^{n-1}),$$

then $U = U'$, $j = \text{Id}$, $J = \text{Id}$.

But at each point $f \in U$, setting $f' = j(f)$, such an isomorphism induces an isomorphism of infinitesimal variations of Hodge structure

$$\begin{array}{ccc} T_{U,f} & \longrightarrow & \bigoplus_p \operatorname{Hom}(H^{p,n-1-p}(Y_f)_{\text{prim}}, H^{p-1,n-p}(Y_f)_{\text{prim}}) \\ \Big\downarrow j_* & & \Big\downarrow J \\ T_{U',f'} & \longrightarrow & \bigoplus_p \operatorname{Hom}(H^{p,n-1-p}(Y_{f'})_{\text{prim}}, H^{p-1,n-p}(Y_{f'})_{\text{prim}}), \end{array} \tag{6.19}$$

i.e. by theorem 6.13 a commutative diagram

$$\begin{array}{ccc} R_f^d & \longrightarrow & \bigoplus_p \operatorname{Hom}\left(R_f^{(n-p)d-n-1}, R_f^{(n-p+1)d-n-1}\right) \\ \downarrow & & \downarrow \\ R_{f'}^d & \longrightarrow & \bigoplus_p \operatorname{Hom}\left(R_{f'}^{(n-p)d-n-1}, R_{f'}^{(n-p+1)d-n-1}\right) \end{array} \tag{6.20}$$

whose vertical arrows are isomorphisms. It thus suffices to prove that the existence of a diagram as above implies that Y_f and Y'_f are isomorphic.

For this, Donagi uses the following lemma.

Lemma 6.31 *Assume that f and f' are homogeneous polynomials of degree d in $n+1$ variables, and that there exists an automorphism $g : S \to S$ of the graded ring S such that $g(J_f) = J_{f'}$. Then there exists an automorphism g' of S such that $g'(f) = f'$.*

Remark 6.32 *One cannot always take $g' = g$, even up to homothety. The simplest counterexample is given by $f = \sum_i X_i^d$, $f' = \sum_i \alpha_i X_i^d$, where the α_i are arbitrary coefficients. Clearly, these polynomials have the same Jacobian ideal, so that we can take $g =$ Id. However, these polynomials are not proportional.*

Proof of lemma 6.31 By replacing f by $g(f)$, we may of course assume that $g = \mathrm{Id}$. Then we know that f and f' have the same associated Jacobian ideal J. But then clearly $f_t = tf + (1-t)f'$ also has the same Jacobian ideal $J_t = J$. Now, $\frac{d}{dt}(f_t) = f - f' \in J = J_t$. As J_t is the tangent space at f_t to the orbit of f_t under the action of $\mathrm{Gl}(n+1)$, we conclude that the trajectory $t \mapsto f_t$ in B projects to a constant in $B' = B/\mathrm{Gl}(n+1)$. Thus, f and f' have the same projection in B', i.e. they are conjugate under the action of $\mathrm{Gl}(n+1)$. □

Donagi's idea was to deduce from the diagram (6.20) a commutative diagram of morphisms of rings

$$\begin{array}{ccc} g : S & \longrightarrow & S \\ \downarrow & & \downarrow \\ R_f & \longrightarrow & R_{f'}, \end{array} \tag{6.21}$$

in which the lower isomorphism is equal to j_* in degree d and is induced by J in the degrees $pd - n - 1$, while the vertical arrows are the natural projections.

This then makes it possible to apply lemma 6.31 to conclude that Y_f and $Y_{f'}$ are isomorphic.

The symmetriser lemma is used to extend the commutative diagram (6.20) as follows. Let k be the smallest non-zero integer of the form $pd - n - 1$. If d does not divide $n+1$, then k is strictly less than d. We know by the symmetriser lemma that given the multiplication

$$R_f^d \times R_f^k \to R_f^{k+d},$$

the set

$$T_f = \left\{\phi \in \mathrm{Hom}\left(R_f^k, R_f^d\right) \mid A\phi(B) = B\phi(A) \in R_f^{k+d}, \quad \forall A,\ B \in R_f^k\right\}$$

can be identified with R_f^{d-k}, and the natural map

$$T_f \times R_f^k \to R_f^d$$

can be identified with the product. Thus, we have constructed a new piece of the Jacobian ring, i.e. the diagram (6.20) also gives a commutative diagram

$$\begin{array}{ccc} R_f^{d-k} & \longrightarrow & \mathrm{Hom}\left(R_f^k, R_f^d\right) \\ \downarrow & & \downarrow J \\ R_{f'}^{d-k} & \longrightarrow & \mathrm{Hom}\left(R_{f'}^k, R_{f'}^d\right), \end{array}$$

where the horizontal arrows are given by multiplication. Iterating this procedure, we easily see that starting from the diagram (6.20), we can construct in this manner a ring isomorphism

$$R'_f \to R'_{f'},$$

where the index $'$ means that we take the subring consisting of the elements divisible by $\delta = \mathrm{GCD}(d, n+1)$. As we know that $\delta < d$ since d does not divide $n+1$, the components of small degree of R'_f are isomorphic to the corresponding components of the ring S, and we can then show that apart from the exceptions listed above, this is sufficient to construct the diagram (6.21). For example, when $\delta = 1$, the symmetriser lemma gives an isomorphism between the Jacobian rings of f and f', and the degree 1 part gives an isomorphism $g_1 : S^1 = R_f^1 \to S^1 = R_{f'}^1$. For $g : S \to S$, we then take the morphism induced by g_1. □

Donagi's theorem was generalised by Green (1984c) to families of sufficiently ample hypersurfaces of a smooth projective variety.

Exercises

1. In this problem, we propose to construct a smooth projective variety X of dimension 3, and a class $0 \neq \alpha \in H^3(X, \mathbb{Q}) \cap F^1H^3(X)$ which is not contained in

$$j_*H_3(Y, \mathbb{Q}) \subset H_3(X, \mathbb{Q}) \cong H^3(X, \mathbb{Q})$$

for any hypersurface $Y \overset{j}{\hookrightarrow} X$. This is thus a counterexample to the naive Hodge conjecture (see vI.11.3.2). The starting point is the observation that for every hypersurface Y, $j_*H_3(Y, \mathbb{Q}) \subset H^3(X, \mathbb{Q})$ is a Hodge substructure of $H^3(X, \mathbb{Q})$ contained in $F^1H^3(X, \mathbb{Q})$. Hence it suffices to construct such a pair (X, α) which moreover satisfies the property:

There exists no non-zero Hodge substructure L_t of $H^3(X, \mathbb{Q})$ contained in $F^1H^3(X, \mathbb{Q})$.

Such a Hodge structure is given by a vector subspace $L_{\mathbb{Q}}$ of $H^3(X, \mathbb{Q})$ contained in $F^1H^3(X)$, such that we have an induced decomposition

$$L_{\mathbb{C}} = L^{2,1} \oplus L^{1,2},$$

$$L^{2,1} := L_{\mathbb{C}} \cap F^2H^3(X), \quad L^{1,2} = \overline{L^{2,1}} \subset F^1H^3(X).$$

The varieties we will consider are smooth quintic hypersurfaces of $\mathbb{P}^4$. Such a hypersurface X satisfies $K_X \cong \mathcal{O}_X$ by the adjunction formula, so that $h^{0,3}(X) = 1$. Choose a connected and simply connected open set

$$U \subset H^0(\mathbb{P}^4, \mathcal{O}_{\mathbb{P}^4}(5)) = S^5$$

containing 0 and parametrising smooth hypersurfaces. For $\alpha \in H^3(X_0, \mathbb{Q})$, let $\alpha_t \in H^3(X_t, \mathbb{Q})$, $t \in U$ denote the class deduced from α by the canonical isomorphism $H^3(X_0, \mathbb{Q}) \cong H^3(X_t, \mathbb{Q})$ given by the simple connectedness of U.

(a) Adapting the proofs of propositions 5.14 and 5.20, show that for $0 \neq \alpha \in H^3(X_0, \mathbb{Q})$, the sets

$$U_\alpha := \{t \in H^3(X_0, \mathbb{Q}) \mid \alpha_t \in F^1H^3(X_t, \mathbb{Q})\}$$

are hypersurfaces of U, and the union of the U_α for $\alpha \neq 0$ is dense in U.

Fix α such that $U_\alpha \neq \emptyset$. We propose to show that the generic point t of U_α is such that there exists no non-zero Hodge substructure L of $H^3(X_t, \mathbb{Q})$ contained in $F^1H^3(X_t)$. Assume, on the contrary, that such a

Hodge substructure L_t exists, and let $L_t \subset H^3(X_t, \mathbb{Q})$ be its underlying $\mathbb{Q}$-vector space. By a countability argument, we may assume that L_t does not depend on t (or more precisely, that it is locally constant).

(b) Show that for $t \in U_\alpha$ and $u \in T_{U_\alpha,t}$, and for every

$$\lambda \in L_t^{1,2} \cong L_{\mathbb{C}}/L_t^{2,1} \subset F^1/F^2H^3(X_t),$$

we have

$$\overline{\nabla}\lambda(u) = 0 \text{ in } H^{0,3}(X_t).$$

(c) Deduce from theorems 6.13 and 6.19, together with the fact that $T_{U_\alpha,t}$ is a hyperplane of S^5, that $\dim L_t^{1,2} \leq 1$.

(d) Assuming that L is non-zero, let λ be a generator of $L_t^{2,1}$. Show that

$$\overline{\nabla}(\lambda)(u) \in L_t^{1,2} \subset F^1/F^2H^3(X_t)$$

for $u \in T_{U_\alpha,t}$.

(e) Let H be a hyperplane of S^5 with no base point (i.e. such that there exists no point of $\mathbb{P}^4$ on which all the elements of H vanish). Show that H generates S^6, i.e. that

$$S^1 \cdot H = S^6.$$

Similarly, let $K \subset S^5$ be a subspace of codimension 2 with no base point. Show that $K \cdot S^2 = S^7$.

(f) Now let H be the hyperplane of S^5 given by $T_{U_\alpha,t}$. Show that H has no base point, and that $\overline{\nabla}(\lambda)(H)$ is contained in $L^{1,2}$. Let $K \subset S^5$ be the subspace of H defined by

$$K = \operatorname{Ker}\left(\overline{\nabla}(\lambda) : H \to L^{1,2}\right).$$

Show that K is of codimension 2, with no base point.

(g) Let $P \in R^5_{f_t}$ be the polynomial representing λ under the isomorphism of corollary 6.12. Show that

$$P \cdot K = 0 \text{ in } R^{10}_{f_t}.$$

(h) Deduce from (e) and (g) that $P \cdot R^7_{f_t} = 0$ in $R^{12}_{f_t}$. Deduce from theorem 6.19 that $P = 0$, so that $L_t = 0$. This contradicts the assumption that L_t is non-zero.

2. *Components of small codimension of the Noether–Lefschetz locus.* In this problem, we propose to prove the theorem of Voisin (1988) and Green (1989a) in the case where $d = 5$.

Let U denote a connected and simply connected open subset of the open set of $S^5 := H^0(\mathbb{P}^3, \mathcal{O}_{\mathbb{P}^3}(5))$ parametrising smooth surfaces. Let $0 \in U$, and let $0 \neq \lambda \in H^2(S_0, \mathbb{Z})_{\text{prim}}$. Consider the component U_λ of the Noether–Lefschetz locus given by

$$U_\lambda = \{t \in U \mid \lambda_t \in F^1 H^2(S_t)\}.$$

Assume that $0 \in U_\lambda$.

(a) Let $P_\lambda \in R^6_{f_0}$ be the polynomial representing $\lambda \in H^{1,1}(S)_{\text{prim}}$ under the isomorphism of corollary 6.12. Using theorem 6.13, show that

$$T_{U_\lambda,0} = \text{Ker}\left(\mu^5_{P_\lambda} : S^5 \to R^{11}_{f_0}\right)$$

where $\mu^i_{P_\lambda}$ denotes multiplication by P_λ, mapping S^i to $R^{6+i}_{f_0}$.

(b) Let $A := \text{Ker}\,(\mu^1_{P_\lambda} : S^1 \to R^7_{f_0})$. By Macaulay's theorem 6.19, A can be identified with the dual of

$$\text{Coker}\left(\mu^5_{P_\lambda} : S^5 \to R^{11}_{f_0}\right).$$

Show that $A \cdot R^5_{f_0} \subset P_\lambda^\perp$, where $\perp$ is relative to the intersection form on $R^6_{f_0}$ given by Macaulay's theorem.

(c) Show that if $B \subset S^1$ is a vector subspace of rank 3, then $B \cdot R^5_{f_0} = R^6_{f_0}$. Deduce that $\text{codim}\, A \geq 2$, hence that $\text{codim}\, U_\lambda \geq 2$.

Assume from now on that $\text{codim}\, A = 2$. By (b), this is equivalent to assuming that $\text{codim}\, T_{U_\lambda,0} = 2$.

(d) Let $\Delta \subset \mathbb{P}^3$ be the line defined by $A \subset S^1 = H^0(\mathbb{P}^3, \mathcal{O}_{\mathbb{P}^3}(1))$. Deduce from (b) that $I_\Delta(6) + J^6_{f_0} \neq S^6$, where $I_\Delta(6) \subset S^6$ is the set of polynomials of degree 6 vanishing on Δ.

(e) Deduce that $\text{rank}\, J^4_{f_0|\Delta} = 2$. (Use the fact that $J^4_{f_0|\Delta} \subset H^0(\mathcal{O}_\Delta(4)$ has no base point.) Show that $I_\Delta(6) + J^6_{f_0}$ is a hyperplane of S^6.

We have thus proved the following result:

Every component U_λ of the Noether–Lefschetz locus is of codimension at least 2, and if U_λ is of codimension equal to 2, then for every point $0 \in U_\lambda$, there exists a line Δ satisfying the following properties:

(i) $\text{rank}\, J^4_{f_0|\Delta} = 2$.

(ii) $I_\Delta(5) \subset T_{U_\lambda}$.

(iii) *The hyperplane* $I_\Delta(6) + J^6_{f_0}$ modulo $J^6_{f_0}$ *is equal to the hyperplane* $P_\lambda^\perp \subset R^6_{f_0}$.

(f) Let $G = G(2,4)$ be the Grassmannian of lines of $\mathbb{P}^3$. Let $Z \subset G \times U$ be the algebraic subset defined by

$$Z = \left\{(\Delta, f) \mid \operatorname{rank} J^4_{f|\Delta} = 2\right\}.$$

Show that the projection of $T_{Z,(\Delta,f)}$ onto $T_{U,f}$ does not contain $I_\Delta(5)$ if Δ is not contained in the surface of equation f.

(g) Deduce from (f) that the line Δ of (e) must be contained in the surface S_0. Deduce from condition (iii) of (e) that λ is proportional to $h - 5[\Delta]$, $h = c_1(\mathcal{O}_{S_0}(1))$. (Hint: Use proposition 5.19 to show that the class $h - 5[\Delta]$ also satisfies condition (iii)).

We have thus proved the following result:

Every component U_λ of the Noether–Lefschetz locus for surfaces of degree 5 in $\mathbb{P}^3$ is of codimension at least 2, and the only component of codimension 2 is the family of surfaces containing a line Δ.

7
Normal Functions and Infinitesimal Invariants

The main theme of this chapter is the Abel–Jacobi map of hypersurfaces of a given projective variety of even dimension $2n$. In two different ways, it constitutes the preparation for the following chapter, devoted to the work of Nori. Firstly we give the following application of the symmetriser lemma, i.e. the exactness of the complex $\mathcal{K}_{p,q}$, $p+q = 2n-1$ in degree ≤ 1 for $q < 2n-2$, for hypersurfaces of projective space of sufficiently large degree.

Theorem 7.1 *If $Y \subset \mathbb{P}^{2n}$ is a general hypersurface of sufficiently large degree, then the Abel–Jacobi map of Y is trivial modulo torsion.*

Secondly, we introduce the notion of the infinitesimal invariant of a normal function, and give a geometric description of this invariant in the case of the normal function associated to a cycle. This is used to give an infinitesimal proof of the following result.

Theorem 7.2 *Let $Z \subset X$ be a cycle of codimension n which is not cohomologous to 0 modulo torsion in a $2n$-dimensional variety, and let L be a sufficiently ample line bundle on X. Then, if $Y \in |L|$ is general and $Z_{|Y}$ is cohomologous to 0, $Z_{|Y}$ is not annihilated by the Abel–Jacobi map of Y.*

This theorem is a consequence of the equality of the infinitesimal invariant of the normal function ν_Z with that of the Hodge class $[Z]$ for a normal function associated to an algebraic cycle Z. Indeed, the infinitesimal invariant of the normal function ν_Z is obtained by differentiating the normal function ν_Z, and its non-triviality thus implies that ν_Z and all of its multiples are non-trivial. To show that ν_Z is non-trivial, it thus suffices to show that the infinitesimal invariant of the class $[Z]$ is non-zero if $[Z]$ is not torsion. We obtain theorem 7.2 by applying this reasoning to the universal family $\mathcal{X} \to B$ of smooth hypersurfaces of the system $|L|$, and to the cycle $p^{-1}(Z)$ where $p : \mathcal{X} \to X$ is the natural map. For this, we first use Nori's theorem, which will be proved in the following chapter, to show that $p^*[Z]$ is non-zero in $H^n(\mathcal{X}, \Omega^n_{\mathcal{X}})$.

A different argument, more global in nature, is given in the following chapter (cf. the proof of theorem 8.21(ii)).

The starting point of the proof of theorem 7.1 is the horizontality property of normal functions associated to algebraic cycles. Let $X \to B$ be a family of smooth projective varieties, and let $Z \subset X$ be a cycle of codimension k cohomologous to 0 on the fibres X_b. The normal function ν_Z is the section of the intermediate Jacobian fibration $J^{2k-1} \to B$ defined by

$$\nu_Z(b) = \Phi_{X_b}(Z_b) \in J^{2k-1}(X_b).$$

The horizontality property is a differential equation analogous to the transversality property satisfied by ν_Z.

The proof of theorem 7.1 uses the following statement.

Proposition 7.3 *Assume that the family $X \to B$ satisfies the property that the complexes $\mathcal{K}_{p,q}$, $p+q = 2k-1$ are exact in degree 0 and 1 for $q \leq k-1$. Then a horizontal section of the Jacobian fibration is locally the projection of a locally constant section of the local system $H^{2k-1}_{\mathbb{C}}$, uniquely defined up to a section of $H^{2k-1}_{\mathbb{Z}}$.*

We conclude the proof of the theorem by applying Macaulay's theorem and the symmetriser lemma, which provide the desired exactness for universal families of hypersurfaces of sufficiently large degree, together with a monodromy argument.

The first section of this chapter is devoted to generalities on Jacobian fibrations and horizontality. We define the infinitesimal invariant of a normal function. In the second section, we prove the horizontality property and compare the infinitesimal invariant of the normal function ν_Z with that of the Hodge class $[Z]$ for a normal function associated to an algebraic cycle Z. Theorem 7.1 is proved in the third section.

7.1 The Jacobian fibration

7.1.1 Holomorphic structure

Let $(H^{2k-1}_{\mathbb{Z}}, F^{\cdot}H^{2k-1}_{\mathbb{C}})$ be an integral Hodge structure of weight $2k-1$. In vI.12.1.7, we introduced the corresponding intermediate Jacobian J^{2k-1}, which is a complex torus constructed using the Hodge structure on $H^{2k-1}_{\mathbb{Z}}$ by setting

$$J^{2k-1} = \frac{H^{2k-1}_{\mathbb{C}}}{F^k H^{2k-1} \oplus H^{2k-1}_{\mathbb{Z}}}.$$

As the Hodge filtration satisfies

$$F^k H^{2k-1} \oplus \overline{F^k H^{2k-1}} = H^{2k-1}_{\mathbb{C}},$$

the composed map

$$H^{2k-1}_{\mathbb{R}} \to H^{2k-1}_{\mathbb{C}} \to H^{2k-1}_{\mathbb{C}} / F^k H^{2k-1}$$

is an isomorphism of real vector spaces, so J^{2k-1} is a torus, which is isomorphic as a real torus to

$$H^{2k-1}_{\mathbb{R}} / H^{2k-1}_{\mathbb{Z}} = H^{2k-1}_{\mathbb{Z}} \otimes \mathbb{R}/\mathbb{Z}.$$

If X is now an n-dimensional compact Kähler manifold, we can apply this construction to the Hodge structure on the integral cohomology modulo torsion $H^{2k-1}(X, \mathbb{Z})$. Poincaré duality, together with the fact that the pairing

$$H^{2k-1}(X, \mathbb{C}) \times H^{2n-2k+1}(X, \mathbb{C}) \to \mathbb{C}$$

identifies $F^k H^{2k-1}(X)$ with the orthogonal complement of $F^{n-k+1} H^{2n-2k+1}(X)$, shows that we also have

$$J^{2k-1}(X) = F^{n-k+1} H^{2n-2k+1}(X)^* / H_{2n-2k+1}(X, \mathbb{Z}),$$

where the map

$$H_{2n-2k+1}(X, \mathbb{Z}) \to F^{n-k+1} H^{2n-2k+1}(X)^*$$

is given by integration. (The image of this map is called the set of periods of X.)

More generally, assume that we have an integral variation of Hodge structure $(H^{2k-1}_{\mathbb{Z}}, F^{\cdot}\mathcal{H}^{2k-1})$ of weight $2k-1$ on a complex manifold Y. We have the holomorphic vector bundle (or the sheaf of free $\mathcal{O}_X$-modules)

$$\mathcal{E} = \mathcal{H}^{2k-1} / F^k \mathcal{H}^{2k-1},$$

and a natural inclusion

$$H^{2k-1}_{\mathbb{Z}} \to \mathcal{E}.$$

The quotient $\mathcal{J}$ can then be viewed as a sheaf of sections of the fibration in Jacobians

$$J \to Y$$

with fibre

$$J_y = J_y^{2k-1} = \mathcal{E}_y / H_{\mathbb{Z},y}^{2k-1}, \quad y \in Y.$$

The fibration in Jacobians is thus equipped with a natural complex structure, the quotient of the complex structure of the holomorphic vector bundle $\mathcal{E}$ by the action by translation of the sheaf of discrete groups $H_{\mathbb{Z}}^{2k-1}$. These translations are indeed biholomorphic transformations, since the local sections of $H_{\mathbb{Z}}^{2k-1}$ are holomorphic sections of $\mathcal{E}$. By the construction given above, the sheaf $\mathcal{J}$ is then the sheaf of holomorphic sections of the fibration $J \to Y$ equipped with this complex structure.

7.1.2 Normal functions

As above, let $J \to Y$ be the fibration in Jacobians associated to a variation of Hodge structure of odd weight $2k-1$, and let $\nu \in \mathcal{J}$ be a holomorphic section of J. By definition, $\mathcal{J}$ is a quotient of the holomorphic bundle $\mathcal{H}^{2k-1}$, which means that the holomorphic sections of $J \to Y$ admit local liftings which are holomorphic sections of the vector bundle $\mathcal{H}^{2k-1}$ on Y. If $\tilde{\nu}$ is a local lifting of ν in $\mathcal{H}^{2k-1}$, consider

$$\nabla \tilde{\nu} \in \mathcal{H}^{2k-1} \otimes \Omega_Y.$$

Note that the lifting $\tilde{\nu}$ is defined up to a section of the form $\eta_F + \eta_{\mathbb{Z}}$, where η_F is a section of $F^k \mathcal{H}^{2k-1}$ and $\eta_{\mathbb{Z}}$ is a section of $H_{\mathbb{Z}}^{2k-1}$. Now, we have $\nabla H_{\mathbb{Z}}^{2k-1} = 0$, and by transversality, we have

$$\nabla F^k \mathcal{H}^{2k-1} \subset F^{k-1} \mathcal{H}^{2k-1} \otimes \Omega_Y.$$

Thus, $\nabla \tilde{\nu}$ is defined modulo $F^{k-1} \mathcal{H}^{2k-1} \otimes \Omega_Y$.

Definition 7.4 *We say that ν is horizontal if ν satisfies the condition*

$$\nabla \tilde{\nu} \in F^{k-1} \mathcal{H}^{2k-1} \otimes \Omega_Y$$

for a local lifting $\tilde{\nu}$ of ν in $\mathcal{H}^{2k-1}$.

By the arguments above, this condition is then satisfied for every local lifting.

Definition 7.5 *The section ν of $J \to Y$ is a normal function if ν is holomorphic and horizontal.*

Remark 7.6 *This is a purely local definition. If we work over a base which is quasi-projective, the definition of a normal function also needs to include conditions on the behaviour at infinity (see El Zein & Zucker 1984).*

7.1.3 Infinitesimal invariants

As above, let $(H_{\mathbb{Z}}^{2k-1}, F^{\cdot}\mathcal{H}^{2k-1})$ be a variation of Hodge structure on Y, and let $\nu \in \mathcal{J}$ be a normal function on Y. For every local lifting $\tilde{\nu}$ of ν in $\mathcal{H}^{2k-1}$, we then have

$$\nabla\tilde{\nu} \in F^{k-1}\mathcal{H}^{2k-1} \otimes \Omega_Y.$$

Consider the projection $\overline{\nabla\tilde{\nu}}$ of $\nabla\tilde{\nu}$ in the quotient

$$(F^{k-1}\mathcal{H}^{2k-1}/F^k\mathcal{H}^{2k-1}) \otimes \Omega_Y = \mathcal{H}^{k-1,k} \otimes \Omega_Y.$$

Lemma 7.7 *$\overline{\nabla\tilde{\nu}}$ is killed by the differential*

$$\overline{\nabla} : \mathcal{H}^{k-1,k} \otimes \Omega_Y \to \mathcal{H}^{k-2,k+1} \otimes \Omega_Y^2$$

of the complex $\mathcal{K}_{k,k-1}$. Moreover, the image of $\overline{\nabla\tilde{\nu}}$ modulo

$$\operatorname{Im}\overline{\nabla} : \mathcal{H}^{k,k-1} \to \mathcal{H}^{k-1,k} \otimes \Omega_Y$$

depends only on ν and not on the choice of lifting.

Proof By definition, for a section α of $\mathcal{H}^{k-1,k} \otimes \Omega_Y$ admitting the local lifting $\tilde{\alpha} \in F^{k-1}\mathcal{H}^{2k-1} \otimes \Omega_Y$, we have

$$\overline{\nabla}(\alpha) = \nabla\tilde{\alpha} \mod F^{k-2}\mathcal{H}^{2k-1} \otimes \Omega_Y^2.$$

As $\nabla\tilde{\nu} \in F^{k-1}\mathcal{H}^{2k-1} \otimes \Omega_Y$ is a lifting of $\overline{\nabla\tilde{\nu}}$, we thus have

$$\overline{\nabla}(\overline{\nabla\tilde{\nu}}) = \nabla \circ \nabla\tilde{\nu} \mod F^{k-2}\mathcal{H}^{2k-1} \otimes \Omega_Y^2$$

and this vanishes in $\mathcal{H}^{k-2,k+1} \otimes \Omega_Y^2$ since $\nabla^2 = 0$.

Moreover, another lifting $\tilde{\nu}'$ is of the form

$$\tilde{\nu}' = \tilde{\nu} + \eta_F + \eta_{\mathbb{Z}},$$

where η_F is a section of $F^k\mathcal{H}^{2k-1}$ and $\eta_{\mathbb{Z}}$ is a section of $H_{\mathbb{Z}}^{2k-1}$. As $\nabla\eta_{\mathbb{Z}} = 0$, we obtain

$$\nabla\tilde{\nu}' = \nabla\tilde{\nu} + \nabla\eta_F.$$

Now, we know that

$$\overline{\nabla}(\overline{\eta}) = \nabla\eta_F \mod F^k\mathcal{H}^{2k-1} \otimes \Omega_Y,$$

where $\overline{\eta}$ is the projection of η_F to $\mathcal{H}^{k,k-1}$. Therefore

$$\overline{\nabla\tilde{\nu}'} - \overline{\nabla\tilde{\nu}} = \overline{\nabla}(\overline{\eta}),$$

which proves the second statement. □

We can thus make the following definition.

Definition 7.8 (Griffiths 1983) *The infinitesimal invariant $\delta\nu$ is the class of*

$$\overline{\nabla\tilde{\nu}} \in \mathrm{Ker}\,\overline{\nabla} \subset \mathcal{H}^{k-1,k} \otimes \Omega_Y$$

in $H^1(\mathcal{K}_{k,k-1})$. For $y \in Y$, the infinitesimal invariant $\delta\nu_y$ is the class of

$$\overline{\nabla\tilde{\nu}}_y \in \mathcal{H}_y^{k-1,k} \otimes \Omega_{Y,y}$$

in

$$H^1(\mathcal{K}_{k,k-1,y}) = \frac{\mathrm{Ker}\left(\overline{\nabla} : \mathcal{H}_y^{k-1,k} \otimes \Omega_{Y,y} \to \mathcal{H}_y^{k-1,k} \otimes \Omega_{Y,y}^2\right)}{\mathrm{Im}\left(\overline{\nabla} : \mathcal{H}_y^{k,k-1} \to \mathcal{H}_y^{k-1,k} \otimes \Omega_{Y,y}\right)}.$$

7.2 The Abel–Jacobi map

7.2.1 General properties

Let $\pi : X \to Y$ be a holomorphic projective fibration of relative dimension n, and let $\mathcal{J}$ be the sheaf of holomorphic sections of the intermediate Jacobian fibration $J^{2k-1} \to Y$, with fibre $J^{2k-1}(X_y)$ at the point y. Let $Z \subset X$ be a relative cycle of codimension k, i.e. a combination with integral coefficients $Z = \sum_i n_i Z_i$, where each $Z_i \subset X$ is a complex analytic subset of codimension k of X, which is reduced, irreducible and flat over Y. Assume that for $y \in Y$, the cycle $Z_y := \sum_i n_i Z_{i,y}$, $Z_{i,y} = Z \cap X_y$ is homologous to 0. (By flatness, if Y is connected and if this condition is satisfied at a point y, it is also satisfied for every $y \in Y$.) We can then define the Abel–Jacobi invariant

$$\Phi_{X_y}^k(Z_y) \in J^{2k-1}(X_y), \quad y \in Y$$

in the following way (see vI.12.1.2): as Z_y is homologous to 0, there exists a real chain $\Gamma \subset X_y$ of dimension $2n-2k-1$ such that $\partial\Gamma = Z_y$. Moreover, Γ is defined up to a cycle. We then show using Hodge theory that integration over Γ acting on the closed forms of type $(2n-2k+1, 0) + \cdots + (n-k+1, n-k)$ defines a linear form on $F^{n-k+1}H^{2n-2k+1}(X_y)$, even though Γ is not closed. For this, we use the isomorphism

$$F^{n-k+1}H^{2n-2k+1}(X_y) = F^{n-k+1}A^{2n-2k+1}(X_y)^c/dF^{n-k+1}A^{2n-2k}(X_y) \quad (7.1)$$

given by the degeneracy at E_1 of the Frölicher spectral sequence. For a class $\eta \in F^{n-k+1}H^{2n-2k+1}(X_y)$, we then set

$$\int_\Gamma \eta = \int_\Gamma \tilde{\eta},$$

where $\tilde{\eta}$ is a closed representative of η in $F^{n-k+1}A^{2n-2k+1}(X_y)$. Stokes' formula and the isomorphism (7.1) show that this is well-defined: indeed, if $\eta \in F^{n-k+1}A^{2n-2k}(X_y)$, then

$$\int_\Gamma d\eta = \int_{Z_y} \eta = 0$$

since Z_y is an algebraic cycle, which implies that the forms of type $(2n-2k, 0) + \cdots + (n-k+1, n-k-1)$, i.e. those in $F^{n-k+1}A^{2n-2k}(X_y)$, vanish on Z_y.

The Abel–Jacobi invariant of Z_y is then defined by

$$\Phi^k_{X_y}(Z_y) = \int_\Gamma \in F^{n-k+1}H^{2n-2k+1}(X_y)^*/H_{2n-2k+1}(X_y, \mathbb{Z}) = J^{2k-1}(X_y).$$

Returning to the family of cycles $(Z_y)_{y\in Y}$, we thus obtain a section ν_Z of the fibration in intermediate Jacobians $J \to Y$, by setting

$$\nu_Z(y) = \Phi_{X_y}(Z_y).$$

Theorem 7.9 (Griffiths) *The section ν_Z is holomorphic and horizontal.*

Proof Assume first that each Z_i is smooth and relatively smooth over Y, and that the components Z_i of Z do not intersect. Then, locally in the neighbourhood of each point $0 \in Y$, we can find a $\mathcal{C}^\infty$ trivialisation

$$T = (T_0, \pi) : X \cong X_0 \times Y$$

of the fibration $\pi : X \to Y$, which induces a trivialisation of each $\pi_{|Z_i} : Z_i \to Y$:

$$T = (T_0, \pi) : Z_i \cong Z_{i,0} \times Y.$$

Let $\Gamma_0 \subset X_0$ be a real chain such that $\partial\Gamma_0 = Z_0$. Then let $\Gamma_y := T^{-1}(\Gamma_0 \times y) \subset X_y$. We have $\partial\Gamma_y = Z_y$.

By definition of the Abel–Jacobi map, the section

$$y \mapsto \int_{\Gamma_y} \in F^{n-k+1}H^{2n-2k+1}(X_y)^*$$

is a lifting $\tilde{\nu}_Z$ of ν_Z to a section which is clearly a $\mathcal{C}^\infty$ section of the vector bundle $(F^{n-k+1}\mathcal{H}^{2n-2k+1})^*$. We thus need to show that $\tilde{\nu}_Z$ is holomorphic, or that for every holomorphic section η of $F^{n-k+1}\mathcal{H}^{2n-2k+1}$ defined in the neighbourhood of 0, the function

$$\tilde{\nu}_Z(\eta)(y) = \int_{\Gamma_y} \eta_y$$

is holomorphic in the neighbourhood of 0. We know (see vI.9.1.1) that we can take the trivialisation T to be such that the subvarieties $T^{-1}(z \times Y)$, $z \in X_0$ are complex subvarieties of X. The horizontal tangent bundle $T_*^{-1}(0 \times T_Y) \subset T_X$ is then a $\mathcal{C}^\infty$ complex subbundle. The section η is represented by a $d_{X/Y}$-closed section ω of the bundle $\pi_* F^{n-k+1}\mathcal{A}^{2n-2k+1}_{X/Y}$, where $F^\cdot$ is the Hodge filtration on the relative differential forms. Under the hypotheses on the trivialisation, the form Ω on X defined by the conditions that $\Omega_{|X_y} = \omega_y$ and $\mathrm{int}(u)(\Omega) = 0$ for a horizontal tangent vector u, also lies in $F^{n-k+1}\mathcal{A}^{2n-2k+1}_X$. Moreover, we know that for $v \in T_{Y,0}$, we have

$$\nabla_v \eta = [\mathrm{int}(\tilde{v})(d\Omega)_{|X_0}],$$

where $\tilde{v}$ is the horizontal lifting of v (Cartan–Lie formula).

Now assume that v is of type $(0, 1)$. Under the hypotheses on the trivialisation, $\tilde{v}$ is also of type $(0, 1)$, so the form $\mathrm{int}(\tilde{v})(d\Omega)_{|X_0}$ lies in $F^{n-k+1}\mathcal{A}^{2n-2k+1}_{X_0}$. As η is holomorphic, we also have $\nabla_v \eta = 0$, so this form is exact. Now, we have already seen that the integral over Γ_0 of an exact form which is in $F^{n-k+1}\mathcal{A}^{2n-2k+1}_{X_0}$ is equal to zero. Thus,

$$\int_{\Gamma_0} \mathrm{int}(\tilde{v})(d\Omega) = 0.$$

Finally, using the fact that the form Ω vanishes on $Z = \partial\Gamma = \partial\Gamma_0 \times Y$, it can be immediately proved that

$$d_v(\tilde{\nu}_Z(\eta)) = d_v \int_{\Gamma_y} \omega_y = \int_{\Gamma_0} \mathrm{int}(\tilde{v})(d\Omega).$$

So this is 0, which shows that ν_Z is holomorphic.

It remains to prove the horizontality. Note that for a section ν' of $\mathcal{H}^{2k-1}$, and $v \in T_{Y,0}$, the condition $\nabla_v \nu' \in F^{k-1}H^{2k-1}(X_0)$ is equivalent to

$$\langle \nabla_v \nu', \eta \rangle = 0, \quad \forall \eta \in F^{n-k+2}H^{2n-2k+1}(X_0). \tag{7.2}$$

But the compatibility of ∇ with Poincaré duality shows that for every section η of $F^{n-k+2}\mathcal{H}^{2k-1}$ defined in the neighbourhood of 0, we have

$$d_v\langle \nu', \eta\rangle - \langle \nu', \nabla_v \eta\rangle = \langle \nabla_v \nu', \eta\rangle.$$

Thus, (7.2) is equivalent to

$$d_v\langle \nu', \eta\rangle = \langle \nu', \nabla_v \eta\rangle, \quad \forall \eta \in F^{n-k+2}\mathcal{H}^{2n-2k+1}(X_0). \tag{7.3}$$

Let us apply this to a lifting ν' of $\tilde{\nu}$ in $\mathcal{H}^{2k-1}$. (The lifting $\tilde{\nu} \in (F^{n-k+1}\mathcal{H}^{2n-2k+1})^*$ considered at the beginning of the proof is only a lifting to $\mathcal{H}^{2k-1}/F^k\mathcal{H}^{2k-1}$.) We need to prove the equality (7.3) for every section η of $F^{n-k+2}\mathcal{H}^{2k-1}$ defined in the neighbourhood of 0. Now, as $\nabla_v\eta \in F^{n-k+1}\mathcal{H}^{2k-1}$ by transversality, we have

$$\langle \nu', \nabla_v \eta\rangle = \tilde{\nu}(\nabla_v \eta) = \int_{\Gamma_0} \nabla_v \eta, \tag{7.4}$$

so we need to prove that

$$d_v\left(\int_{\Gamma_y} \eta_y\right) = \int_{\Gamma_0} \nabla_v \eta. \tag{7.5}$$

This can be shown exactly as above. Indeed, with the same notation, we have the formula

$$d_v\left(\int_{\Gamma_y} \eta_y\right) = \int_{\Gamma_0} \mathrm{int}(\tilde{v})(d\Omega).$$

But under the hypotheses on the trivialisation, the closed form $\mathrm{int}(\tilde{v})(d\Omega)_{|X_0}$ lies in $F^{n-k+1}\mathcal{A}^{2n-2k+1}_{X_0}$ and represents $\nabla_v\eta$. Thus, by the definition of $\int_{\Gamma_0}$, we have

$$\int_{\Gamma_0} \mathrm{int}(\tilde{v})(d\Omega) = \int_{\Gamma_0} \nabla_v \eta. \tag{7.6}$$

The general case can be deduced from this as follows. In the preceding argument, it is possible to replace the hypothesis that the support $\mathrm{Supp}\, Z_y := \bigcup_i Z_{i,y}$ of Z is smooth by the hypothesis that it varies equisingularly above Y. This means that locally along $\mathrm{Supp}\, Z$, we have a holomorphic trivialisation $\mathrm{Supp}\, Z \cong \mathrm{Supp}\, Z_y \times Y$. Indeed, for such an equisingular family, we have a global trivialisation $\mathcal{C}^\infty$ of $X \to Y$ similar to the one introduced at the beginning of the proof, i.e. one which induces a trivialisation of $\mathrm{Supp}\, Z \to Y$.

Moreover, since the section is continuous, to prove that it is holomorphic, it suffices to prove it on a non-empty Zariski open analytic subset of Y, i.e. on the complement of a proper closed analytic subset of Y. Similarly, it suffices by continuity to prove the horizontality on a Zariski open analytic subset. But we know that there exists such an open set above which Supp $Z \to Y$ is equisingular. It thus suffices to work in this open set and apply the arguments above. □

Remark 7.10 *It is important to note that the equalities (7.4) and (7.6) are not tautologies, since the contour Γ_0 is not closed. The integral of a class along this contour is thus not actually defined in general. It is defined only for the classes in $F^{n-k+1}H^{2n-2k+1}(X_0)$, and even then the equality above holds on the condition that the chosen representative lies in $F^{n-k+1}\mathcal{A}^{2n-2k+1}_{X_0}$.*

7.2.2 Geometric interpretation of the infinitesimal invariant

In the preceding situation, the relative cycle $Z = \sum_i n_i Z_i$, where each $Z_i \subset X$ is an irreducible algebraic subvariety of codimension k, admits a class in $H^k(X, \Omega_X^k)$. We will only consider the image of this class in $H^0(Y, R^k\pi_*\Omega_X^k)$, which we still denote by $[Z]$. Its value $[Z]_y$ at the point $y \in Y$ is by definition the image of $[Z]$ in $H^k(X_y, \Omega^k_{X|X_y})$. By definition, the class $[Z]$ is given by

$$[Z] = \sum_i n_i [Z_i],$$

and the class $[Z_i]_y$ is described as follows. To simplify, assume that Z_i is smooth in X and that π remains a submersion on Z_i. (In fact, the following description remains valid in the case where Z_i is a local complete intersection and $\pi_{|Z_i}$ is flat.) This implies that Z_i, resp. $Z_{i,y}$, has invertible canonical bundle $K_{Z_i} = \Omega_{Z_i}^{N-k}$ (with $N = \dim X$), resp. $K_{Z_{i,y}}$, and we have the adjunction formula

$$K_{Z_i\,|Z_{i,y}} = K_{Z_{i,y}}, \tag{7.7}$$

since the normal bundle of $Z_{i,y}$ in Z_i is trivial.

Serre duality then gives an isomorphism

$$H^k\big(X_y, \Omega^k_{X|X_y}\big) \cong \big(H^{n-k}\big(X_y, \Omega_X^{N-k}{}_{|X_y}\big)\big)^*,$$

again with $N = \dim X$, using the fact that

$$\Omega_X^{N-k}{}_{|X_y} = \mathrm{Hom}\,\big(\Omega^k_{X|X_y}, K_{X|X_y}\big) = \mathrm{Hom}\,\big(\Omega^k_{X|X_y}, K_{X_y}\big).$$

Here, the last equality is obtained by the adjunction formula, noting that the normal bundle of X_y in X is trivial. The class $[Z_i]_y \in (H^{n-k}(X_y, \Omega_X^{N-k}{}_{|X_y}))^*$

is then obtained as the linear form on $H^{n-k}(X_y, \Omega_X^{N-k}{}_{|X_y})$ given by the composition

$$H^{n-k}\big(X_y, \Omega_X^{N-k}{}_{|X_y}\big) \to H^{n-k}\big(Z_{i,y}, \Omega_X^{N-k}{}_{|Z_{i,y}}\big) \to H^{n-k}\big(Z_{i,y}, \Omega_{Z_i}^{N-k}{}_{|Z_{i,y}}\big)$$
$$\to H^{n-k}\big(Z_{i,y}, K_{Z_i|Z_{i,y}}\big) \to H^{n-k}\big(Z_{i,y}, K_{Z_{i,y}}\big) \stackrel{\int_{Z_{i,y}}}{\to} \mathbb{C},$$

where the two first arrows are the restriction arrows, the third arrow is given by the natural morphism

$$\Omega_{Z_i}^{N-k} \to K_{Z_i}$$

and the fourth arrow is given by the adjunction isomorphism (7.7).

Remark 7.11 *A very beautiful definition of the cohomology class of a local complete intersection subscheme can be found in Bloch (1972); it does not use Serre duality, and is thus valid for a closed subscheme with no compactness or projectivity hypotheses.*

Remark 7.12 *Note that the definition above shows that the class $[Z_i]_y$ depends only on the subscheme*

$$Z_{i,y}^\epsilon \subset X_y^\epsilon,$$

where $X_y^\epsilon = \pi^{-1}(Y_y^\epsilon)$, Y_y^ϵ is the first-order infinitesimal neighbourhood of y in Y, and

$$Z_{i,y}^\epsilon = X_y^\epsilon \cap Z_i.$$

Remark 7.13 *We used the triviality of the normal bundle of X_y in X (or rather of its maximal exterior power) twice, a trivialisation being given by the choice of a basis of the vector space $T_{Y,y}$. The first time was in the equality $K_{X|X_y} = K_{X_y}$, and the second time was in the equality $K_{Z_i|Z_{i,y}} = K_{Z_{i,y}}$. It is essential to use the same trivialisation in the two cases, in order to obtain a result which is independent of the choice of trivialisation.*

Let us check that the restriction to $H^k(X_y, \Omega_{X_y}^k)$ of the class $[Z]_y \in H^k(X_y, \Omega_{X|X_y}^k)$ defined above is the class $[Z_y]$ of Z_y. We see this by noting that the Serre dual of the restriction map

$$H^k\big(X_y, \Omega_{X|X_y}^k\big) \to H^k\big(X_y, \Omega_{X_y}^k\big)$$

is the map

$$H^{n-k}(X_y, \Omega_{X_y}^{n-k}) \to H^{n-k}(X_y, \Omega_X^{N-k}{}_{|X_y})$$

induced by the map

$$\bigwedge \pi^*\kappa_y : \Omega_{X_y}^{n-k} \to \Omega_X^{N-k}{}_{|X_y},$$

where κ_y is a generator of $K_{Y,y}$ and $\pi^*\kappa_y$ is thus a trivialisation of $\bigwedge^{N-n} N^*_{X_y/X}$. (Here, to apply Serre duality, we use the fact that

$$K_{X|X_y} \cong K_{X_y} \tag{7.8}$$

in order to identify $(\Omega^k_{X|X_y})^* \otimes K_{X_y}$ with $\Omega_X^{N-k}{}_{|X_y}$; the isomorphism (7.8) must then be induced by the trivialisation $\pi^*\kappa_y$ of $\bigwedge^{N-n} N^*_{X_y/X}$.)

The equality $[Z]_{y|X_y} = [Z_y]$ in $H^k(X_y, \Omega^k_{X_y})$ is then equivalent to the statement that for every form $\alpha \in H^{n-k}(X_y, \Omega_{X_y}^{n-k})$, we have

$$\langle [Z_y], \alpha \rangle = \langle [Z]_{y|X_y}, \alpha \rangle = \langle [Z]_y, \pi^*\kappa_y \wedge \alpha \rangle. \tag{7.9}$$

But

$$\langle [Z_y], \alpha \rangle = \sum_i n_i \int_{Z_{i,y}} \alpha,$$

whereas by the description of $\langle [Z]_y, \beta \rangle$ given at the beginning of 7.2.2, we find that

$$\langle [Z]_y, f^*\kappa_y \wedge \alpha \rangle = \sum_i n_i \int_{Z_{i,y}} \alpha_i,$$

where $\alpha_i \in H^{n-k}(Z_{i,y}, \Omega^{n-k}_{Z_{i,y}})$ is deduced from the image of $\pi^*\kappa_y \wedge \alpha$ by the restriction to $H^{n-k}(Z_{i,y}, \Omega_{Z_i}^{N-k}{}_{|Z_{i,y}})$, under the isomorphism

$$\Omega_{Z_i}^{N-k}{}_{|Z_{i,y}} \cong \Omega^{n-k}_{Z_{i,y}}$$

given by a trivialisation of $\bigwedge^{N-n} N_{Z_{i,y}/Z_i}$. As we must take the restriction of κ_y^{-1} for our trivialisation (see remark 7.13), it is clear that $\alpha_i = \alpha_{|Z_{i,y}}$, so that we obtain the equality (7.9).

Now assume that Z_y is homologous to 0 for every $y \in Y$. For $0 \in Y$, the class $[Z]_0$ is then an element of

$$\mathrm{Ker}\,(H^k(X_0, \Omega^k_{X|X_0}) \to H^k(X_0, \Omega^k_{X_0})),$$

so it has an infinitesimal invariant

$$\delta[Z]_0 \in H^1(\mathcal{K}_{k,k-1,0}) = \frac{\operatorname{Ker}\left(\overline{\nabla}: \mathcal{H}_0^{k-1,k} \otimes \Omega_{Y,0} \to \mathcal{H}_0^{k-1,k} \otimes \Omega^2_{Y,0}\right)}{\operatorname{Im}\left(\overline{\nabla}: \mathcal{H}_0^{k,k-1} \to \mathcal{H}_0^{k-1,k} \otimes \Omega_{Y,0}\right)}$$

(see definition 5.10).

Furthermore, we have the normal function ν_Z associated to the cycle Z by theorem 7.9, and its infinitesimal invariant

$$\delta\nu_{Z,0} \in \frac{\operatorname{Ker}\left(\overline{\nabla}: \mathcal{H}_0^{k-1,k} \otimes \Omega_{Y,0} \to \mathcal{H}_0^{k-1,k} \otimes \Omega^2_{Y,0}\right)}{\operatorname{Im}\left(\overline{\nabla}: \mathcal{H}_0^{k,k-1} \to \mathcal{H}_0^{k-1,k} \otimes \Omega_{Y,0}\right)}$$

(see definition 7.8.)

Theorem 7.14 (Voisin 1988) *The infinitesimal invariants $\delta\nu_{Z,0}$ and $\delta[Z]_0$ coincide.*

Proof By Serre duality, the dual of $\frac{\mathcal{H}_0^{k-1,k}\otimes\Omega_{Y,0}}{\operatorname{Im}(\overline{\nabla}:\mathcal{H}_0^{k,k-1}\to\mathcal{H}_0^{k-1,k}\otimes\Omega_{Y,0})}$ is given by

$$\operatorname{Ker}\left({}^t\overline{\nabla}: \mathcal{H}_0^{n-k+1,n-k} \otimes T_{Y,0} \to \mathcal{H}_0^{n-k,n-k+1}\right),$$

where ${}^t\overline{\nabla}$ is the transpose of $\overline{\nabla}$. Using the fact that

$$\langle\overline{\nabla}_u(\lambda), \mu\rangle = \langle\overline{\nabla}_u(\mu), \lambda\rangle \tag{7.10}$$

for $u \in T_{Y,0}$, $\lambda \in \mathcal{H}_0^{k,k-1}$, $\mu \in \mathcal{H}_0^{n-k+1,n-k}$, we then find that

$${}^t\overline{\nabla}(\mu \otimes u) = \overline{\nabla}_u(\mu).$$

Here, the equality (7.10) is obtained by noting that if $\tilde{\lambda}$ and $\tilde{\mu}$ are sections of $F^k\mathcal{H}^{2k-1}$ and $F^{n-k+1}\mathcal{H}^{2n-2k+1}$ respectively, we have $\langle\lambda, \mu\rangle = 0$ for the Poincaré duality. Differentiating this relation at the point 0, we obtain the equality (7.10), where we use the Serre duality between $\mathcal{H}_0^{k-1,k}$ and $\mathcal{H}_0^{n-k+1,n-k}$ on the left, and the Serre duality between $\mathcal{H}_0^{n-k,n-k+1}$ and $\mathcal{H}_0^{k,k-1}$ on the right.

By Serre duality, the infinitesimal invariants $\delta\nu_{Z,0}$ and $\delta[Z]_0$ both give linear forms on $\operatorname{Ker}({}^t\overline{\nabla}: \mathcal{H}_0^{n-k+1,n-k} \otimes T_{Y,0} \to \mathcal{H}_0^{n-k,n-k+1})$, for which we use the same notation; we need to show that these linear forms coincide.

Let us first compute $\delta\nu_{Z,0}$. If ν' is a lifting of ν in $\mathcal{H}^{2k-1}$, we know that $\delta\nu_{Z,0}$ is the projection of $\nabla\nu' \in F^{k-1}\mathcal{H}^{2k-1} \otimes \Omega_{Y,0}$. Thus, for

$$\eta = \sum_j \eta_j \otimes u_j \in \operatorname{Ker}{}^t\overline{\nabla} \subset \mathcal{H}_0^{n-k+1,n-k} \otimes T_{Y,0},$$

we have

$$\delta\nu_{Z,0}(\eta) = \left\langle \sum_j \nabla_{u_j}\nu', \eta_j{}' \right\rangle, \tag{7.11}$$

where $\eta_j{}'$ is any lifting of η_j to $F^{n-k+1}\mathcal{H}_0^{2n-2k+1}$. Let $\tilde{\eta}_j$ be a section of $F^{n-k+1}\mathcal{H}^{2n-2k+1}$ such that $\eta_j(0) = \eta'_j$. We have the equality

$$\begin{aligned}\left\langle \sum_j \nabla_{u_j}\nu', \eta_j{}' \right\rangle &= \sum_j d_{u_j}\langle \nu', \tilde{\eta}_j\rangle - \left\langle \nu', \sum_j \nabla_{u_j}\tilde{\eta}_j \right\rangle \\ &= \sum_j d_{u_j}\int_{\Gamma_y} \tilde{\eta}_j - \int_{\Gamma_0}\sum_j \nabla_{u_j}\tilde{\eta}_j,\end{aligned} \tag{7.12}$$

where, to compute the last term, we used the fact that

$$\sum_j \nabla_{u_j}\tilde{\eta}_j \in F^{n-k+1}\mathcal{H}_0^{2n-2k+1}$$

since $\sum_j \eta_j \otimes u_j \in \operatorname{Ker}{}^t\overline{\nabla}$. This means that, in the integral, we need to choose a representative of $\sum_j \nabla_{u_j}\tilde{\eta}_j$ which is a closed form in $F^{n-k+1}A_{X_0}^{2n-2k+1}$.

Let us now use the notation and the computations introduced in the proof of theorem 7.9. Thus, $(\omega_{j,y})_{y\in Y}$ is a $d_{X/Y}$-closed section of $\pi_* F^{n-k+1}\mathcal{A}_{X/Y}^{2n-2k+1}$ representing $\tilde{\eta}_j$, $\Omega_j \in F^{n-k+1}\mathcal{A}_X^{2n-2k+1}$ is the form associated to ω_j by the trivialisation, and we have

$$d_{u_j}\int_{\Gamma_y} \tilde{\eta}_j = \int_{\Gamma_0} \operatorname{int}(\tilde{u}_j)(d\Omega_j), \tag{7.13}$$

where $\tilde{u}_j$ is the lifting of u_j in $T_{X|X_0}$ given by the trivialisation. Furthermore, the form $\sum_j \operatorname{int}(\tilde{u}_j)(d\Omega_j)$ which represents $\sum_j \nabla_{u_j}\tilde{\eta}_j$ lies in $F^{n-k}A_{X_0}^{2n-2k+1}$, and its cohomology class lies in $F^{n-k+1}H_{X_0}^{2n-2k+1}$ by hypothesis. Thus, there exist forms

$$\alpha \in F^{n-k}\mathcal{A}_{X_0}^{2n-2k}, \quad \beta \in F^{n-k+1}\mathcal{A}_{X_0}^{2n-2k+1}, \tag{7.14}$$

such that

$$\sum_j \operatorname{int}(\tilde{u}_j)(d\Omega_j)_{|X_0} = \beta + d\alpha. \tag{7.15}$$

By Stokes' formula, we then have

$$\int_{\Gamma_0}\sum_j \nabla_{u_j}\tilde{\eta}_j := \int_{\Gamma_0}\beta = \int_{\Gamma_0}\sum_j \operatorname{int}(\tilde{u}_j)(d\Omega_j) - \sum_i n_i \int_{Z_i,0}\alpha, \tag{7.16}$$

so by (7.11), (7.12), (7.13), and (7.16), we have

$$\delta\nu_{Z,0}(\eta) = \sum_j d_{u_j} \int_{\Gamma_y} \tilde{\eta}_j - \int_{\Gamma_0} \sum_j \nabla_{u_j} \tilde{\eta}_j = \sum_i n_i \int_{Z_i,0} \alpha. \tag{7.17}$$

Now consider $\delta[Z]_0$. We have the exact sequence on X_0 given by

$$0 \to \Omega_{X_0}^{k-1} \otimes \pi^*\Omega_{Y,0} \to L^0/L^2\Omega^k_{X\,|X_0} \to \Omega^k_{X_0} \to 0, \tag{7.18}$$

where L is the Leray filtration on $\Omega^k_{X\,|X_0}$, which induces an isomorphism

$$\operatorname{Ker}\left(H^k(X_0, L^0\Omega^k_X/L^2\Omega^k_{X\,|X_0}) \to H^k(X_0, \Omega^k_{X_0})\right) \cong \frac{H^k(X_0, \Omega^{k-1}_{X_0}) \otimes \Omega_{Y,0}}{\overline{\nabla} H^{k-1}(X_0, \Omega^k_{X_0})}.$$

Then $\delta[Z]_0$ is the image under this isomorphism of the projection of $[Z]_0$ to

$$H^k(X_0, L^0/L^2\Omega^k_{X\,|X_0}).$$

Indeed, this follows from the definition of $\delta[Z]_0$ via the cohomology spectral sequence of the sheaf $\Omega^k_{X\,|X_0}$ for the Leray filtration, together with the identification of the complex $(\mathcal{K}^p_{k,k-1}, \overline{\nabla})$ with $(E_1^{p,k-1}, d_1)$ (see proposition 5.9).

Note now that $(L^0/L^2\Omega^k_{X\,|X_0})^* \otimes K_{X_0}$ is isomorphic to $L^{b-1}\Omega^{N-k}_{X\quad|X_0}$, where $b = \dim Y = N - n$, via the exterior product

$$L^0/L^2\Omega^k_{X\,|X_0} \otimes L^{b-1}\Omega^{N-k}_{X\quad|X_0} \to K_{X\,|X_0}$$

followed by the identification $K_{X_0} \cong K_{X\,|X_0}$ given by a generator κ_0 of $K_{Y,0}$. It follows that the Serre dual of the space $H^k(X_0, L^0/L^2\Omega^k_{X\,|X_0})$ is the space $H^{n-k}(X_0, L^{b-1}\Omega^{N-k}_{X\quad|X_0})$. Now, we have the exact sequence (which can be identified with the exact sequence obtained taking the dual of (7.18) and tensoring with K_{X_0}):

$$0 \to \Omega^{n-k}_{X_0} \to L^{b-1}\Omega^{N-k}_{X\quad|X_0} \to \Omega^{n-k+1}_{X_0} \otimes \pi^* T_{Y,0} \to 0, \tag{7.19}$$

where the first map is given by $\alpha \mapsto \alpha \wedge \pi^*\kappa_0$, and the last map can be identified with the composed map

$$L^{b-1}\Omega^{N-k}_{X\quad|X_0} \to L^{b-1}/L^b\Omega^{N-k}_{X\quad|X_0} \cong \Omega^{n-k+1}_{X_0} \otimes \pi^*\Omega^{b-1}_{Y,0}$$

via the isomorphism

$$\operatorname{int}()\kappa_0 : T_{Y,0} \cong \Omega^{b-1}_{Y,0}.$$

The exact sequence (7.19) gives the long exact sequence

$$H^{n-k}(X_0, L^{b-1}\Omega^{N-k}_{X\quad|X_0}) \xrightarrow{r} T_{Y,0} \otimes H^{n-k}(X_0, \Omega^{n-k+1}_{X_0}) \xrightarrow{{}^t\overline{\nabla}} H^{n-k+1}(X_0, \Omega^{n-k}_{X_0}).$$

Here, the map r is the transposed map with respect to Serre duality of the map

$$H^k(X_0, \Omega_{X_0}^{k-1}) \otimes \Omega_{Y,0} \to H^k(X_0, L^0/L^2\Omega^k_{X\,|X_0})$$

induced by (7.18). It follows that for

$$\eta = r(\psi) = \sum_i \eta_i \otimes u_i \in \operatorname{Im} r = \operatorname{Ker}{}^t\overline{\nabla} \subset \mathcal{H}_0^{n-k+1,n-k} \otimes T_{Y,0},$$

we have

$$\delta[Z]_0(\eta) = \langle [Z]_0, \psi \rangle,$$

where the pairing $\langle\,,\,\rangle$ is given by the Serre duality between $H^k(X_0, \Omega^k_{X\,|X_0})$ and $H^{n-k}(X_0, \Omega_X^{N-k}{}_{|X_0})$.

To compute $\delta[Z]_0(\eta)$, it thus remains to explicitly compute a lifting

$$\psi \in H^{n-k}(X_0, L^{b-1}\Omega_X^{N-k}{}_{|X_0})$$

of $\eta \in \operatorname{Ker}{}^t\overline{\nabla}$. With notation as above, let $\epsilon_i = \operatorname{int}(u_i)(\kappa_0)$. Writing $\omega_i^{n-k+1,n-k}$ for the component of type $(n-k+1, n-k)$ of ω_i, we can view $\omega_i^{n-k+1,n-k}$ as a form of type $(0, n-k)$ with values in $\Omega_{X_0}^{n-k+1}$, and the form

$$\sum_i \Omega_i^{n-k+1,n-k}{}_{|X_0} \wedge \epsilon_i$$

is a $(0, n-k)$-form with values in $L^{b-1}\Omega_X^{n-k+1}{}_{|X_0}$, which is a lifting of

$$\sum_i \omega_i \otimes u_i \in A_{X_0}^{0,n-k}(\Omega_{X_0}^{n-k+1}) \otimes T_{Y,0}.$$

This form is not $\overline{\partial}$-closed, but denoting by $\alpha^{n-k,n-k}$ the component of type $(n-k, n-k)$ of the form α introduced in (7.14), and taking the components of type $(n-k, n-k+1)$, the relation (7.15) yields the relation

$$\sum_i \operatorname{int}(\tilde{u}_i)(\overline{\partial}\Omega_i^{n-k+1,n-k})_{|X_0} = \overline{\partial}\alpha^{n-k,n-k}.$$

Thus, the form

$$\tilde{\psi} = \sum_i \Omega_i^{n-k+1,n-k} \wedge \pi^*\epsilon_i + \alpha^{n-k,n-k} \wedge \pi^*\kappa_0 \in A_{X_0}^{0,n-k}(L^{b-1}\Omega_X^{N-k}{}_{|X_0}) \tag{7.20}$$

is $\overline{\partial}$-closed. Therefore, the Dolbeault cohomology class ψ of $\tilde{\psi}$ is a lifting of $\sum_i \eta_i \otimes u_i$ to $H^{n-k}(X_0, L^{b-1}\Omega_X^{N-k}{}_{|X_0})$, i.e. it satisfies

$$r(\psi) = \eta.$$

Thus, we have

$$\delta[Z]_0(\eta) = \langle [Z]_0, \tilde{\psi} \rangle = \sum_i n_i \int_{Z_i,0} \alpha^{n-k,n-k} = \sum_i n_i \int_{Z_i,0} \alpha,$$

where the second equality follows from formula (7.20) for $\tilde{\psi}$ and the description of the class $[Z]_0$ given at the beginning of this section. Comparing with (7.17), we find that we have proved the equality

$$\delta[Z]_0(\eta) = \delta\nu_{Z,0}(\eta).$$

□

This result has the following application (see Green & Müller-Stach 1996).

Theorem 7.15 *Let $Z \subset X$ be a cycle of codimension n in a smooth projective $2n$-dimensional variety, and let L be a sufficiently ample line bundle on X. Assume that the class $[Z]$ is not torsion, and that its restriction to X_t for $X_t \in |L|$ vanishes in $H^{2n}(X_t, \mathbb{Z})$. Then if $X_t \in |L|$ is general, $Z_t := Z_{|X_t}$ is not annihilated by the Abel–Jacobi map of X_t.*

Proof Let $U \subset H^0(X, L)$ be the open set parametrising the smooth members of $|L|$. Let $\pi : \mathcal{X} \to U$ be the universal family of hypersurfaces parametrised by U. Let $p : \mathcal{X} \to X$ be the natural map. Consider the cycle $p^*Z =: \mathcal{Z}$. By hypothesis, its class $[\mathcal{Z}]$ vanishes under restriction to the fibres X_y of π. Consider the associated normal function $\nu_{\mathcal{Z}}$, defined by

$$\nu_{\mathcal{Z}}(y) = \Phi_{X_y}\left(\mathcal{Z}_{|X_y}\right) = \Phi_{X_y}(Z_y).$$

We want to show that $\nu_{\mathcal{Z}}$ is non-zero. (This will imply the result, as the set of points y such that $\nu_{\mathcal{Z}}(y) = \Phi_{X_y}(Z_y) = 0$ is then a proper analytic subset.)

Assume, then, that $\nu_{\mathcal{Z}} = 0$. Then in particular, the infinitesimal invariant $\delta\nu_{\mathcal{Z}}$ is zero. Now, we have just shown that it is also the first infinitesimal invariant of the class

$$[\mathcal{Z}] \in L^1 H^n\left(\mathcal{X}, \Omega^n_{\mathcal{X}}\right).$$

Consider the spectral sequences associated to the sheaves $\Omega^n_{\mathcal{X}}$ and $\Omega^n_{X \times U}$, to the Leray filtration and to the functors π_*, pr_{2*} respectively where pr_2 is the second projection onto U. The terms $E_1^{p,q}$ are respectively equal to $R^{p+q}\pi_*\Omega^{n-p}_{\mathcal{X}/U} \otimes \Omega^p_U$ and to $R^{p+q}\mathrm{pr}_{2*}(p^*\Omega^{n-p}_X) \otimes \Omega^p_U$. By the Lefschetz theorem 1.23, the restriction map

$$H^q\left(X, \Omega^p_X\right) \to H^q\left(X_y, \Omega^p_{X_y}\right)$$

is an isomorphism for $p+q < 2n-1$, so that the restriction induces an isomorphism of spectral sequences on the $E_1^{p,q}$ for $p+q=n,\ p>1$ and for $p+q=n+1,\ p>2$. It follows that the restriction induces a surjection

$$L^2R^n\mathrm{pr}_{2*}\Omega^n_{X\times U} \to L^2R^n\pi_*\Omega^n_{\mathcal{X}}.$$

Moreover, by definition, the kernel of the 'infinitesimal invariant' map

$$L^1R^n\pi_*\Omega^n_{\mathcal{X}} \to E_2^{1,n-1}$$

is equal to $L^2R^n\pi_*\Omega^n_{\mathcal{X}}$. We thus conclude that the image of $[\mathcal{Z}]$ in $H^0(U, R^n\pi_*\Omega^n_{\mathcal{X}})$ is the restriction of a class $\alpha \in H^0(U, L^2R^n\mathrm{pr}_{2*}\Omega^n_{X\times U})$. Finally, as U is affine, the cohomology groups $H^i(U,\mathcal{F})$ are zero for every algebraic coherent sheaf on U and every $i>0$ (see Hartshorne 1977, III.3). Thus, by the Leray spectral sequence

$$H^n(\mathcal{X},\Omega^n_{\mathcal{X}}) = H^0(X, R^n\pi_*\Omega^n_{\mathcal{X}}),$$

we have

$$H^n(X\times U,\Omega^n_{X\times U}) = H^0(X, R^n\mathrm{pr}_{2*}\Omega^n_{X\times U}).$$

It follows that there exists a class $\alpha \in L^2H^n(X\times U,\Omega^n_{X\times U})$ such that

$$\alpha_{|\mathcal{X}} = [\mathcal{Z}] = \mathrm{pr}_1^*[Z]_{|\mathcal{X}} \text{ in } H^n(\mathcal{X},\Omega^n_{\mathcal{X}}),$$

where pr_1 is the first projection onto U. Thus, $(\alpha-\mathrm{pr}_1^*[Z])$ vanishes on $\mathcal{X}$. But as $\mathrm{pr}_1^*[Z] \notin L^2H^n(X\times U,\Omega^n_{X\times U})$, we have $\alpha-\mathrm{pr}_1^*[Z]\neq 0$ in $H^n(X\times U,\Omega^n_{X\times U})$, and this contradicts the following result, which is a consequence of the proof of Nori's theorem (see the next chapter).

Proposition 7.16 *The restriction map*

$$H^n(X\times U,\Omega^n_{X\times U}) \to H^n(\mathcal{X},\Omega^n_{\mathcal{X}})$$

is injective. □

7.3 The case of hypersurfaces of high degree in $\mathbb{P}^n$

7.3.1 Application of the symmetriser lemma

Let $\pi : \mathcal{X} \to B$ be the universal family of smooth hypersurfaces of degree d in $\mathbb{P}^n$. Recall that we have the complexes $\mathcal{K}_{p,q},\ p+q=n-1$ given by

$$0 \to \mathcal{H}^{p,q} \xrightarrow{\overline{\nabla}} \mathcal{H}^{p-1,q+1}\otimes\Omega_B \xrightarrow{\overline{\nabla}^2} \mathcal{H}^{p-2,q+2}\otimes\Omega_B^2 \to \cdots$$

coming from the variation of Hodge structure on the vanishing cohomology (or the primitive cohomology, these two notions being equivalent for hypersurfaces of projective space) of the hypersurfaces parametrised by B. Macaulay's theorem and theorem 6.17 first show that the complex $\mathcal{K}_{p,q}$ is exact in degree 0 whenever $\mathcal{H}^{p-1,q+1} \neq 0$. Indeed, at every point $f \in B$,

$$\overline{\nabla} : \mathcal{H}_f^{p,q} \to \mathcal{H}_f^{p-1,q+1} \otimes \Omega_{B,f}$$

can be identified with the map

$$R_f^{(n-p)d-n-1} \to \operatorname{Hom}\left(S^d, R_f^{(n-p+1)d-n-1}\right)$$

given by the product. Now, by Macaulay's theorem 6.19, this map is injective whenever $R_f^{(n-p+1)d-n-1} \neq 0$. Furthermore, we have the following consequence of the symmetriser lemma (proposition 6.21).

Proposition 7.17 *The complex $\mathcal{K}_{p,q}$ is exact in degree 1 if $(n-p+2)d-n-1 < N = (n+1)(d-2)$.*

Proof By theorem 6.17, this is equivalent to the statement that for every $f \in B$, the sequence

$$R_f^{(n-p)d-n-1} \to \operatorname{Hom}\left(S^d, R_f^{(n-p+1)d-n-1}\right) \to \operatorname{Hom}\left(\bigwedge^2 S^d, R_f^{(n-p+2)d-n-1}\right),$$

where the maps are given by the product, is exact at the middle.

The symmetriser lemma says that this is satisfied if

$$d + (n-p+1)d - n - 1 = (n-p+2)d - n - 1 < N,$$

$$d - 1 + (n-p+1)d - n - 1 \leq N.$$

But clearly the first condition implies the second. □

In particular, applying this to the family of hypersurfaces of degree d in $\mathbb{P}^{2n}$, we obtain the following result.

Corollary 7.18 *If $(n+2)d - (2n+1) < (2n+1)(d-2)$, then the complexes*

$$\mathcal{K}_{p,q},\ p+q = 2n-1,\ p \geq n,$$

are exact in degree 0 and 1.

Proof By Macaulay's theorem, we know that the complex $\mathcal{K}_{p,q}$, $p+q = 2n-1$ is exact in degree 0 when $\mathcal{K}^1_{p,q} = \mathcal{H}^{p-1,q+1} \otimes \Omega_B$ is non-zero, i.e. when

$$(2n-p)d - 2n - 1 \leq N = (d-2)(2n+1).$$

But this inequality is satisfied if $(n+2)d - (2n+1) < (2n+1)(d-2)$ and $p \geq n$.

Similarly, the preceding proposition says that the complex $\mathcal{K}_{p,q}$, $p+q = 2n-1$ is exact in degree 1 if $(2n-p+2)d - 2n - 1 < N = (2n+1)(d-2)$. But this inequality is satisfied if $(n+2)d - (2n+1) < N = (2n+1)(d-2)$ and $p \geq n$. □

7.3.2 Generic triviality of the Abel–Jacobi map

The preceding result and the analysis of the infinitesimal invariants of normal functions now yield the following result.

Theorem 7.19 *Consider the family $\pi : \mathcal{X} \to B$ of smooth hypersurfaces of degree d in $\mathbb{P}^{2n}$. Then if $(n+2)d - (2n+1) < (2n+1)(d-2)$, for a general point $f \in B$ the image of the Abel–Jacobi map*

$$\Phi^n_{Y_f} : \mathcal{Z}^n(Y_f)_{\text{hom}} \to J^{2n-1}(Y_f)$$

is contained in the torsion of $J^{2n-1}(Y_f)$.

This theorem was proved independently by Green (1989b) in every dimension, and by the author (unpublished) for 3-dimensional hypersurfaces. The first step of the proof consists in proving the following result.

Proposition 7.20 *Let $U \subset B$ be a simply connected open set, and let*

$$\alpha \in \mathcal{H}^{2n-1}/F^n\mathcal{H}^{2n-1}(U)$$

be a section satisfying the horizontality condition

$$\nabla\tilde{\alpha} \in F^{n-1}\mathcal{H}^{2n-1} \otimes \Omega_U,$$

where $\tilde{\alpha}$ is a lifting of α to $\mathcal{H}^{2n-1}$.

Then if $(n+2)d - (2n+1) < (2n+1)(d-2)$, there exists a unique section β of $H^{2n-1}_{\mathbb{C}}$ on U which projects to α.

Here, one uses the inclusion

$$H^{2n-1}_{\mathbb{C}} \subset \mathcal{H}^{2n-1}$$

of the sheaf of locally constant sections (annihilated by the Gauss–Manin connection) into the sheaf of holomorphic sections.

Proof of proposition 7.20 We begin with the uniqueness. Let β and β' be two such sections; then the difference $\lambda = \beta - \beta'$ is a section of $H^{2n-1}_{\mathbb{C}}$ on U which lies in $F^n\mathcal{H}^{2n-1}$. By the generalised Noether–Lefschetz theorem 6.24, we thus have $\lambda = 0$, since the numerical hypotheses are such that $\mathcal{H}^{n-1,n} \neq 0$.

To prove the existence, note that as U is simply connected, it suffices to prove the existence locally on U. We will show by induction on $l \geq n$ that there exists a local lifting β_l of α in $\mathcal{H}^{2n-1}$ satisfying the condition

$$\nabla\beta_l \in F^{l-1}\mathcal{H}^{2n-1} \otimes \Omega_U. \tag{7.21}$$

For $l = n$, this is the hypothesis of the proposition, and for $l > 2n$, it is the conclusion of the proposition since then $F^{l-1}\mathcal{H}^{2n-1} \otimes \Omega_U = 0$.

Assume that condition (7.21) is satisfied for l. We then have $\nabla\beta_l \in F^{l-1}\mathcal{H}^{2n-1} \otimes \Omega_U$, and as $\nabla \circ \nabla = 0$, its projection $\overline{\nabla\beta_l}$ in $\mathcal{H}^{l-1,2n-l} \otimes \Omega_U$ satisfies

$$\overline{\nabla}(\overline{\nabla\beta_l}) = 0 \text{ in } \mathcal{H}^{l-2,2n-l+1} \otimes \Omega_U^2.$$

Now, by corollary 7.18 together with the fact that $l \geq n$, the complex

$$0 \to \mathcal{H}^{l,2n-l-1} \xrightarrow{\overline{\nabla}} \mathcal{H}^{l-1,2n-l} \otimes \Omega_U \xrightarrow{\overline{\nabla}} \mathcal{H}^{l-2,2n-l+1} \otimes \Omega_U^2$$

is exact at the middle, so locally, there exists a section η of $\mathcal{H}^{l,2n-l-1}$ such that $\overline{\nabla}\eta = \overline{\nabla\beta_l}$. Let $\tilde{\eta}$ be a local lifting of η to $F^l\mathcal{H}^{2n-1}$. Then $\beta_{l+1} = \beta_l - \tilde{\eta}$ satisfies

$$\nabla\beta_{l+1} = 0 \text{ mod } F^l\mathcal{H}^{2n-1} \otimes \Omega_U,$$

since by definition of $\overline{\nabla}$, we have

$$\overline{\nabla}\eta = \nabla\tilde{\eta} \text{ mod } F^l\mathcal{H}^{2n-1} \otimes \Omega_U.$$

Thus, we have $\nabla\beta_{l+1} \in F^l\mathcal{H}^{2n-1} \otimes \Omega_U$, and as $\eta \in F^l\mathcal{H}^{2n-1}$ with $l \geq n$, β_{l+1} also projects to α.

□

The following result can be deduced from this.

Corollary 7.21 *Under the same numerical hypotheses, if U is a simply connected open set of B and ν is a normal function on U, i.e. a holomorphic and horizontal section of the family of intermediate Jacobians $(J^{2n-1}(Y_f))_{f\in B}$, then there exists a section β of $H^{2n-1}_{\mathbb{C}}$ on U such that β projects to ν under the* *composition*

$$H_{\mathbb{C}}^{2n-1} \to \mathcal{H}^{2n-1} \to \mathcal{H}^{2n-1}/(F^n\mathcal{H}^{2n-1} \oplus H_{\mathbb{Z}}^{2n-1}) = \mathcal{J}.$$

Moreover, β is unique up to a section of $H_{\mathbb{Z}}^{2n-1}$.

Proof As U is simply connected, every section of $H_{\mathbb{C}}^{2n-1}$ on an open set of U extends to U, so it suffices to prove the existence locally. Now by definition, every local lifting α of ν in $\mathcal{H}^{2n-1}/F^n\mathcal{H}^{2n-1}$ satisfies the horizontality condition. Proposition 7.20 thus yields the existence of β.

Finally, if β and β' are two sections $H_{\mathbb{C}}^{2n-1}$ on U which project to ν, the difference $\eta = \beta - \beta'$ can be written

$$\eta = \eta_{\mathbb{Z}} + \eta_F,$$

where $\eta_{\mathbb{Z}}$ is a section of $H_{\mathbb{Z}}^{2n-1}$ and η_F is a section of $F^n\mathcal{H}^{2n-1}$.

Then we have $\nabla\eta_F = 0$, and we conclude that $\eta_F = 0$ as in the proof of proposition 7.20. □

Theorem 7.22 *Still under the hypothesis $(n+2)d-(2n+1) < (2n+1)(d-2)$, let $r : V \to B$ be an étale morphism, where V is a quasi-projective variety, and let ν be a normal function on V, i.e. a holomorphic and horizontal section of the family of intermediate Jacobians $(J^{2n-1}(Y_{r(t)}))_{t\in V}$. Then ν is torsion.*

Proof As V is locally isomorphic to B, since r is étale, we can apply corollary 7.21 to conclude that locally on V, there exists a section $\beta \in r^{-1}H_{\mathbb{C}}^{2n-1}$ which is well-defined up to a section of $r^{-1}H_{\mathbb{Z}}^{2n-1}$ and projects to ν. These sections β thus provide a global section β of the local system $r^{-1}(H_{\mathbb{C}}^{2n-1}/H_{\mathbb{Z}}^{2n-1})$.

Using a monodromy argument, we will now show that such a section is necessarily torsion, i.e. is a section of $r^{-1}(H_{\mathbb{Q}}^{2n-1}/H_{\mathbb{Z}}^{2n-1})$. Let $v \in V$, and let

$$\beta_v \in H^{2n-1}(X_{r(v)}, \mathbb{C})/H^{2n-1}(X_{r(v)}, \mathbb{Z})$$

be the value of β at the point v. Then β_v is invariant under the monodromy action on the cohomology with coefficients $\mathbb{C}/\mathbb{Z}$

$$\overline{\rho} : \pi_1(V, v) \stackrel{r_*}{\to} \pi_1(B, r(v)) \to \mathrm{Aut}\,(H^{2n-1}(X_{r(v)}, \mathbb{C})/H^{2n-1}(X_{r(v)}, \mathbb{Z})).$$

If $\tilde{\beta}_v$ is a lifting of β_v to $H^{2n-1}(X_{r(v)}, \mathbb{C})$, this means that for every $\gamma \in \pi_1(V, v)$, we have

$$\rho(\gamma)(\tilde{\beta}_v) - \tilde{\beta}_v \in H^{2n-1}(X_{r(v)}, \mathbb{Z}), \qquad (7.22)$$

where ρ is the monodromy action on $H^{2n-1}(X_{r(v)}, \mathbb{Z})$. Now, the image of the map $r_* : \pi_1(V, v) \to \pi_1(B, r(v))$ is of finite index M in $\pi_1(B, r(v))$.

Moreover, by lemma 2.26 and theorem 3.16, there exist vanishing cycles $\delta_i \in H^{2n-1}(X_{r(v)}, \mathbb{Z})$, and loops $\gamma_i \in \pi_1(B, r(v))$, such that for every $\eta \in H^{2n-1}(X_{r(v)}, \mathbb{Z})$, we have

$$\rho(\gamma_i)(\eta) = \eta + \langle \eta, \delta_i \rangle \delta_i.$$

Finally, the δ_i generate $H^{2n-1}(X_{r(v)}, \mathbb{Z})$, since the whole cohomology $H^{2n-1}(X_{r(v)}, \mathbb{Z})$ is vanishing, as the cohomology of projective space is zero in odd degree.

Now let $\gamma_i' \in \pi_1(V, v)$ be such that $r_*(\gamma_i') = \gamma_i^M$. Then for every

$$\eta \in H^{2n-1}(X_{r(v)}, \mathbb{C}),$$

we have

$$\rho(\gamma_i')(\eta) = \eta + M\langle \eta, \delta_i \rangle \delta_i.$$

Let us apply this to $\tilde{\beta}_v$. We obtain

$$\rho(\gamma_i')(\tilde{\beta}_v) - \tilde{\beta}_v = M\langle \tilde{\beta}_v, \delta_i \rangle \delta_i$$

and by (7.22), this has to belong to $H^{2n-1}(X_{r(v)}, \mathbb{Z})$ for every i. Thus, we have $\langle \tilde{\beta}_v, \delta_i \rangle \in \mathbb{Q}$ for all i, and as the intersection form is non-degenerate on $H^{2n-1}(X_{r(v)}, \mathbb{Z})$, which is generated by the δ_i, it follows that

$$\tilde{\beta}_v \in H^{2n-1}(X_{r(v)}, \mathbb{Q}).$$

Thus, the section β is in fact a section of the local system $r^{-1}(H_{\mathbb{Q}}^{2n-1}/H_{\mathbb{Z}}^{2n-1})$, so it is torsion, as is its projection $\nu \in r^*\mathcal{J}$. □

Proof of theorem 7.19 It now remains simply to explain how theorem 7.22 implies theorem 7.19. As in the proof of the Noether–Lefschetz theorem, this follows from the existence of a countable set of relative Hilbert schemes $p_i : H_i \to B$, where the morphisms p_i are projective, parametrising the pairs (Z, f) consisting of a point $f \in B$ parametrising a smooth hypersurface Y_f, and an $(n-1)$-dimensional subscheme of Y_f. As $p_i(H_i)$ is a closed algebraic subset of B, if $f \in B$ is general and $f \in p_i(H_i)$, then $p_i(H_i) = B$.

Now let $f \in B$ be a general point, and let $Z_1, \ldots, Z_k \subset Y_f$ be subschemes of codimension n such that $Z = \sum_j n_j Z_j$ is homologous to 0 in Y_f. We want to prove that $\Phi_{Y_f}^n(Z)$ is a torsion point of $J^{2n-1}(Y_f)$. Each (Z_j, f) is an element of a relative Hilbert scheme H_j, which by the generality of f has to dominate B via $\mathrm{pr}_2 = p_j$, so the variety

$$W = H_1 \times_B H_2 \times_B \cdots \times_B H_k \xrightarrow{p} B$$

also dominates B, and we have a relative cycle

$$\mathcal{Z} \subset \mathcal{Y}_W,$$

where $\mathcal{Y}_W$ is the pullback via p of the universal hypersurface $\pi : \mathcal{Y} \to B$, $\mathcal{Z} = \sum_j n_j \mathcal{Z}_j$, and each $\mathcal{Z}_j$ is the pullback by the jth projection $\mathrm{pr}_j : W \to H_j$ of the universal subscheme parametrised by H_j (see Kollár 1996). By construction, the cycle $Z \subset Y_f$ is one of the cycles parametrised by this family. Now, up to replacing W by a quasi-projective subvariety of W, using standard arguments, we may assume that $p : W \to B$ is étale and that $(Z, f) \in W$. As the fibres $\mathcal{Z}_w \subset Y_{p(w)}$ of the relative cycle $\mathcal{Z} \subset \mathcal{X}_W$ are homologous to 0, theorem 7.9 yields a normal function $\nu_{\mathcal{Z}}$ on W such that

$$\nu_{\mathcal{Z}}(Z, f) = \Phi^n_{Y_f}(Z) \in J^{2n-1}(Y_f).$$

By theorem 7.22, $\nu_{\mathcal{Z}}$ is torsion, so $\Phi^n_{Y_f}(Z)$ is torsion. □

Theorem 7.19 was generalised by Green & Müller-Stach (1996) to the case of the family of sufficiently ample hypersurfaces of any $2n$-dimensional smooth projective variety X. Let $Y \subset X$ be such a hypersurface. By restriction, we obtain a map

$$J^{2n-1}(X) \to J^{2n-1}(Y),$$

which is injective by the Lefschetz theorem 1.23; we write $J^{2n-1}(Y)_0$ for the cokernel of this map. If $Z \subset X$ is a primitive cycle of codimension n, i.e. such that the restriction

$$[Z]_{|Y} = 0 \text{ in } H^{2n}(Y, \mathbb{Z}),$$

then we have a point

$$\Phi^n_Y(Z_{|Y}) \in J^{2n-1}(Y)$$

and its projection

$$\Phi^n_{Y,0}(Z_{|Y}) \in J^{2n-1}(Y)_0.$$

The result of Green & Müller-Stach is as follows.

Theorem 7.23 *If Y is a sufficiently ample general hypersurface of X, then the image of the primitive part of the Abel–Jacobi map*

$$\Phi^n_{Y,0} : \mathcal{Z}^n_{\mathrm{hom}}(Y) \stackrel{\Phi^n_Y}{\to} J^{2n-1}(Y) \to J^{2n-1}(Y)_0$$

is generated by the $\Phi^n_{Y,0}(Z_{|Y})$ *up to torsion, where* $Z \subset X$ *runs through the set of primitive cycles of codimension* n.

The original proof of this theorem was obtained by refining the proof of theorem 7.19, noting in particular that the cohomology of the complexes $\mathcal{K}_{p,q}$ for $p+q = 2n-1$ in degree 1, which are constructed using variations of Hodge structure on the vanishing cohomology of the hypersurfaces Y, is precisely equal to $H^n(X, \Omega^n_X)_{\text{prim}}$ at every point.

This proof was absorbed by the work of Nori which will be described in the next chapter.

Exercises

1. *Mixed Hodge structures and intermediate Jacobians* (see Carlson 1987). Let $(H_{\mathbb{Z}}, W, F)$ be a mixed Hodge structure (of weight 0) defined over $\mathbb{Z}$. Assume that for some integer k, we have

$$W_{2k-2}H_{\mathbb{Z}} = 0, \quad W_{2k}H_{\mathbb{Z}} = H_{\mathbb{Z}}.$$

Then $H_{\mathbb{Z}}$ is an extension

$$0 \to H^{2k-1}_{\mathbb{Z}} \to H_{\mathbb{Z}} \stackrel{q}{\to} H^{2k}_{\mathbb{Z}} \to 0, \tag{7.23}$$

where $H^{2k-1}_{\mathbb{Z}} = W_{2k-1}H_{\mathbb{Z}}$, the quotient $H^{2k}_{\mathbb{Z}} = W_{2k}/W_{2k-1}H_{\mathbb{Z}}$ is assumed to be torsion-free, and $H^{2k-1}_{\mathbb{Z}}$ and $H^{2k}_{\mathbb{Z}}$ are equipped with pure Hodge structures induced by F, of weight $2k-1$ and $2k$ respectively.

(a) Show that there exists a section $\sigma_F : H^{2k}_{\mathbb{C}} \to H_{\mathbb{C}}$ (i.e. a right inverse for $q_{\mathbb{C}}$) compatible with the Hodge filtrations, i.e. satisfying

$$\sigma_F\left(F^i H^{2k}_{\mathbb{C}}\right) \subset F^i H_{\mathbb{C}}.$$

(b) Let $\sigma_{\mathbb{Z}} : H^{2k}_{\mathbb{Z}} \to H_{\mathbb{Z}}$ be a right inverse for q defined over $\mathbb{Z}$. Show that

$$\sigma_{\mathbb{Z}} \otimes \mathbb{C} - \sigma_F \in \operatorname{Hom}_{\mathbb{C}}\left(H^{2k}_{\mathbb{C}}, H^{2k-1}_{\mathbb{C}}\right).$$

Show that σ_F is defined up to an element of

$$\operatorname{Hom}_F\left(H^{2k}_{\mathbb{C}}, H^{2k-1}_{\mathbb{C}}\right) := \left\{\phi \in \operatorname{Hom}_{\mathbb{C}}\left(H^{2k}_{\mathbb{C}}, H^{2k-1}_{\mathbb{C}}\right) \mid \phi\left(F^i H^{2k}_{\mathbb{C}}\right) \subset F^i H^{2k-1}_{\mathbb{C}},\ \forall i\right\}.$$

Deduce that $\sigma_{\mathbb{Z}} \otimes \mathbb{C} - \sigma_F$ is determined by the mixed Hodge structure on $H_{\mathbb{Z}}$, up to an element of

$$\operatorname{Hom}_F\left(H^{2k}_{\mathbb{C}}, H^{2k-1}_{\mathbb{C}}\right) \oplus \operatorname{Hom}_{\mathbb{Z}}\left(H^{2k}_{\mathbb{Z}}, H^{2k-1}_{\mathbb{Z}}\right).$$

Deduce that the mixed Hodge structure on $H_{\mathbb{Z}}$ determines an element

$$e \in \frac{\operatorname{Hom}_{\mathbb{C}}\left(H_{\mathbb{C}}^{2k}, H_{\mathbb{C}}^{2k-1}\right)}{\operatorname{Hom}_F\left(H_{\mathbb{C}}^{2k}, H_{\mathbb{C}}^{2k-1}\right) \oplus \operatorname{Hom}_{\mathbb{Z}}\left(H_{\mathbb{Z}}^{2k}, H_{\mathbb{Z}}^{2k-1}\right)}.$$

This element is called the extension class associated to the exact sequence (7.23) of mixed Hodge structures.

(c) Show that $\frac{\operatorname{Hom}_{\mathbb{C}}(H_{\mathbb{C}}^{2k}, H_{\mathbb{C}}^{2k-1})}{\operatorname{Hom}_F(H_{\mathbb{C}}^{2k}, H_{\mathbb{C}}^{2k-1}) \oplus \operatorname{Hom}_{\mathbb{Z}}(H_{\mathbb{Z}}^{2k}, H_{\mathbb{Z}}^{2k-1})}$ is a complex torus. (It is the intermediate Jacobian associated to the induced Hodge structure of odd weight on $\operatorname{Hom}(H_{\mathbb{Z}}^{2k}, H_{\mathbb{Z}}^{2k-1})$.) Show that if the Hodge structure on $H_{\mathbb{Z}}^{2k}$ is trivial, i.e. if

$$H_{\mathbb{Z}}^{2k} = \mathbb{Z}, \quad F^{k+1} H_{\mathbb{C}}^{2k} = 0, \quad F^k H_{\mathbb{C}}^{2k} = H_{\mathbb{C}}^{2k},$$

then the complex torus $\frac{\operatorname{Hom}_{\mathbb{C}}(H_{\mathbb{C}}^{2k}, H_{\mathbb{C}}^{2k-1})}{\operatorname{Hom}_F(H_{\mathbb{C}}^{2k}, H_{\mathbb{C}}^{2k-1}) \oplus \operatorname{Hom}_{\mathbb{Z}}(H_{\mathbb{Z}}^{2k}, H_{\mathbb{Z}}^{2k-1})}$ is isomorphic to the intermediate Jacobian J^{2k-1} associated to the Hodge structure of weight $2k-1$ on $H_{\mathbb{Z}}^{2k-1}$.

(d) Conversely, let

$$e \in \frac{\operatorname{Hom}_{\mathbb{C}}\left(H_{\mathbb{C}}^{2k}, H_{\mathbb{C}}^{2k-1}\right)}{\operatorname{Hom}_F\left(H_{\mathbb{C}}^{2k}, H_{\mathbb{C}}^{2k-1}\right) \oplus \operatorname{Hom}_{\mathbb{Z}}\left(H_{\mathbb{Z}}^{2k}, H_{\mathbb{Z}}^{2k-1}\right)},$$

and let $\tilde{e} \in \operatorname{Hom}_{\mathbb{C}}(H_{\mathbb{C}}^{2k}, H_{\mathbb{C}}^{2k-1})$ be a lifting of e. Put a mixed Hodge structure on $H_{\mathbb{Z}} = H_{\mathbb{Z}}^{2k-1} \oplus H_{\mathbb{Z}}^{2k}$, by setting

$$\begin{aligned} W_{2k-1} H_{\mathbb{Z}} &= H_{\mathbb{Z}}^{2k-1}, \quad W_{2k} H_{\mathbb{Z}} = H_{\mathbb{Z}}, \\ F^i H_{\mathbb{C}} &= F^i H_{\mathbb{C}}^{2k-1} \oplus (\mathrm{Id} - \tilde{e})(F^i H_{\mathbb{C}}^{2k}). \end{aligned}$$

Show that the extension class of this mixed Hodge structure is equal to e.

(e) Show that two mixed Hodge structures on $H_{\mathbb{Z}}$ inducing given Hodge structures on $H_{\mathbb{Z}}^{2k-1}$ and $H_{\mathbb{Z}}^{2k}$ are isomorphic, via an isomorphism which induces the identity on $H_{\mathbb{Z}}^{2k-1}$ and $H_{\mathbb{Z}}^{2k}$, if and only if the associated extension classes are equal.

2. *Horizontality and transversality.* Let $(H_{\mathbb{Z}}^{2k-1}, F^i \mathcal{H}^{2k-1})$ be a variation of pure Hodge structures of weight $2k-1$ over a base B. (In particular, the transversality condition

$$\nabla F^i \mathcal{H}^{2k-1} \subset F^{i-1} \mathcal{H}^{2k-1} \otimes \Omega_B$$

is satisfied.) Let ν be a holomorphic section of the associated family of intermediate Jacobians

$$\nu(b) \in \frac{\mathcal{H}_b^{2k-1}}{F^k\mathcal{H}_b^{2k-1} \oplus H_{\mathbb{Z},b}^{2k-1}}.$$

(a) Using the preceding exercise, construct using ν a holomorphic filtration F on $\mathcal{H} := (H_{\mathbb{Z}}^{2k-1} \oplus \mathbb{Z}) \otimes \mathcal{O}_B$ which induces the Hodge filtration on $\mathcal{H}^{2k-1}$ and the trivial filtration of weight $2k$, given by

$$F^{k+1}\mathcal{O}_B = 0, \quad F^k\mathcal{O}_B = \mathcal{O}_B,$$

on the quotient $\mathcal{O}_B$.

(b) Show that the section ν is horizontal if and only if the filtration F thus constructed satisfies the transversality condition.

8

Nori's Work

In this chapter, we finally give the proof of Nori's connectivity theorem, which is a strengthened Lefschetz theorem for locally complete families of sufficiently ample hypersurfaces or complete intersections. This statement, which brings us back to our starting point, i.e. the Lefschetz theorem on hyperplane sections, enables us to evaluate the contribution of the theory of infinitesimal variations of Hodge structure.

Theorem 8.1 (Nori) *Let $j : \mathcal{Y}_B \subset X \times B$ be the universal family of complete intersections of r sufficiently ample hypersurfaces in a smooth $(n+r)$-dimensional variety X. Then for every smooth quasi-projective variety T and any submersive morphism $\phi : T \to B$, the restriction*

$$j^* : H^k(X \times T, \mathbb{Q}) \to H^k(\mathcal{Y}_T, \mathbb{Q})$$

is an isomorphism for $k < 2n$, and is injective for $k \leq 2n$, where $\mathcal{Y}_T$ is the product bundle $T \times_B \mathcal{Y}_B$.

Apart from an argument using mixed Hodge structures, which reduces this statement to a comparison statement on the Dolbeault cohomology in degree $\leq n$, the main ingredients of the proof are: the computation of the cohomology of the complexes $\mathcal{K}_{p,q}$, $p+q=n$ associated to the variations of Hodge structure on the vanishing cohomology of the complete intersections Y_b, $b \in B$, and the interpretation of the complexes $\mathcal{K}_{p,q}$ in terms of the Leray spectral sequences of the sheaves $\Omega^q_{\mathcal{Y}_T}$ for the Leray filtration and the functors $R^p\phi_*$ given in proposition 5.9.

The remainder of this chapter is devoted to a remarkable application of this theorem to the theory of algebraic cycles. This application concerns the Griffiths group of algebraic varieties, and we first devote an entire section to various results of Griffiths.

We begin with the proof of Griffiths' theorem, which states that the Abel–Jacobi map for general quintic hypersurfaces of $\mathbb{P}^4$ is non-trivial modulo torsion. The most important consequence of this theorem concerns the Griffiths group

$$\mathrm{Griff}^2(Y) = \mathcal{Z}^2(Y)_{\mathrm{hom}}/\mathcal{Z}^2(Y)_{\mathrm{alg}}$$

for such a quintic Y, where $\mathcal{Z}^2(Y)_{\mathrm{alg}}$ is the subgroup of the cycles algebraically equivalent to 0.

Indeed, Griffiths also proved the following result, which is a consequence of the horizontality of the normal functions associated to algebraic cycles.

Theorem 8.2 *For every projective smooth variety Y, the Abel–Jacobi map Φ_Y^k sends the group $\mathcal{Z}^2(Y)_{\mathrm{alg}}$ to the 'algebraic part' of the intermediate Jacobian $J^{2k-1}(Y)$.*

The generalised Noether–Lefschetz theorem for hypersurfaces then allows us to show that the algebraic part of the intermediate Jacobian $J^3(Y)$ is trivial if Y is a general quintic of $\mathbb{P}^4$, and combining these statements, we obtain the following theorem.

Theorem 8.3 *The Griffiths group* $\mathrm{Griff}^2(X)$ *is not a torsion group if Y is a general quintic of $\mathbb{P}^4$.*

Griffiths thus proved the existence of cycles homologous to 0, none of whose multiples is algebraically equivalent to 0. The cycles discovered by Griffiths can, however, be detected by an invariant of Hodge theory, namely their Abel–Jacobi invariant modulo the algebraic part of the intermediate Jacobian.

As an extremely ingenious consequence of his connectivity theorem, Nori proved the existence of cycles homologous to 0, which are annihilated by the Abel–Jacobi map and no multiple of which is algebraically equivalent to 0. He also defined a filtration on the Griffiths group and showed in a precise way its non-triviality.

The first section of this chapter is devoted to the connectivity theorem. We have restricted the last part of the proof to the case of hypersurfaces in projective space, because the ground was prepared starting in chapter 6 in order for the proof to appear easy in this case. The proof we give is Nori's original proof in the case of hypersurfaces in projective space. His later proof (Nori 1993) is valid in all cases, but the role played by the infinitesimal variations of Hodge structure of the varieties does not stand out as clearly. Fakhrudin (1996) studies along the same lines the cohomology of locally complete quasi-projective families of abelian varieties.

The second section is devoted to Griffiths' theorem, and particularly, to the notion of the Hodge class of a normal function. In the last section, we present the Nori filtration and Nori's theorem.

8.1 The connectivity theorem

8.1.1 Statement of the theorem

Let X be an $(n+r)$-dimensional projective variety, and let $L_1, \dots, L_r$ be ample line bundles on X. If σ_i, $i = 1, \dots, r$ are sections of L_i, and if

$$Y = \bigcap_i V(\sigma_i) \stackrel{l}{\hookrightarrow} X$$

is smooth of dimension n, then by the Lefschetz theorem 1.23, we know that the restriction map

$$l^* : H^k(X, \mathbb{Z}) \to H^k(Y, \mathbb{Z})$$

is an isomorphism for $k < n$ and is injective for $k = n$.

Let $B \subset \prod_i H^0(X, L_i)$ be the open set consisting of the r-tuples $b = (\sigma_1, \dots, \sigma_r)$ such that $Y_b = \bigcap_i V(\sigma_i)$ is smooth of dimension n. We have the universal complete intersection

$$\mathcal{Y}_B \stackrel{l}{\hookrightarrow} B \times X, \quad \mathcal{Y}_B = \{(b, x) \mid x \in Y_b\}.$$

For every variety T and every morphism $\phi : T \to B$, consider the universal family of complete intersections parametrised by T, i.e. the inclusion

$$\mathcal{Y}_T \stackrel{l}{\hookrightarrow} T \times X, \quad \mathcal{Y}_T = \{(t, x) \mid x \in Y_b,\ b = \phi(t)\}.$$

Writing π_Y and π_X for the projections to T of $\mathcal{Y}_T$ and $T \times X$ respectively, the Lefschetz theorem gives an isomorphism of local systems $l^* : R^k\pi_{X*}\mathbb{Z} \cong R^k\pi_{Y*}\mathbb{Z}$ for $k < n$. Moreover, l^* is injective for $k = n$. The Leray spectral sequences (see subsection 4.1.3) of π_X and π_Y thus coincide starting from E_2 in bidegree (p, q), $q < n$, i.e.

$$l^* : E_{2,X}^{p,q} = H^p(T, R^q\pi_{X*}\mathbb{Z}) \to E_{2,Y}^{p,q} = H^p(T, R^q\pi_{Y*}\mathbb{Z})$$

is an isomorphism for $q < n$. Moreover, for $p = 0, q = n$ we have an injection

$$l^* : E_{2,X}^{0,n} = H^0(T, R^n\pi_{X*}\mathbb{Z}) \hookrightarrow E_{2,Y}^{0,n} = H^0(T, R^n\pi_{Y*}\mathbb{Z}).$$

It follows immediately that $l^* : H^k(T \times X, \mathbb{Z}) \to H^k(\mathcal{Y}_T, \mathbb{Z})$ is an isomorphism for $k < n$, and is injective for $k = n$. Nori's theorem shows that the

Lefschetz theorem above can be radically improved if one considers submersive morphisms ϕ from a smooth quasi-projective variety, i.e. Zariski open in a projective variety, so that the $\mathcal{Y}_T$ considered are locally complete algebraic families of complete intersections. We can thus view this theorem as a statement concerning the 'cohomology of the generic complete intersection'.

Theorem 8.4 *In the preceding situation, assume that T is smooth and quasi-projective, and that the L_i are sufficiently ample. Then for every submersive morphism $\phi : T \to B$, the map*

$$l^* : H^k(T \times X, \mathbb{Q}) \to H^k(\mathcal{Y}_T, \mathbb{Q})$$

is an isomorphism for $k < 2n$, and is injective for $k = 2n$.

Remark 8.5 *Note that we have passed to the rational cohomology here; this statement does not hold in integral cohomology.*

Remark 8.6 *An important difference from the usual Lefschetz theorem is the need to consider sufficiently ample hypersurfaces. We can easily give counter-examples for hypersurfaces of low degree: if we take for B the family of smooth quadrics Y_q in $X = \mathbb{P}^3$, and for T the family parametrising the lines contained in these quadrics (every smooth quadric $Y_q \subset \mathbb{P}^3$ contains two families of lines, each parametrised by a $\mathbb{P}^1$), then the hypersurface $\mathcal{Y}_T \subset T \times \mathbb{P}^3$ contains a tautological divisor*

$$D = \{((\Delta, q), x) \in T \times \mathbb{P}^3 \mid x \in \Delta,\ q \in B,\ \Delta \subset q\},$$

and it is obvious that its cohomology class in $H^2(\mathcal{Y}_T, \mathbb{Q})$ does not come from a class in $H^2(T \times \mathbb{P}^3, \mathbb{Q})$.

8.1.2 Algebraic translation

Let us explain how theorem 8.4 follows from an algebraic statement concerning the Dolbeault cohomology of $T \times X$ and $\mathcal{Y}_T$.

Proposition 8.7 (Nori 1993) *Let $U \stackrel{l}{\hookrightarrow} V$ be two smooth quasi-projective varieties, with U closed in V. Assume that for some integer n, the restriction map*

$$l^* : H^p\big(V, \Omega_V^q\big) \to H^p\big(U, \Omega_U^q\big)$$

is an isomorphism for $p+q<2n$, $p<n$, and is injective for $p+q\leq 2n$, $p\leq n$. Then the restriction map

$$l^*: H^k(V,\mathbb{Q}) \to H^k(U,\mathbb{Q})$$

is an isomorphism for $k<2n$, and is injective for $k=2n$.

Proof The holomorphic de Rham complexes $\Omega_U^\bullet$ and $\Omega_V^\bullet$ are resolutions of constant sheaves $\mathbb{C}$ on U and V respectively. Equivalently, the inclusions

$$\Omega_U^\bullet \subset \mathcal{A}_U^\bullet, \quad \Omega_V^\bullet \subset \mathcal{A}_V^\bullet$$

are quasi-isomorphisms.

Let us introduce the following complex of sheaves on V:

$$\Omega_{V,U}^\bullet := \mathrm{Ker}\,(l^*: \Omega_V^\bullet \to \Omega_U^\bullet). \tag{8.1}$$

We have a natural inclusion

$$\Omega_{V,U}^\bullet \subset \mathcal{A}_{V,U}^\bullet, \tag{8.2}$$

where

$$\mathcal{A}_{V,U}^\bullet := \mathrm{Ker}\,(l^*: \mathcal{A}_V^\bullet \to \mathcal{A}_U^\bullet).$$

The diagram of exact sequences

$$\begin{array}{ccccccccc}
0 & \longrightarrow & \Omega_{V,U}^\bullet & \longrightarrow & \Omega_V^\bullet & \longrightarrow & \Omega_U^\bullet & \longrightarrow & 0 \\
 & & \downarrow & & \downarrow & & \downarrow & & \\
0 & \longrightarrow & \mathcal{A}_{V,U}^\bullet & \longrightarrow & \mathcal{A}_V^\bullet & \longrightarrow & \mathcal{A}_U^\bullet & \longrightarrow & 0,
\end{array}$$

where the last two vertical arrows are quasi-isomorphisms, shows that the first arrow is also a quasi isomorphism. Now, we know that the cohomology of the complex of global sections $A_{V,U}^\bullet$ of $\mathcal{A}_{V,U}^\bullet$ is equal to the relative cohomology of the pair (V,U). As this complex is a complex of fine sheaves, we thus have a composed isomorphism

$$H^k(V,U,\mathbb{C}) \cong H^k(A_{V,U}^\bullet) \cong \mathbb{H}^k(\mathcal{A}_{V,U}^\bullet) \cong \mathbb{H}^k(V,\Omega_{V,U}^\bullet). \tag{8.3}$$

The isomorphism (8.3) enables us to put a decreasing filtration $G^\cdot$ on the relative cohomology $H^k(V,U,\mathbb{C})$ by setting

$$G^q H^k(V,U,\mathbb{C}) = \mathrm{Im}\,(\mathbb{H}^k(V, G^q\Omega_{V,U}^\bullet) \to \mathbb{H}^k(V,\Omega_{V,U}^\bullet)),$$

where

$$G^q\Omega_{V,U}^\bullet = 0 \to \Omega_{V,U}^q \to \Omega_{V,U}^{q+1} \to \cdots.$$

Now assume that

$$l^* : H^p(V, \Omega_V^q) \to H^p(U, \Omega_U^q)$$

is an isomorphism for $p + q < 2n,\ p < n$ and is injective for $p + q \leq 2n$, $p \leq n$. Then the long exact sequence associated to the exact sequence

$$0 \to \Omega_{V,U}^\bullet \to \Omega_V^\bullet \to \Omega_U^\bullet \to 0$$

shows that

$$H^p(V, \Omega_{V,U}^q) = 0 \ \text{ for } \ p + q \leq 2n, \quad p \leq n.$$

The spectral sequence associated to the filtration $G^\bullet$ then shows immediately that

$$G^q H^k(V, U, \mathbb{C}) = 0 \ \text{ for } \ k \leq 2n, \quad q \geq k - n.$$

It remains to deduce that $H^k(V, U, \mathbb{Q}) = 0$ for $k \leq 2n$, which is equivalent to the conclusion of the proposition. But this is a consequence of the following result.

Theorem 8.8 (Deligne 1975) *There exists a mixed Hodge structure of weight k on $H^k(V, U, \mathbb{Q})$, for which the weight filtration W satisfies $W_{-2} = 0$ and the Hodge filtration F satisfies*

$$F^q H^k(V, U, \mathbb{C}) \subset G^q H^k(V, U, \mathbb{C}).$$

Later, we will sketch the proof of this theorem. The proof of proposition 8.7 can now be concluded as follows. We have

$$F^q H^k(V, U, \mathbb{C}) \subset G^q H^k(V, U, \mathbb{C}) = 0 \ \text{ for } \ k \leq 2n, \quad q \geq k - n.$$

Now by lemma 4.21, we have a decomposition

$$W_i H^k(V, U, \mathbb{C}) = \bigoplus_{p+q \leq k+i} L^{p,q}, \tag{8.4}$$

$$F^i H^k(V, U, \mathbb{C}) = \bigoplus_{p \geq i} L^{p,q}, \tag{8.5}$$

where $L^{p,q}$ can be identified with

$$\mathrm{Gr}_F^p \mathrm{Gr}_{p+q-k}^W H^k(V, U, \mathbb{C}) = H^{p,q}\big(\mathrm{Gr}_{p+q-k}^W H^k(V, U, \mathbb{C})\big).$$

By Hodge symmetry for the Hodge structures on the $\mathrm{Gr}_{p+q-k}^W H^k(V, U, \mathbb{C})$, we have in particular

$$L^{p,q} = 0 \Leftrightarrow L^{q,p} = 0. \tag{8.6}$$

Furthermore, the equality (8.4) and the fact that $W_{-2} = 0$ show that $L^{p,q} = 0$ for $p+q < k-1$. Moreover, the equality (8.5) and the fact that $F^i H^k(V, U, \mathbb{C}) = 0$ for $i \geq k-n$ also show that we have $L^{p,q} = 0$ for $p \geq k-n$, so by the symmetry (8.6), also for $q \geq k-n$. But if $p+q \geq k-1$ and $k \leq 2n$, then $p \geq k-n$ or $q \geq k-n$. Thus, $L^{p,q} = 0$ for all p, q and $H^k(V, U, \mathbb{C}) = 0$ for $k \leq 2n$. This of course implies that $H^k(V, U, \mathbb{Q}) = 0$ for $k \leq 2n$, and concludes the proof of the proposition. □

Proof of theorem 8.8 To construct the Hodge filtration and the weight filtration on $H^k(V, U, \mathbb{C})$, we represent this cohomology group as follows. By Hironaka (1964), there exists a smooth projective variety $\overline{V}$ containing V as a dense Zariski open set and such that $D = \overline{V} - V$ is a normal crossing divisor. We write $j : V \hookrightarrow \overline{V}$ for the inclusion. Up to blowing up along D, we may also assume that the Zariski closure $\overline{U}$ of U in $\overline{V}$ is smooth and meets D and all the intersections of the components of D transversally. In particular, $D' := \overline{U} - U$ is a normal crossing divisor in $\overline{U}$. We write $j : U \hookrightarrow \overline{U}$ for the restriction of j to U, and $l : \overline{U} \hookrightarrow \overline{V}$ for the inclusion, which extends the initial inclusion $l : U \hookrightarrow V$.

We then have the logarithmic de Rham complexes (see vI.8.2.2)

$$\Omega^\bullet_{\overline{V}}(\log D), \quad \Omega^\bullet_{\overline{U}}(\log D')$$

and a surjective restriction arrow

$$l^* : \Omega^\bullet_{\overline{V}}(\log D) \to \Omega^\bullet_{\overline{U}}(\log D'),$$

which enables us to define $\Omega^\bullet_{\overline{V,U}}(\log D)$ by the exact sequence

$$0 \to \Omega^\bullet_{\overline{V,U}}(\log D) \to \Omega^\bullet_{\overline{V}}(\log D) \to \Omega^\bullet_{\overline{U}}(\log D') \to 0. \tag{8.7}$$

We have the obvious inclusion of $\Omega^\bullet_{\overline{V,U}}(\log D)$ in $j_*\mathcal{A}^\bullet_{V,U}$, which for every k gives a map

$$\mathbb{H}^k(\overline{V}, \Omega^\bullet_{\overline{V,U}}(\log D)) \to H^k(V, U, \mathbb{C}). \tag{8.8}$$

This inclusion is compatible with the inclusions

$$\Omega^\bullet_{\overline{V}}(\log D) \subset j_*\mathcal{A}^\bullet_V, \quad \Omega^\bullet_{\overline{U}}(\log D') \subset j_*\mathcal{A}^\bullet_U,$$

which are quasi-isomorphisms (see vI.8.2.3) and thus induce isomorphisms

$$\mathbb{H}^k(\overline{V}, \Omega^\bullet_{\overline{V}}(\log D)) \cong H^k(V, \mathbb{C}), \quad \mathbb{H}^k(\overline{U}, \Omega^\bullet_{\overline{U}}(\log D')) \cong H^k(U, \mathbb{C})$$

for every k. Together, the long exact sequence associated to the exact sequence (8.7) and the arrows (8.8) yield a commutative diagram

$$\begin{array}{ccccccc}
\mathbb{H}^k(\overline{V}, \Omega^\bullet_{\overline{V,U}}(\log D)) & \longrightarrow & \mathbb{H}^k(\overline{V}, \Omega^\bullet_{\overline{V}}(\log D)) & \longrightarrow & \mathbb{H}^k(\overline{U}, \Omega^\bullet_{\overline{U}}(\log D')) & \longrightarrow & \mathbb{H}^{k+1}(\overline{V}, \Omega^\bullet_{\overline{V,U}}(\log D)) \\
\downarrow & & \downarrow & & \downarrow & & \downarrow \\
H^k(V, U, \mathbb{C}) & \longrightarrow & H^k(V, \mathbb{C}) & \longrightarrow & H^k(U, \mathbb{C}) & \longrightarrow & H^{k+1}(V, U, \mathbb{C}),
\end{array}$$

and the five lemma implies that the arrows (8.8) are isomorphisms.

The Hodge filtration on $H^k(V, U, \mathbb{C}) = \mathbb{H}^k(\overline{V}, \Omega^\bullet_{\overline{V,U}}(\log D))$ is then defined by

$$\begin{aligned}
&F^p\mathbb{H}^k(\overline{V}, \Omega^\bullet_{\overline{V,U}}(\log D)) \\
&\quad = \mathrm{Im}\,(\mathbb{H}^k(\overline{V}, F^p\Omega^\bullet_{\overline{V,U}}(\log D)) \to \mathbb{H}^k(\overline{V}, \Omega^\bullet_{\overline{V,U}}(\log D))),
\end{aligned}$$

where as usual, we define

$$F^p\Omega^\bullet_{\overline{V,U}}(\log D) = 0 \to \Omega^p_{\overline{V,U}}(\log D) \to \Omega^{p+1}_{\overline{V,U}}(\log D) \to \cdots.$$

The natural map

$$\mathbb{H}^k\big(\overline{V}, F^p\Omega^\bullet_{\overline{V,U}}(\log D)\big) \to \mathbb{H}^k\big(\overline{V}, \Omega^\bullet_{\overline{V,U}}(\log D)\big)$$

is in fact injective, by the degeneration at E_1 of the spectral sequence associated to the filtration F.

Recall finally that for the absolute cohomology $H^*(V, \mathbb{C})$, the weight filtration W is induced in cohomology by the filtration W on the complex $\Omega^\bullet_{\overline{V}}(\log D)$, whose description (see vI.8.4.1) we recall is as follows. The divisor D is locally defined by an equation of the form $z_1 \ldots z_r = 0$, and $W_k\Omega^\bullet_{\overline{V}}(\log D)$ is locally generated by the forms $\sum_{i_1,\ldots,i_k \le r} \frac{dz_{i_1}}{z_{i_1}} \wedge \cdots \wedge \frac{dz_{i_k}}{z_{i_k}} \wedge \alpha_{i_1\ldots i_k}$, where the $\alpha_{i_1\ldots i_k}$ are holomorphic; the filtration W on the complex $\Omega^\bullet_{\overline{U}}(\log D')$ is defined similarly. Note finally that the complex $\Omega^\bullet_{\overline{V,U}}(\log D)$ is quasi-isomorphic to the cone $C^*(l^*)$ of the restriction map

$$l^* : \Omega^\bullet_{\overline{V}}(\log D) \to \Omega^\bullet_{\overline{U}}(\log D'),$$

i.e.

$$\begin{aligned}
C^k(l^*) &= \Omega^k_{\overline{V}}(\log D) \oplus \Omega^{k-1}_{\overline{U}}(\log D') \\
d &= (d_V \pm l^*, d_U) : C^k(l^*) \to C^{k+1}(l^*).
\end{aligned}$$

We then define the weight filtration on $C^*(l^*)$ by

$$W_l C^k(l^*) = W_l\Omega^k_{\overline{V}}(\log D) \oplus W_{l+1}\Omega^{k-1}_{\overline{U}}(\log D').$$

Having defined the filtrations F and W, one shows, essentially using formal arguments concerning the degeneration at E_1 of the Frölicher spectral sequence for smooth projective varieties and the strictness of the morphisms of Hodge structures, that they define a mixed Hodge structure of weight k on $H^k(V, U, \mathbb{C})$. By definition, we have $W_{-2}H^k(V, U, \mathbb{C}) = 0$, since the weight filtration W on the complexes $\Omega^\bullet_{\overline{V}}(\log D)$ and $\Omega^\bullet_{\overline{U}}(\log D')$ satisfies $W_{-1} = 0$. To conclude, it remains only to prove the inclusion $F^q H^k(V, U, \mathbb{C}) \subset G^q H^k(V, U, \mathbb{C})$. But this is obvious, since the inclusion $\Omega^\bullet_{\overline{V,U}}(\log D) \subset j_*\Omega^\bullet_{V,U}$, which induces the isomorphism

$$\mathbb{H}^k\big(\overline{V}, \Omega^\bullet_{\overline{V,U}}(\log D)\big) \cong \mathbb{H}^k(\overline{V}, j_*\Omega^\bullet_{V,U})$$
$$= \mathbb{H}^k(V, \Omega^\bullet_{V,U}) = H^k(V, U, \mathbb{C}),$$

sends $F^p\Omega^\bullet_{\overline{V,U}}(\log D)$ to $G^p j_*\Omega^\bullet_{V,U} = j_*G^p\Omega^\bullet_{V,U}$, and by the definition of the filtration G, we have

$$G^p H^k(V, U, \mathbb{C}) = \operatorname{Im}(\mathbb{H}^k(V, G^p\Omega^\bullet_{V,U}) \to \mathbb{H}^k(V, \Omega^\bullet_{V,U})).$$

But via the isomorphism

$$\mathbb{H}^k(V, \Omega^\bullet_{V,U}) \simeq \mathbb{H}^k(\overline{V}, j_*\Omega^\bullet_{V,U}),$$

this image is contained in a

$$\operatorname{Im}(\mathbb{H}^k(\overline{V}, j_*G^p\Omega^\bullet_{V,U}) \to \mathbb{H}^k(\overline{V}, j_*\Omega^\bullet_{V,U})).$$

□

8.1.3 The case of hypersurfaces of projective space

The results of the preceding section show that theorem 8.4 is a consequence of the following theorem.

Theorem 8.9 *With the notation of theorem 8.4 and under the same hypotheses, the restriction map*

$$l^* : H^p\big(T \times X, \Omega^q_{T\times X}\big) \to H^p\big(\mathcal{Y}_T, \Omega^q_{\mathcal{Y}_T}\big)$$

is an isomorphism for $p + q < 2n$, $p < n$, and is injective for $p + q \leq 2n$, $p \leq n$.

More precisely, if we consider the Leray spectral sequences for the cohomology of the sheaves $\Omega^q_{T\times X}$ relative to π_X and $\Omega^q_{\mathcal{Y}_T}$ relative to π_Y, we see that this theorem can actually be deduced from the following statement, for which we recall that $\pi_X : T \times X \to T$ is the first projection, and $\pi_Y : \mathcal{Y}_T \to T$ is equal to $\pi_X \circ l$.

Theorem 8.10 *With the notation of theorem 8.4, and under the same hypotheses,*

$$l^* : R^p \pi_{X*} \Omega^q_{T\times X} \to R^p \pi_{Y*} \Omega^q_{\mathcal{Y}_T}$$

is an isomorphism for $p+q < 2n$, $p < n$, and is injective for $p+q \leq 2n$, $p \leq n$.

We restrict ourselves to proving theorem 8.10 in the case where $X = \mathbb{P}^{n+1}$ and $r = 1$, i.e. the case of hypersurfaces of high degree in projective space.

Proof Consider the family $\pi_Y : \mathcal{Y}_T \to T$ of smooth hypersurfaces of degree d parametrised by T, and consider the variation of Hodge structure on the local system $H^n_{\mathbb{C}} = R^n \pi_{Y*}\mathbb{C}_{\text{prim}}$. Recall that we have the Hodge bundles $\mathcal{H}^n = H^n_{\mathbb{C}} \otimes \mathcal{O}_T$ on T, equipped with the filtration $F^p\mathcal{H}^n$ satisfying the transversality condition for the Gauss–Manin connection ∇, the bundles $\mathcal{H}^{p,q} = F^p\mathcal{H}^n / F^{p+1}\mathcal{H}^n$, and the complexes (see subsection 5.1.3)

$$\begin{aligned}\mathcal{K}_{p,q} &= \mathrm{Gr}^p_F(DR(H^n_{\mathbb{C}}))\\ &= 0 \to \mathcal{H}^{p,q} \overset{\overline{\nabla}}{\to} \mathcal{H}^{p-1,q+1} \otimes \Omega_T \overset{\overline{\nabla}}{\to} \mathcal{H}^{p-2,q+2} \otimes \Omega^2_T \to \cdots.\end{aligned}$$

We shall prove

Theorem 8.11 *If d is sufficiently large, the complexes $\mathcal{K}_{p,q}$ are exact in degree $k < \inf(p, n)$.*

In the case of hypersurfaces of projective space, the proof will actually provide an explicit bound for d as a function of n.

Admitting theorem 8.11, let us show how to conclude the proof of theorem 8.10 (for hypersurfaces of projective space). We need to show that

$$l^* : R^p \pi_{X*} \Omega^q_{T\times X} \to R^p \pi_{Y*} \Omega^q_{\mathcal{Y}_T}$$

is an isomorphism for $p+q < 2n$, $p < n$ and is injective for $p+q \leq 2n$, $p \leq n$. Consider the Leray filtrations L on $\Omega^q_{T\times X}$ and $\Omega^q_{\mathcal{Y}_T}$:

$$L^r \Omega^q_{T\times X} = \pi_X^* \Omega^r_T \wedge \Omega^{q-r}_{T\times X}, \quad L^r \Omega^q_{\mathcal{Y}_T} = \pi_Y^* \Omega^r_T \wedge \Omega^{q-r}_{\mathcal{Y}_T}.$$

The restriction morphism $l^* : \Omega^q_{T\times X} \to \Omega^q_{\mathcal{Y}_T}$ is compatible with the Leray filtrations, and it suffices to show that the restriction induces an isomorphism at

E_2 of the corresponding spectral sequences for every bidegree (r, s) such that $s + r + q < 2n$, $s + r < n$, and an injection if $s + r + q \leq 2n$, $s + r \leq n$.

For the first sheaf, the term $E_1^{r,s}$ of the spectral sequence is equal to

$$\Omega_T^r \otimes H^{s+r}\left(\mathbb{P}^{n+1}, \Omega_{\mathbb{P}^{n+1}}^{q-r}\right),$$

as the differential d_1 is zero. For the second sheaf, the term $E_1^{r,s}$ can be identified by proposition 5.9 with $\mathcal{K}'^r_{q,s}$, where the complex $\mathcal{K}'_{q,s}$ is the complex constructed by considering the whole variation of Hodge structure of the hypersurfaces $(Y_t)_{t\in T}$, instead of only considering its primitive part, i.e.

$$\mathcal{K}'^r_{q,s} = \Omega_T^r \otimes R^{r+s}\pi_{Y*}\Omega_{Y_T/T}^{q-r}.$$

Moreover, we know that d_1 can be identified with $\overline{\nabla}$.

As we have a decomposition as a direct sum

$$H^*(Y_t, \mathbb{C}) = H^*(\mathbb{P}^{n+1}, \mathbb{C}) \oplus H^*(Y_t, \mathbb{C})_{\text{prim}}$$

in degree $\leq 2n$, we conclude that for $s + q \leq 2n$, the complexes $(\mathcal{K}'^r_{q,s}, \overline{\nabla})$ are (actually split) extensions

$$0 \to \left(\Omega_T^r \otimes H^{s+r}\left(\mathbb{P}^{n+1}, \Omega_{\mathbb{P}^{n+1}}^{q-r}\right), d = 0\right) \to \left(\mathcal{K}'^r_{q,s}, \overline{\nabla}\right) \to \left(\mathcal{K}^r_{q,s}, \overline{\nabla}\right) \to 0.$$

The exactness of the complexes $(\mathcal{K}_{q,s}, \overline{\nabla})$ in degree $k < \inf(n, q)$ for $q + r = n$, and at every degree for $q + r \neq n$, then shows that for $q + s \leq 2n$, the complex $(\mathcal{K}'_{q,s}, \overline{\nabla})$ has cohomology given by $\Omega_T^r \otimes H^{s+r}(\mathbb{P}^{n+1}, \Omega_{\mathbb{P}^{n+1}}^{q-r})$ in degree $r < \inf(n, q)$ for $q + s = n$, and in every degree for $q + s \neq n$.

Thus, for $q + s \leq 2n$, the restriction arrow induces an isomorphism on the E_2 of the Leray spectral sequences for the sheaves $\Omega_{T\times X}^q$ and $\Omega_{\mathcal{Y}_T}^q$ in the bidegrees (r, s) such that $r < \inf(n, q)$ or $q + s \neq n$.

Now assume that $s + r < n$ and $s + r + q < 2n$; then $q + s \leq 2n$. If $q + s \neq n$, we have already seen that the restriction arrow induces an isomorphism on the $E_2^{r,s}$ of the Leray spectral sequences. Moreover, if $q + s = n$, we have the implications

$$s + r + q < 2n \Rightarrow r < n,$$
$$s + r < n \Rightarrow r < q.$$

Thus, we also have an isomorphism on the E_2 of the Leray spectral sequences in bidegree (r, s) in this case. We show similarly that l^* is injective on the terms $E_2^{r,s}$ for $s + r \leq n$, $s + r + q \leq 2n$. □

Proof of theorem 8.11 The proof is the same as for the symmetriser lemma (proposition 6.21). One first checks that it suffices to prove the exactness for

$T = B$. Then, in this case, the exactness of the complex $\mathcal{K}_{p,q}$ in degree $l < \inf(p, n)$ is equivalent to the exactness at the middle of the sequence

$$\mathcal{H}^{p-l+1,q+l-1} \otimes \Omega_B^{l-1} \xrightarrow{\overline{\nabla}} \mathcal{H}^{p-l,q+l} \otimes \Omega_B^{l} \xrightarrow{\overline{\nabla}} \mathcal{H}^{p-l-1,q+l+1} \otimes \Omega_B^{l+1}$$

for $p - l - 1 \geq 0$, $l \leq n - 1$, or also to the exactness at the middle of the sequence

$$\mathcal{H}^{p,q} \otimes \Omega_B^{l-1} \xrightarrow{\overline{\nabla}} \mathcal{H}^{p-1,q+1} \otimes \Omega_B^{l} \xrightarrow{\overline{\nabla}} \mathcal{H}^{p-2,q+2} \otimes \Omega_B^{l+1}$$

for $p - 2 \geq 0$ and $l \leq n - 1$.

By theorem 6.17, for $f \in B$, we have isomorphisms

$$R_f^{kd-n-2} \cong H^{n+1-k,k-1}(Y_f)_{\text{prim}}$$

identifying the map

$$\overline{\nabla} : H^{n+1-k,k-1}(Y_f)_{\text{prim}} \to \text{Hom}\,(T_{B,b}, H^{n-k,k}(Y_f)_{\text{prim}})$$

with the map

$$R_f^{kd-n-2} \to \text{Hom}\left(S^d, R_f^{(k+1)d-n-2}\right)$$

given by the product, where R_f is the Jacobian ring of f. Using lemma 5.8, we thus see that for every $f \in B$, we need to prove the exactness at the middle of the sequence

$$\left(\bigwedge^{l-1} S^d\right)^* \otimes R_f^{(k-1)d-n-2} \to \left(\bigwedge^{l} S^d\right)^* \otimes R_f^{kd-n-2}$$
$$\to \left(\bigwedge^{l+1} S^d\right)^* \otimes R_f^{(k+1)d-n-2}$$

for $k + 1 \leq n + 1$, $l \leq n - 1$, where the differentials d are given by

$$d\phi(A_1 \wedge \cdots \wedge A_{a+1}) = \sum_i (-1)^i A_i \phi(A_1 \wedge \cdots \wedge \hat{A}_i \wedge \cdots \wedge A_{a+1})$$

for $\phi \in (\bigwedge^a S^d)^* \otimes R_f^l = \text{Hom}\,(\bigwedge^a S^d, R_f^l)$.

Applying Macaulay's theorem 6.19, this exact sequence dualises to

$$\bigwedge^{l+1} S^d \otimes R_f^{(n-k+1)d-n-2} \xrightarrow{\delta} \bigwedge^{l} S^d \otimes R_f^{(n-k+2)d-n-2}$$
$$\xrightarrow{\delta} \bigwedge^{l-1} S^d \otimes R_f^{(n-k+3)d-n-2}, \tag{8.9}$$

where the differentials δ (called the Koszul differentials) are given by

$$\delta(A_1 \wedge \cdots \wedge A_{k+1} \otimes B) = \sum_i (-1)^i A_1 \wedge \cdots \wedge \hat{A}_i \wedge \cdots \wedge A_{k+1} \otimes A_i B. \tag{8.10}$$

We now need the following result, due to M. Green (1984b).

Theorem 8.12 *The sequence*

$$\bigwedge^{l+1} S^d \otimes S^a \to \bigwedge^{l} S^d \otimes S^{a+d} \to \bigwedge^{l-1} S^d \otimes S^{a+2d},$$

where the differentials are given by (8.10), is exact at the middle if $a \geq l$.

In particular, we find that the sequence

$$\bigwedge^{l+1} S^d \otimes S^{(n-k+1)d-n-2} \to \bigwedge^{l} S^d \otimes S^{(n-k+2)d-n-2} \\ \to \bigwedge^{l-1} S^d \otimes S^{(n-k+3)d-n-2} \tag{8.11}$$

is exact at the middle if $(n-k+1)d-n-2 \geq l$.

Now, if $k+1 \leq n+1$, then $(n-k+1)d-n-2 \geq d-n-2$, and for $d \geq 2n+1$, we find that $(n-k+1)d-n-2 \geq n-1$. Thus, under the hypotheses $k+1 \leq n+1, l \leq n-1$, and $d \geq 2n+1$, we find that the sequence (8.11) is exact at the middle. An additional argument, which makes use of the resolution of $R_f^{\cdot}$ as an $S^{\cdot}$-module, then shows that this implies the exactness at the middle of the sequence (8.9) for $d \geq 2n+1$, and for $k+1 \leq n+1$, $l \leq k-1$. □

Remark 8.13 *The estimate on d is optimal: for $d \leq 2n-1$, the generic hypersurfaces of degree d in $\mathbb{P}^{n+1}$ contain lines, as shown by an argument of dimension-counting similar to the one given in the proof of theorem 8.21. Taking for T the variety*

$$T = \{(\Delta, b) \in \mathrm{Grass}(1, n+1) \times B \mid \Delta \subset Y_b\},$$

the projection $T \to B$ is dominant, so generically submersive. But the universal hypersurface $\mathcal{Y}_T$ contains the locus

$$\mathcal{D}_T = \left\{(x, \Delta, b) \in \mathbb{P}^{n+1} \times \mathrm{Grass}(1, n+1) \times B \mid x \in \Delta \subset Y_b\right\}.$$

The cohomology class of this locus is of degree $\leq 2n-2$, and we can show that it is not the restriction of a cohomology class on $T \times \mathbb{P}^{n+1}$. For $d = 2n$, we can also construct a counterexample to theorem 8.4 by using the existence of a 1-dimensional family of osculating lines to order d for a generic hypersurface of degree d in $\mathbb{P}^{n+1}$ and the notion of higher Chow groups (see Voisin 2002).

8.2 Algebraic equivalence

8.2.1 General properties

When X is a smooth projective variety, we introduced in vI.11.1.2 the group $\mathcal{Z}^k(X)$ of algebraic cycles of codimension k on X, i.e. the group of combinations $\sum_i n_i Z_i$, where the n_i are integers and each $Z_i \subset X$ is a closed, irreducible and reduced algebraic subset of codimension k of X. If C is a smooth curve and $Z \subset C \times X$ is a closed algebraic subset of codimension k each of whose components dominates C, then for each point c, we can consider the cycle $Z_c = Z \cap (c \times X)$ of X. This is the cycle associated to the schematic intersection of Z with $c \times X$ (see vI.11.1.2).

Definition 8.14 *The subgroup $\mathcal{Z}^k(X)_{\text{alg}}$ of $\mathcal{Z}^k(X)$ is the subgroup generated by the cycles of the form $Z_c - Z_{c'}$, for every smooth connected curve C, for every pair of points $c, c' \in C$, and for every cycle Z of codimension k in $C \times X$ each of whose components dominates C.*

It follows from the compatibility of the 'cycle class map' with correspondences (see section 9.2.4) that the group $\mathcal{Z}^k(X)_{\text{alg}}$ is contained in the group $\mathcal{Z}^k(X)_{\text{hom}}$ of cycles which are homologically equivalent to 0. When $k = 1$, we know by the theory of the Picard group that the divisors homologous to 0 are parametrised up to rational equivalence by a connected algebraic variety $\text{Pic}^0(X)$, so they are algebraically equivalent to 0. If $k = n = \dim X$ and X is connected, a 0-cycle is always supported on a smooth connected curve contained in X, and since the cycles homologous to 0 are the 0-cycles of degree 0, we also conclude that

$$\mathcal{Z}^n(X)_{\text{alg}} = \mathcal{Z}^n(X)_{\text{hom}}.$$

The following result is one step in the proof of Griffiths' theorem, which shows that in general, we do not have

$$\mathcal{Z}^k(X)_{\text{alg}} = \mathcal{Z}^k(X)_{\text{hom}}$$

for $2 \leq k < n$. Let $J^{2k-1}(X)$ be the kth intermediate Jacobian of X (see section 7.1), and let

$$J^{2k-1}(X)_{\text{alg}} \subset J^{2k-1}(X)$$

be the largest complex subtorus satisfying the property that its tangent space at 0 is contained in $H^{k-1,k}(X)$. (Here, we use the natural identification of $T_{J^{2k-1}(X),0}$ with $H^{2k-1}(X, \mathbb{C})/F^k H^{2k-1}(X)$.) Note that $J^{2k-1}(X)_{\text{alg}}$ is also the complex torus associated to the largest Hodge substructure of $H^{2k-1}(X,\mathbb{Z})$ contained in $F^{k-1}H^{2k-1}(X)$. The following theorem was proved in vI.12.2.1.

Theorem 8.15 *If Z is algebraically equivalent to 0 in X, then the image $\Phi_X^k(Z)$ of Z under the Abel–Jacobi map of X lies in $J^{2k-1}(X)_{\mathrm{alg}}$.*

Writing $\mathrm{Griff}^k(X) := \mathcal{Z}^k(X)_{\mathrm{hom}}/\mathcal{Z}^k(X)_{\mathrm{alg}}$ and

$$J^{2k-1}(X)_{\mathrm{tr}} := J^{2k-1}(X)/J^{2k-1}(X)_{\mathrm{alg}},$$

we easily deduce the following corollary.

Corollary 8.16 *The Abel–Jacobi map Φ_X^k gives a commutative diagram*

$$\begin{array}{lccc} \Phi_X^k : & \mathcal{Z}^k(X)_{\mathrm{hom}} & \longrightarrow & J^{2k-1}(X) \\ & \downarrow & & \downarrow \\ \overline{\Phi}_X^k : & \mathrm{Griff}^k(X) & \longrightarrow & J^{2k-1}(X)_{\mathrm{tr}}. \end{array}$$

In particular, if $J^{2k-1}(X)_{\mathrm{alg}} = 0$, then Φ_X^k factors through $\mathrm{Griff}^k(X)$.

Remark 8.17 *We write* tr *as subscript to indicate 'transcendental' rather than 'algebraic'. Indeed, we can show that the tori $J^{2k-1}(X)$ are not in general algebraic, whereas the quotients $J^{2k-1}(X)_{\mathrm{alg}}$ are algebraic, thanks to the Hodge index theorem (vI.6.3.2).*

Theorem 6.24 implies the following result.

Lemma 8.18 *The condition $J^{2k-1}(X)_{\mathrm{alg}} = 0$ is satisfied for general hypersurfaces of sufficiently large degree in $\mathbb{P}^{2k}$ for $k \geq 2$.*

Proof The condition $J^{2k-1}(X)_{\mathrm{alg}} \neq 0$ is equivalent to the statement that $H^{2k-1}(X,\mathbb{Z})$ has a non-trivial Hodge substructure, and a fortiori at least one non-zero integral class contained in $F^{k-1}H^{2k-1}(X)$. If $k \geq 2$ and d is sufficiently large, then $F^{k-1}H^{2k-1}(X) \neq H^{2k-1}(X)$ for X of degree d in $\mathbb{P}^{2k}$. Theorem 6.24 then says that the Hodge loci U_λ^{k-1} (defined locally in B) are proper analytic subsets of U for $\lambda \neq 0$. Thus, when X lies outside the countable union of the U_λ^{k-1} for integral λ, we have $J^{2k-1}(X)_{\mathrm{alg}} = 0$. □

8.2.2 The Hodge class of a normal function

Let $\pi : \mathcal{X} \to B$ be a proper, submersive and projective morphism between two complex manifolds, and let $\nu \in H^0(B, \mathcal{J}^{2k-1})$ be a holomorphic section of the intermediate Jacobian fibration $J^{2k-1}(X_b)_{b\in B}$ (see section 7.1). Recall that $\mathcal{J}^{2k-1}$ is defined by the exact sequence

$$0 \to H_{\mathbb{Z}}^{2k-1} \to \mathcal{H}^{2k-1}/F^k\mathcal{H}^{2k-1} \to \mathcal{J}^{2k-1} \to 0, \tag{8.12}$$

where $H_{\mathbb{Z}}^{2k-1} = R^{2k-1}\pi_*\mathbb{Z}/\text{torsion}$ and $\mathcal{H}^{2k-1} = H_{\mathbb{Z}}^{2k-1} \otimes \mathcal{O}_B$.

The exact sequence (8.12) gives a connecting map

$$\alpha : H^0(B, \mathcal{J}^{2k-1}) \to H^1(B, R^{2k-1}\pi_*\mathbb{Z}/\text{torsion}).$$

Definition 8.19 *The Hodge class $[\nu]$ of ν is the element*

$$\alpha(\nu) \in H^1(B, R^{2k-1}\pi_*\mathbb{Z}/\text{torsion}).$$

This terminology can be explained as follows. Assume that B is projective. Then we can show that the group $H^1(B, R^{2k-1}\pi_*\mathbb{Z})$ modulo its torsion is equipped with a Hodge structure of weight $2k$. Moreover, the morphism given by

$$\beta : L^1H^{2k}(\mathcal{X}, \mathbb{Z}) = \operatorname{Ker}(H^{2k}(\mathcal{X}, \mathbb{Z}) \to H^{2k}(X_b, \mathbb{Z})) \to H^1(B, R^{2k-1}\pi_*\mathbb{Z}),$$

where L is the Leray filtration, is a morphism of Hodge structures. Then $[\nu]$ is a Hodge class of $H^1(B, R^{2k-1}\pi_*\mathbb{Z}/\text{torsion})$, by the fact that $[\nu]$ vanishes in

$$H^1(B, \mathcal{H}^{2k-1}/F^k\mathcal{H}^{2k-1}),$$

which admits

$$H^1(B, R^{2k-1}\pi_*\mathbb{C})/F^kH^1(B, R^{2k-1}\pi_*\mathbb{C})$$

as a quotient.

Now assume that we have a relative cycle $\mathcal{Z} \subset \mathcal{X}$ of codimension k such that the restrictions $Z_b = \mathcal{Z} \cap X_b$ are homologous to 0. By theorem 7.9, we then have a normal function $\nu_{\mathcal{Z}}$, defined by

$$\nu_{\mathcal{Z}}(b) = \Phi_{X_b}(Z_b) \in J^{2k-1}(X_b).$$

Moreover, the cohomology class $[\mathcal{Z}]$ of the cycle $\mathcal{Z}$ vanishes by hypothesis on the fibres X_b, and thus it lies in $L^1H^{2k}(\mathcal{X}, \mathbb{Z})$.

Lemma 8.20 *The classes $\beta([\mathcal{Z}])$ and $[\nu_{\mathcal{Z}}]$ coincide in $H^1(B, R^{2k-1}\pi_*\mathbb{Z}/$ torsion).*

Proof In this proof, we write $\nu_{\mathcal{Z}} = \nu$. Let U_i be an open covering of B such that there exist real differentiable chains $\Gamma_i \subset \mathcal{X}_{U_i}$ satisfying $\partial\Gamma_i = \mathcal{Z}_{U_i}$, where $\mathcal{Z}_{U_i} := \mathcal{Z} \cap \mathcal{X}_{U_i}$, $\mathcal{X}_{U_i} = \pi^{-1}(U_i)$, modulo a chain supported on the boundary of $\mathcal{X}_{U_i}$ (i.e. $\mathcal{X}_{\partial U_i}$).

We obtain such a chain by writing $Z_b = \partial\gamma_{i,b}$ locally, i.e. for $b \in U_i$, where the chain $\gamma_{i,b} \subset X_b$ varies continuously with b. Set

$$\Gamma_i = \bigcup_{b\in U_i} \gamma_{i,b}.$$

Then, by definition of the Abel–Jacobi map, over each U_i there is a lifting $\tilde{\nu}_i$ of ν to a holomorphic section of

$$\mathcal{H}^{2k-1}/F^k\mathcal{H}^{2k-1} \cong (F^{n-k+1}\mathcal{H}^{2n-2k+1})^*$$

given by $\tilde{\nu}_i(\eta)(b) = \int_{\gamma_{i,b}} \eta_b$, where η is a local section of $F^{n-k+1}\mathcal{H}^{2n-2k+1}$.

By the definition of the connecting map α, we thus see that $\alpha(\nu)$ is represented by the Čech cocycle $T_{ij} \in \Gamma(U_i \cap U_j, R^{2k-1}\pi^*\mathbb{Z}/\text{torsion})$ with values in the sheaf $R^{2k-1}\pi^*\mathbb{Z}/\text{torsion}$, and relative to the covering U_i of B, given by

$$T_{ij}(b) = \text{Image of } \gamma_{j,b} - \gamma_{i,b} \in H_{2n-2k+1}(X_b, \mathbb{Z}) \text{ in } H^{2k-1}(X_b, \mathbb{Z})/\text{torsion}, \tag{8.13}$$

where the map is given by Poincaré duality (modulo torsion).

It remains to see that $\beta([\mathcal{Z}])$ can be represented similarly.

As we are interested in the cohomology modulo torsion, we will prove this inside $H^1(B, R^{2k-1}\pi_*\mathbb{R})$. For this we introduce the complex $(\mathcal{D}_\bullet, \partial)$ of currents on $\mathcal{X}$; here $\mathcal{D}_l$ is the sheaf associated to the presheaf

$$U \mapsto \Gamma_c(U, \mathcal{A}^k_{\mathcal{X}})^*,$$

where Γ_c denotes the set of sections with compact support. The differential $\partial : \mathcal{D}_l \to \mathcal{D}_{l-1}$ is the transpose of the exterior differential $d : \mathcal{A}^{l-1}_{\mathcal{X}} \to \mathcal{A}^l_{\mathcal{X}}$. We have an obvious inclusion

$$i : \mathcal{A}^{N-k}_{\mathcal{X}} \hookrightarrow \mathcal{D}_k,$$

where $N = \dim_{\mathbb{R}}\mathcal{X}$, given by $\alpha \mapsto \int_{\mathcal{X}} \alpha\wedge$. Moreover, $\mathcal{D}_l$ also contains the differentiable l-chains, via their currents of integration.

We can show that i is a quasi-isomorphism. This implies that the complex

$$\mathcal{D}'^l := \mathcal{D}_{N-l},$$

equipped with the differential ∂, is an acyclic resolution (since the $\mathcal{D}'^l$ are sheaves of $\mathcal{C}^\infty$-modules) of the constant sheaf $\mathbb{R}$ on $\mathcal{X}$. In particular, for $U \subset \mathcal{X}$, we have $H^k(U, \mathbb{R}) = H^k(\Gamma(U, \mathcal{D}'^\bullet))$.

By definition, the Leray filtration on $H^*(X, \mathbb{R})$ can be computed as follows. Let $(\mathcal{M}^{\bullet,\bullet}, \partial, \delta)$ be the double complex of sheaves on B defined by $\mathcal{M}^{p,q} = \check{\mathcal{C}}^q(\pi_*\mathcal{D}'^p)$. Here, $(\check{\mathcal{C}}^q(\,\cdot\,), \delta)$ denotes the Čech resolution of $(\,\cdot\,)$. Then the simple complex associated to the double complex of global sections of $\mathcal{M}^{\bullet,\bullet}$

has cohomology given by $H^*(X, \mathbb{R})$, and the Leray filtration is given by the filtration by the second index. Now let $\mu \in \Gamma(\mathcal{X}, \mathcal{D}'^{2k})$ be a ∂-closed section defining a class in $L^1H^{2k}(X, \mathbb{R})$. Then there exists an open covering U_i of B such that

$$\mu_{|\mathcal{X}_{U_i}} = \partial \mu_i, \quad \mu_i \in \Gamma(\mathcal{X}_{U_i}, \mathcal{D}'^{2k-1}).$$

Then we have $(\mu_i) \in \check{C}^0(\pi_*\mathcal{D}'^{2k-1})$ and

$$(\partial + \delta)(\mu_i) = (\mu_{|\mathcal{X}_{U_i}}, (\mu_i - \mu_j)) \in \check{C}^0(\pi_*\mathcal{D}'^{2k}) \oplus \check{C}^1(\pi_*\mathcal{D}'^{2k-1}).$$

Thus, $(\mu_{|U_i}) \in \check{C}^0(\pi_*\mathcal{D}'^{2k})$ is cohomologous to the ∂-closed element

$$(\mu_j - \mu_i) \in \check{C}^1(\pi_*\mathcal{D}'^{2k-1})$$

in the double complex $\Gamma(\mathcal{M}^{\bullet,\bullet})$. We thus have a representative $(\mu_j - \mu_i)$ of μ in $L^1\Gamma(\mathcal{M}^{\bullet,\bullet}) = \Gamma(\mathcal{M}^{\bullet,\geq 1})$. As every $\mu_j - \mu_i$ is ∂-closed on $\mathcal{X}_{U_{ij}}$, it also defines a class $[\mu_j - \mu_i]$ in $R^{2k-1}\pi_*\mathbb{R}_{|U_{ij}}$, and we now note that

$$\beta(\mu) \in H^1(B, R^{2k-1}\pi_*\mathbb{R})$$

is represented by the cocycle $([\mu_j - \mu_i]) \in \check{C}^1(R^{2k-1}\pi_*\mathbb{R})$. Indeed, β is the natural composed map

$$L^1H^{2k}(\mathcal{X}, \mathbb{R}) \to E_\infty^{1,k-1} \subset E_2^{1,2k-1} = H^1(B, R^{2k-1}\pi_*\mathbb{R})$$

deduced from the Leray spectral sequence. But the analysis of the spectral sequence of a double complex shows that this composed map sends the class of a $(\partial \pm \delta)$-closed element in $L^1\Gamma(\mathcal{M}^{\bullet,\bullet})$ to the class relative to the differential δ of the class relative to the differential ∂ of its projection in $L^1/L^2\Gamma(\mathcal{M}^{\bullet,\bullet}) = \Gamma(\mathcal{M}^{\bullet,1})$.

Let us now return to the class $[\mathcal{Z}] \in L^1H^{2k}(\mathcal{X}, \mathbb{Z})$. We can view $\mathcal{Z}$ as a current μ on $\mathcal{X}$, and similarly, the Γ_i introduced above are currents μ_i such that $\partial \mu_i = \mu_{|\mathcal{X}_{U_i}}$. Thus, we have

$$\beta([\mathcal{Z}]) = \text{class}(\mu_j - \mu_i) \in H^1(B, R^{2k-1}\pi_*\mathbb{R}),$$

which together with (8.13) proves the equality

$$\alpha(\nu) = \beta([\mathcal{Z}]),$$

since $[\mu_i - \mu_j] = \gamma_{j,b} - \gamma_{i,b} \in R^{2k-1}\pi_*\mathbb{R}$. □

8.2.3 Griffiths' theorem

The proof of lemma 8.18 can be used to show that if $Y \subset \mathbb{P}^4$ is a general quintic hypersurface, then $J^3(Y)_{\text{alg}} = 0$. On the other hand, Griffiths proved the following.

Theorem 8.21

(i) If Y is as above, then Y contains a finite number $N > 1$ of lines $\Delta_1, \ldots, \Delta_N$.
(ii) For $i \neq j$, $\Phi_Y^2(\Delta_i - \Delta_j)$ is not a torsion point in $J^3(Y)$.

Corollary 8.16 shows that for such a variety, the Abel–Jacobi map Φ_Y^2 factors through the Griffiths group $\text{Griff}^2(Y)$. Thus, we have the following corollary of theorem 8.21.

Theorem 8.22 *For Y as above, the group* $\text{Griff}^2(Y)$ *is not a torsion group.*

Proof of theorem 8.21(i) Let $G = \text{Grass}(2, 5)$ be the Grassmannian of lines in $\mathbb{P}^4$. We have $\dim G = 6$. Let $S^5 = H^0(\mathbb{P}^4, \mathcal{O}_{\mathbb{P}^4}(5))$, and let $W \subset G \times S^5$ be the closed algebraic subset defined by

$$W = \{(\Delta, f) \mid f_{|\Delta} = 0\}.$$

As the first projection makes W into a vector bundle on G of rank equal to $\dim S^5 - \dim H^0(\mathcal{O}_{\mathbb{P}^1}(5)) = \dim S^5 - 6$, we see that W is smooth and irreducible of dimension equal to $\dim S^5$. It follows that the generic fibre of $\text{pr}_2 : W \to S^5$ is finite, i.e. that the generic hypersurface Y of degree 5 contains at most a finite number $N = \deg \text{pr}_2$ of lines. To see that $N \geq 2$, we reason as in exercise 1 of chapter 3: we note that for fixed generic f, the fibre $\text{pr}_2^{-1}(f)$, or the set of lines contained in Y_f, is the zero locus of the section σ_f of the vector bundle E of rank 6 on G with fibre $E_\Delta = H^0(\Delta, \mathcal{O}_\Delta(5))$ at the point $\Delta \in G$. The number N is thus equal to the Chern class $c_6(E) \in H^{12}(G, \mathbb{Z}) = \mathbb{Z}$, and we check that this number is greater than 2 (in fact, it is equal to 2875; see Clemens 1984). $\square$

Proof of theorem 8.21(ii) Note first that if Y is a generic quintic in $\mathbb{P}^4$ and Δ, Δ' are two lines contained in Y, then there exists a smooth quintic $X \subset \mathbb{P}^5$ such that Y is equal to $X \cap \mathbb{P}^4$, and containing two planes P and P', such that $\Delta = P \cap Y$ and $\Delta' = P' \cap Y$. Indeed, this follows from Bertini's lemma (see exercise 1 of chapter 2), which shows that if the planes P and P' are fixed, and intersect $\mathbb{P}^4$ along the lines Δ and Δ' respectively, the generic quintic hypersurface of $\mathbb{P}^5$ containing Y, P and P' is smooth outside $Y \cup P \cup P'$. As Y is smooth, such a hypersurface X is also smooth along Y. But a local analysis along $P \cup P'$ easily shows that it is also smooth along $P \cup P'$.

Now take a Lefschetz pencil $(Y_t)_{t \in \mathbb{P}^1}$ of hyperplane sections of X, such that $Y_0 = Y$. The planes P and P' are not contained in any member of the pencil, since a quintic in $\mathbb{P}^4$ containing a plane cannot be smooth, nor can it have exactly one ordinary double point. Let $U \subset \mathbb{P}^1$ be the open set parametrising the smooth hyperplane sections Y_t. For $t \in U$, we have the two lines $\Delta_t = Y_t \cap P$ and $\Delta'_t = Y_t \cap P'$. We will show that for general t in U, the point $\Phi^2_{Y_t}(\Delta_t - \Delta'_t) \in J^3(Y_t)$ is not a torsion point. Note that if

$$\begin{array}{ccc} f : \widetilde{X} & \longrightarrow & \mathbb{P}^1 \\ \downarrow \tau & & \\ X & & \end{array}$$

is the blowup of X along the base locus of the pencil, and $\widetilde{P}$, $\widetilde{P}'$ are the strict transforms of P and P', i.e. the irreducible components of $\tau^{-1}(P)$ (resp. $\tau^{-1}(P')$) dominating P (resp. P') via τ, the cycle $\mathcal{Z} = \widetilde{P} - \widetilde{P}'$ satisfies

$$Z_t := \mathcal{Z} \cdot f^{-1}(t) = \Delta_t - \Delta'_t \in \mathcal{Z}^2(Y_t), \quad t \in U.$$

Thus, the associated normal function $\nu_{\mathcal{Z}}$ defined on U satisfies

$$\nu_{\mathcal{Z}}(t) = \Phi^2_{Y_t}(\Delta_t - \Delta'_t), \quad t \in U.$$

Note that for every integer M, the condition

$$M \Phi^2_{Y_t}(\Delta_t - \Delta'_t) = 0 \text{ in } J^3(Y_t)$$

defines a closed analytic subset of U, since $\nu_{\mathcal{Z}}$ is holomorphic by theorem 7.9. It thus suffices to show that each of these analytic subsets is proper.

Assume, on the contrary, that $M \Phi^2_{Y_t}(\Delta_t - \Delta'_t) = 0$ in $J^3(Y_t)$ for $t \in U$. Then we have $M \nu_{\mathcal{Z}} = 0$, so that $M[\nu_{\mathcal{Z}}] = 0$ in $H^1(U, R^3 f_* \mathbb{Z})$. But we know by lemma 8.20 that $[\nu_{\mathcal{Z}}]$ is the image of

$$[\mathcal{Z}]_{|\widetilde{X}_U} \in L^1 H^4(\widetilde{X}_U, \mathbb{Z})$$

in $H^1(U, R^3 f_* \mathbb{Z})$, where $\widetilde{X}_U := f^{-1}(U)$. Now, as U has the homotopy type of a 1-dimensional CW-complex, the cohomology of U with values in any local system is zero in degree > 1, and thus we actually have

$$L^1 H^4(\widetilde{X}_U, \mathbb{Z}) = H^1(U, R^3 f_* \mathbb{Z}).$$

This means that $M[\mathcal{Z}]_{|\widetilde{Y}_U} = 0$ in $H^4(\widetilde{X}_U, \mathbb{Z})$, which implies that there exist classes $\gamma_i \in H_4(Y_{t_i}, \mathbb{Z})$, where the t_i are the critical values of the pencil, such that

$$M[\mathcal{Z}] = \sum_i \mathrm{PD}(j_{t_i *}\gamma_i) \in H^4(\widetilde{X}, \mathbb{Z}),$$

where $\mathrm{PD} : H_4(\widetilde{X}, \mathbb{Z}) \cong H^4(\widetilde{X}, \mathbb{Z})$ is the isomorphism coming from Poincaré duality, and j_{t_i} is the natural inclusion of Y_{t_i} in $\widetilde{X}$. Applying τ_*, we now find that

$$M[P - P'] = \sum_i \mathrm{PD}(i_{t_i *}\gamma_i) \in H^4(X, \mathbb{Z}),$$

where i_{t_i} is the natural inclusion of Y_{t_i} in X. Now, as the Y_{t_i} have a unique ordinary double point as singularity, the image of

$$\mathrm{PD} \circ i_{t_i *} : H_4(Y_{t_i}, \mathbb{Z}) \to H^4(X, \mathbb{Z})$$

is equal to $h^2\mathbb{Z}$, where $h = c_1(\mathcal{O}_X(1)) \in H^2(X, \mathbb{Z})$. Indeed, this follows from the fact that this image is also equal to the image of the Gysin morphism

$$\tilde{i}_{t_i *} : H^2(\widetilde{Y}_{t_i}, \mathbb{Z}) \to H^4(X, \mathbb{Z}),$$

where $\widetilde{Y}_{t_i}$ is the desingularisation of Y_{t_i} obtained by blowing up the singular point. But it is shown in Clemens (1983b) that $H^2(\widetilde{Y}_{t_i}, \mathbb{Z})$ is generated by $\tilde{i}_{t_i}^* h$ and by the class of the exceptional divisor of the blowup, which implies the result easily.

Thus, the hypothesis would imply that $M[P - P']$ is a multiple of h^2. But simply by computing the intersection matrix in $H^4(X, \mathbb{Z})$ of the classes h^2, $[P]$ and $[P']$, it is easy to see that this is false.

The hypothesis was thus absurd, which concludes the proof of theorem 8.21. □

8.3 Application of the connectivity theorem

8.3.1 The Nori equivalence

We can define the group $\mathcal{Z}^k(X)_{\mathrm{alg}}$ as the subgroup of $\mathcal{Z}^k(X)$ generated by cycles of the form

$$\Gamma_*(Z) =: p(q^{-1}(Z)), \quad Z \in \mathcal{Z}_0(Y)_{\mathrm{hom}} := \mathcal{Z}^{\dim Y}(Y)_{\mathrm{hom}}$$

for every smooth projective variety Y, and every diagram

$$\begin{array}{ccc} \Gamma & \xrightarrow{p} & X \\ \downarrow q & & \\ Y, & & \end{array}$$

where we assume that q is flat over Y along $\operatorname{Supp} Z$, and that $\dim \Gamma - \dim Y = \dim X - k$. (The flatness hypothesis guarantees that $q^{-1}(Z)$ is of dimension $\dim X - k$.)

If we work on the level of the Chow groups $\mathrm{CH}^k(X)$ of cycles modulo rational equivalence, we can define $\Gamma_*(Z)$ without any flatness hypothesis (see Fulton (1984) and the next chapter); we will do this below, citing the next chapter for the precise definitions.

This version of algebraic equivalence led Nori to introduce an increasing filtration $\mathcal{N}_r$, $r \geq 0$, on the group $\mathrm{CH}^k(X)_{\mathrm{hom}}$ having the property that $\mathcal{N}_0 = \mathrm{CH}^k(X)_{\mathrm{alg}}$; this is of course equivalent to giving the corresponding increasing filtration on the group

$$\mathrm{Griff}^k(X) = \mathcal{Z}^k(X)_{\mathrm{hom}}/\mathcal{Z}^k(X)_{\mathrm{alg}} = \mathrm{CH}^k(X)_{\mathrm{hom}}/\mathrm{CH}^k(X)_{\mathrm{alg}}.$$

Definition 8.23 *The subgroup $\mathcal{N}_r\mathrm{CH}^k(X)_{\mathrm{hom}}$ is the subgroup of $\mathrm{CH}^k(X)$ generated by cycles of the form*

$$\Gamma_*(Z) = p_*(q^*(Z)), \quad Z \in \mathrm{CH}_r(Y)_{\mathrm{hom}} := \mathrm{CH}^{\dim Y - r}(Y)_{\mathrm{hom}}$$

for every smooth projective variety Y and every diagram

$$\begin{array}{ccc} \Gamma & \xrightarrow{p} & X \\ \downarrow q & & \\ Y & & \end{array}$$

with

$$\dim \Gamma - \dim Y + r = \dim X - k,$$

in other words, for every correspondence $\Gamma \in \mathrm{CH}^{k+r}(Y \times X)$.

Lemma 8.24 *$\mathcal{N}$ is an increasing filtration such that*

$$\mathcal{N}_{\dim X - k}\mathrm{CH}^k(X)_{\mathrm{hom}} = \mathrm{CH}^k(X)_{\mathrm{hom}}, \quad \mathcal{N}_0\mathrm{CH}^k(X)_{\mathrm{hom}} = \mathrm{CH}^k(X)_{\mathrm{alg}}.$$

Proof If $Z \in \mathrm{CH}_r(Y)_{\mathrm{hom}}$, then $Z \times \mathbb{P}^1 \in \mathrm{CH}_{r+1}(Y \times \mathbb{P}^1)_{\mathrm{hom}}$ and satisfies

$$(Z \times \mathbb{P}^1) \cap (Y \times 0) = Z,$$

i.e. $i_0^*(Z \times \mathbb{P}^1) = Z$, where i_0 is the inclusion of $Y = Y \times 0$ into $Z \times \mathbb{P}^1$. Given a correspondence $\Gamma \subset Y \times X$, consider the correspondence

$$\Gamma' = \Gamma \times 0 \subset (Y \times \mathbb{P}^1) \times X.$$

Clearly $\Gamma'_*(Z \times \mathbb{P}^1) = \Gamma_* Z$, which shows that $\mathcal{N}_r\mathrm{CH}^k(X)_{\mathrm{hom}} \subset \mathcal{N}_{r+1}\mathrm{CH}^k(X)_{\mathrm{hom}}$.

Moreover, the second property is clear, since for $Z \in \mathrm{CH}^k(X)_{\mathrm{hom}}$, we have also $Z \in \mathrm{CH}_{\dim X - k}(X)_{\mathrm{hom}}$, so it suffices to take $X = Y$ and Γ the diagonal of X. Then the maps p and q are the identity, so that $p_* \circ q^*$ is the identity, and $Z = p_*(q^*Z)$ with Z homologous to 0 and of dimension $\dim X - k$.

The last equality follows from the definitions. □

8.3.2 Nori's theorem

Griffiths' theorem 8.21 admits the following obvious generalisation. Let X be a smooth variety of even dimension $2r$, equipped with an ample invertible bundle L, and let $Z \in \mathrm{CH}^r(X)$ be a primitive cycle, i.e. a cycle such that $c^1(L) \cup [Z] = 0$ in $H^{2r+2}(X, \mathbb{Z})$. By the Lefschetz theorem on hyperplane sections, it follows that if the hypersurface $Y \in |H^0(X, kL)|$ is smooth, then the restriction of the class $[Z]$ to $H^{2r}(Y, \mathbb{Z})$ is equal to zero. The cycle $Z_Y = Z \cap Y$ then admits an Abel–Jacobi invariant

$$\Phi_Y^r(Z_Y) \in J^{2r-1}(Y)_0,$$

where $J^{2r-1}(Y)_0 := J^{2r-1}(Y)/J^{2r-1}(X)$.

Theorem 8.25 *If kL is very ample (so that $|kL|$ has sufficiently many Lefschetz pencils), and Y is general, then $\Phi_Y^r(Z_Y)$ is not a torsion point of $J^{2r-1}(Y)_0$. Moreover, if kL is sufficiently ample, and $r \geq 2$, the algebraic part $J^{2r-1}(Y)_{0,\mathrm{alg}}$ is reduced to 0, so that by corollary 8.16, Z_Y is not algebraically equivalent to 0, nor is any of its multiples.*

Proof The first part of the statement follows from theorem 7.15. The second part can be proved by an infinitesimal argument, exactly as in lemma 8.18, using a generalised Noether–Lefschetz theorem similar to theorem 6.25. □

Hence it is possible to detect cycles not algebraically equivalent to 0 using the Abel–Jacobi invariant above. Nori (1993) proved the following result.

Theorem 8.26 *Let X be an $(n+k)$-dimensional variety, and let $Z \in \mathcal{Z}^d(X)$. Then if $[Z] \neq 0$ in $H^{2d}(X, \mathbb{Q})$, and if $n - d > r \geq 0$, for sufficiently ample line bundles $L_1, \ldots, L_k$ on X, and for a general complete intersection of hypersurfaces*

$$Y = \bigcap_i Y_i,\ Y_i \in |L_i|,$$

no multiple of the cycle $Z_Y = Z \cap Y \in \mathrm{CH}_{n-d}(Y) = \mathrm{CH}^d(Y)$ lies in $\mathcal{N}_r\mathrm{CH}^d(Y)_{\mathrm{hom}}$.

This is optimal since for $n - d = r$, we have $Z_Y \in \mathcal{N}_r\mathrm{CH}^d(Y)_{\mathrm{hom}}$ once by $Z_Y \in \mathrm{CH}^d(Y)_{\mathrm{hom}}$ by lemma 8.24. The case $k = 1,\ r = 0,\ n + 1 = 2d$ of this theorem is exactly Griffiths' theorem 8.25.

This theorem shows that the filtration $\mathcal{N}$ is non-trivial modulo torsion, i.e. that its graded group $\mathrm{Gr}_r^{\mathcal{N}}$ can be non-zero in any degree. For this, it suffices to start from a primitive r-dimensional cycle Z on a $2r$-dimensional variety X, relative to a polarisation L. Then for every $k < r$, and any general complete intersection Y of k members of L^p with $p \gg 0$, Z_Y does not lie in $\mathcal{N}_{r-k-1}\mathrm{CH}^r(Y)_{\mathrm{hom}}$, nor do any of its multiples, whereas $Z_Y \in \mathcal{N}_{r-k}\mathrm{CH}^r(Y)_{\mathrm{hom}}$.

For example, X can be a $2r$-dimensional quadric in $\mathbb{P}^{2r+1}$, and Z the difference of two projective spaces $\mathbb{P}^r$ contained in X, with distinct cohomology classes (see chapter 2, exercise 2).

Moreover, Nori's theorem implies the existence of cycles which are not algebraically equivalent to 0, even modulo torsion, but which are homologically equivalent to 0 and annihilated by the Abel–Jacobi map. Indeed, in the preceding example, whenever $k > 1$, if X satisfies $J^{2r-1}(X) = 0$, then the varieties Y satisfy $J^{2r-1}(Y) = 0$ by the Lefschetz theorem on hyperplane sections. The cycles Z_Y are thus annihilated by the Abel–Jacobi map. But Nori's theorem shows that if $r \geq k+1$, then no multiple of Z_Y lies in $\mathcal{N}_0\mathrm{CH}^r(Y) = \mathrm{CH}^r(Y)_{\mathrm{alg}}$.

Remark 8.27 *We must have $r > k \geq 2$, so $r \geq 3$, in order for the preceding construction to yield cycles of codimension r which are annihilated by the Abel–Jacobi map but which are not torsion elements of the Griffiths group. It is conjectured that such cycles do not exist in codimension 2 (see Nori 1993).*

Proof of theorem 8.26 We prove it by contradiction. So assume that the L_i are sufficiently ample, and in particular that they satisfy the conclusion of theorem 8.4, whereas for the general intersection Y_b, where b is a general point of the open set $B \subset \prod_i H^0(X, L_i)$ parametrising the smooth complete intersections, we have $mZ_Y \in \mathcal{N}_r\mathrm{CH}^d(Y)_{\mathrm{hom}}$ for a certain integer $m > 0$. We must show that then $[Z] \in H^{2d}(X, \mathbb{Q})$ is equal to 0.

Consider the set consisting of the isomorphism classes of quintuplets

$$(W, S, \Gamma, b, m),$$

where W is a smooth projective variety, S is an r-dimensional cycle homologous to 0 on W, $b \in B$, m is a non-zero integer, and Γ is a cycle in $W \times Y_b$ such that $\Gamma_*(S) = mZ_{Y_b}$ in $\mathrm{CH}^d(Y_b)$. By the theory of Hilbert schemes, this set is a countable union of irreducible algebraic sets T_i (on which the discrete invariants, such as the integer m, the number of components of Γ, their degrees, and the Hilbert polynomial of W are constant), and the hypothesis is that its projection to B contains the complement of a countable union of proper closed algebraic subsets. Thus, by Baire's theorem, at least one of the T_i, which we may assume to be smooth by desingularisation, actually dominates B, and up to replacing T_i by an open subset T, we may thus assume that the projection $T_i \to B$ is submersive. Moreover, there exists a universal smooth variety $\mathcal{W} \to T$ over T, and a relative cycle $\Gamma \subset \mathcal{W} \times_T \mathcal{Y}_T$, where $\mathcal{Y}_T$ is the universal complete intersection parametrised by T via the map $T \to B$; moreover there exists a cycle $\mathcal{S} \subset \mathcal{W}$ of relative dimension r over T and homologous to 0 on the fibres W_t, such that the cycle

$$\mathcal{Z}' := \mathrm{pr}_{2*}(\Gamma \cdot \mathrm{pr}_1^*\mathcal{S}) \in \mathrm{CH}^d(\mathcal{Y}_T)$$

satisfies

$$\mathcal{Z}' \cap Y_t = mZ_{Y_t} \text{ in } \mathrm{CH}^d(Y_t)$$

for $t \in T$. Here,

$$\mathrm{pr}_1 : \mathcal{W} \times_T \mathcal{Y}_T \to \mathcal{W} \text{ and } \mathrm{pr}_2 : \mathcal{W} \times_T \mathcal{Y}_T \to \mathcal{Y}_T$$

denote the projections.

Now, let $\pi : \mathcal{Y}_T \to X$ denote the natural map (induced by the second projection and the inclusion $\mathcal{Y}_T \subset T \times X$). Then $Z_{Y_t} = \pi^*Z \cap Y_t$. The cycles $\mathcal{Z}'$ and $m\pi^*Z$ thus satisfy the property that $(\mathcal{Z}' - m\pi^*Z) \cap Y_t$ is rationally equivalent to 0 for each fibre Y_t of $\mathcal{Y}_T \to T$, and by the Bloch–Srinivas lemma (theorem 10.19), it follows that there exists a Zariski open set T' of T and an integer $m' > 0$ such that $m'(\mathcal{Z}' - m\pi^*Z)$ vanishes in $\mathrm{CH}^d(\mathcal{Y}_{T'})$.

In particular, we have $m'[\mathcal{Z}' - m\pi^*Z] = 0$ in $H^{2d}(\mathcal{Y}_{T'}, \mathbb{Z})$, so

$$[\mathcal{Z}' - m\pi^*Z] = 0 \text{ in } H^{2d}(\mathcal{Y}_{T'}, \mathbb{Q}). \tag{8.14}$$

Now, by the compatibility of the cycle class map with the actions of correspondences (see section 9.2.4), we have

$$[\mathcal{Z}'] = \mathrm{pr}_{2*}([\Gamma] \cup \mathrm{pr}_1^*([\mathcal{S}])) \in H^{2d}(\mathcal{Y}_T, \mathbb{Q}). \tag{8.15}$$

Finally, note that the cycle Γ is of codimension $d+r$ in $\mathcal{W} \times_T \mathcal{Y}_T$. Its class $[\Gamma]$ thus lies in $H^{2d+2r}(\mathcal{W} \times_T \mathcal{Y}_T, \mathbb{Q})$. As $2d+2r < 2n$, we can apply theorem 8.4 to the family of complete intersections

$$\pi \circ \mathrm{pr}_2 : \quad \begin{array}{ccc} \mathcal{W} \times_T \mathcal{Y}_T & \longrightarrow & X \\ {\scriptstyle \mathrm{pr}_1}\downarrow & & \\ \mathcal{W}. & & \end{array}$$

It follows that there exists a class $\delta \in H^{2d+2r}(\mathcal{W} \times X, \mathbb{Q})$ such that

$$(\mathrm{pr}_1, \pi \circ \mathrm{pr}_2)^* \delta = [\Gamma] \in H^{2d+2r}(\mathcal{W} \times_T \mathcal{Y}_T, \mathbb{Q}). \tag{8.16}$$

Let p_1 and p_2 denote the (relative) projections of $\mathcal{W} \times X$ onto $\mathcal{W}$ and $T \times X$ respectively, and let

$$\gamma = p_{2*}(\delta \cup p_1^*[\mathcal{S}]) \text{ in } H^{2d}(T \times X, \mathbb{Q}).$$

The equalities (8.14), (8.15) and (8.16) then show that if l is the inclusion of $\mathcal{Y}_T$ in $T \times X$, we have

$$l^*\gamma = m\pi^*[Z] \text{ in } H^{2d}(\mathcal{Y}_T, \mathbb{Q}). \tag{8.17}$$

As $\pi = q_2 \circ l$, where q_2 is the second projection of $T \times X$ on X, the equality (8.17) can also be written

$$l^*(\gamma - mq_2^*[Z]) = 0 \text{ in } H^{2d}(\mathcal{Y}_T, \mathbb{Q}).$$

As $2d \leq 2n$, it follows by theorem 8.4, which we now apply to the family of complete intersections $\mathcal{Y}_T \overset{l}{\hookrightarrow} T \times X$, that we have

$$\gamma = mq_2^*[Z] \text{ in } H^{2d}(T \times X, \mathbb{Q}).$$

Let us now take $t \in T$, and restrict the last equality to $t \times X$. We obtain

$$m[Z] = \gamma_{|t \times X} = r_{2*}(\delta_t \cup r_1^*[S_t]) \text{ in } H^{2d}(t \times X, \mathbb{Q}),$$

where δ_t is the restriction of δ to $W_t \times X$, and $[S_t]$ is the class of the cycle $S_t = \mathcal{S} \cap W_t$. Here, the maps $r_i = p_{i|W_t \times X}$ are the projections of $W_t \times X$ onto its factors. But by construction, we have $[S_t] = 0$ in $H^*(W_t)$, and thus $m[Z] = 0$ in $H^{2d}(X, \mathbb{Q})$. □

Exercises

1. *Picard groups of surfaces.* Let S be a smooth projective connected surface, and L a very ample line bundle on S. The sections of L thus

embed S into a $\mathbb{P}^N$. For a smooth member $C \in |L|$, the Lefschetz theorem 1.23 thus gives an injection

$$H^1(S, \mathbb{Z}) \overset{j^*}{\hookrightarrow} H^1(C, \mathbb{Z})$$

which is a morphism of Hodge structures.

(a) Using theorem 1.23 for the cohomology with finite coefficients, show that the cokernel of j^* has no torsion.

(b) Deduce that

$$j^* : \mathrm{Pic}^0 S \to \mathrm{Pic}^0 C$$

is injective.

Let $(C_t)_{t \in \mathbb{P}^1}$ be a pencil of elements of $|L|$ whose generic member is smooth. Let $D \in \mathrm{Pic}\, S$ be such that $D \cdot C := \int_S [D] \cup [C] = 0$ for $C \in |L|$.

(c) Show that if $j_t^* D = 0$ in $\mathrm{Pic}\, C_t$, for C_t smooth, then the class $[D]$ is a combination of the classes of irreducible components of the curves C_t (cf. the proof of theorem 8.21(ii)).

(d) Assume from now on that the pencil is a Lefschetz pencil. Show that the singular fibres C_i of the pencil are irreducible, unless the surface S is ruled, i.e. there exists a morphism $\pi : S \to X$ with values in a smooth curve whose fibre is isomorphic to $\mathbb{P}^1$, and the divisor L is of degree 1 on the fibres of π, or the surface is isomorphic to $\mathbb{P}^2$, embedded in $\mathbb{P}^4$ or $\mathbb{P}^5$ by a linear system of conics (this is called the Veronese surface).
(If a singular fibre is reducible, i.e. $C_i = D \cup D'$, then since C_i has one ordinary double point as its only singularity, we must have $D \cdot D' = 1$. Use the Hodge index theorem for surfaces (see volume I, chapter 11, exercise 2) to deduce that either one of the components, say D, has self-intersection equal to 0 and is of degree 1 in the embedding L (so it is a line), or one of the components has self-intersection 1, is of degree 2 and the two components are homologous. Show that in the first case the deformations of D yield such a morphism π, and that the second case gives the Veronese embedding of $\mathbb{P}^2$.)

(e) Deduce that $j_t^* D \neq 0$ in $\mathrm{Pic}\, C_t$, for t generic in $\mathbb{P}^1$, except in the cases described in (d). (Note that in fact the Veronese case does not give an exception since $\mathrm{Pic}\, \mathbb{P}^2 = \mathbb{Z}$.) Deduce from theorem 9.25, proved in the next chapter, that the first case described in (d) is indeed a counterexample to this statement.

(f) Show that if we are not in the cases described in (d), for a general member C of $|L|$, the restriction

$$j^* : \operatorname{Pic} S \to \operatorname{Pic} C$$

is injective. (Use the fact that the Néron–Severi group $\mathrm{NS}(S)$ is countable.)

Part III
Algebraic Cycles

9
Chow Groups

The third and final part of this volume is devoted to exploring the interactions between the Hodge theory and the algebraic cycles of an algebraic variety. We have not observed this interaction so far, except via the Hodge class or the Abel–Jacobi invariant of a cycle, which take into account only algebraic cycles modulo relatively crude equivalence relations such as homological equivalence, Abel–Jacobi equivalence, or the algebraic equivalence related to these by the arguments introduced in the preceding chapter (Griffiths' theorem 8.15).

On the other hand, the relation of rational equivalence which we introduce here is known to be the finest equivalence relation which satisfies a 'Chow moving lemma', and is stable under correspondences. In the present chapter, we give a brief presentation of the construction of Chow groups and their various functoriality properties. It is important to work with the Chow groups of singular or open varieties, even though the good functoriality properties in question are satisfied only for smooth varieties. Indeed, the more general groups appear in the following localisation theorem, which is a valuable tool for computing Chow groups.

Theorem 9.1 *The sequence*

$$\mathrm{CH}(F) \xrightarrow{i_*} \mathrm{CH}(X) \xrightarrow{j^*} \mathrm{CH}(U) \to 0$$

is exact, where i is a closed immersion and j is the open immersion of the complement.

In the first section of this chapter, we define Chow groups and describe their basic properties, following the book by Fulton (1984). We prove the existence of the morphism i^* when i is a flat map or the immersion of a Cartier divisor, and that of i_* when i is a proper map. We establish the localisation exact sequence, and deduce from it the following homotopy invariance property.

Proposition 9.2 *Let $\pi : E \to X$ be a vector bundle of rank r. Then the map*

$$\mathrm{CH}_k(X) \stackrel{\pi^*}{\to} \mathrm{CH}_{k+r}(E)$$

is an isomorphism.

In the second section, we describe Fulton's theory of refined intersection, and use it to show that the direct sum of the Chow groups of a smooth variety is equipped with the structure of a graded ring (graded by the codimension), where the product is the intersection product. This theory also enables us to prove the existence of a morphism $\phi^* : \mathrm{CH}_k(X) \to \mathrm{CH}_{k-r}(Y)$, $r = \dim X - \dim Y$ for every morphism from a variety Y to a smooth variety X.

We also show that the different cycle class maps constructed in vI.11.1.2, vI.12.1.2, and vI.12.3.3, for cycles in a smooth variety – namely the class $[Z] \in H^{2k}(X, \mathbb{Z})$ where $k = \operatorname{codim} Z$, and the Abel–Jacobi invariant of a cycle homologous to 0 when X is projective, or more generally, the Deligne class $[Z]_{\mathrm{D}} \in H^{2k}_{\mathrm{D}}(X, \mathbb{Z}_{\mathrm{D}}(k))$ – factor through the relation of rational equivalence, and are compatible with the intersection product. Finally, we introduce the correspondences $\Gamma \in \mathrm{CH}^{\gamma}(X \times Y)$ for smooth projective varieties X and Y, and their action on the Chow groups

$$\Gamma_* : \mathrm{CH}^k(X) \to \mathrm{CH}^{k+l}(Y), \quad l = -\dim X + \gamma,$$

and show that for two correspondences $\Gamma \in \mathrm{CH}(X \times Y)$, $\Gamma' \in \mathrm{CH}(Y \times W)$, their composition satisfies

$$(\Gamma' \circ \Gamma)_* = \Gamma'_* \circ \Gamma_* : \mathrm{CH}(X) \to \mathrm{CH}(W).$$

These correspondences and their actions on the Chow groups are the version on the level of Chow groups of the action of cohomological correspondences, i.e. of Hodge classes in products of varieties $X \times Y$, whose Künneth components give morphisms of Hodge structures $H^{\alpha}(X) \to H^{\alpha+2l}(Y)$ (see vI.11.3.3).

The last section of this chapter is devoted to some classical examples. For example, we show that the Chow groups of a projective bundle or of a blowup are determined by those of the base (and of the blown-up locus), exactly as for the cohomology (see vI.7.3.3).

Theorem 9.3 *Let $\pi : \mathbb{P}(E) \to X$ be a projective bundle, with* rank $E = r$. *Then the map*

$$\bigoplus_{i=0}^{i=r-1} h^i \pi^* : \bigoplus_{i=0}^{i=r-1} \mathrm{CH}_{k-r+1+i}(X) \to \mathrm{CH}_k(\mathbb{P}(E))$$

is an isomorphism, where $h = \mathrm{cl}(\mathcal{O}_{\mathbb{P}(E)}(1)) \in \operatorname{Pic} \mathbb{P}(E)$.

We apply this result to the computation of the Chow groups of small dimension of the hypersurfaces of degree d of $\mathbb{P}^N$, when d is small with respect to N.

9.1 Construction

9.1.1 *Rational equivalence*

Let X be a quasi-projective scheme over a field K, i.e. a Zariski open set in a projective scheme. Let $\mathcal{Z}_k(X)$ be the group of k-dimensional algebraic cycles of X, i.e. the free abelian group generated by the (reduced and irreducible) closed k-dimensional subvarieties of X. If $Y \subset X$ is a subscheme of dimension $\leq k$, we can associate a cycle $c(Y) \in \mathcal{Z}_k(X)$ to it as follows: set

$$c(Y) = \sum_W n_W W, \tag{9.1}$$

where the sum is taken over the k-dimensional irreducible reduced components of Y, and the multiplicity n_W is equal to the length $l(\mathcal{O}_{Y,W})$ of the Artinian ring $\mathcal{O}_{Y,W}$, the localisation of $\mathcal{O}_Y$ at the point W.

If $\phi : Y \to X$ is a proper morphism between quasi-projective schemes, we can define

$$\phi_* : \mathcal{Z}_k(Y) \to \mathcal{Z}_k(X)$$

by associating to a reduced irreducible subscheme $Z \subset Y$ the cycle

$$\deg[K(Z) : K(Z')]Z', \quad Z' = \phi(Z),$$

if $\phi : Z \to Z'$ is generically finite, and 0 otherwise (in the latter case we have $\dim Z' < k$). The properness is used here in order to guarantee that Z' is closed in X.

If W is a normal algebraic variety, the localised rings at the points of codimension 1 of W are valuation rings, so we can define the divisor $\operatorname{div}(\phi)$ of a non-zero rational function $\phi \in K(W)^*$. If $W \subset X$ is a closed subvariety and $\tau : \widetilde{W} \to X$ is the normalisation of W, we also have the map $\tau_* : \mathcal{Z}_k(\widetilde{W}) \to \mathcal{Z}_k(X)$. This allows us to give the following definition.

Definition 9.4 *The subgroup $\mathcal{Z}_k(X)_{\text{rat}}$ of cycles rationally equivalent to 0 is the subgroup of $\mathcal{Z}_k(X)$ generated by the cycles of the form*

$$\tau_* \operatorname{div}(\phi), \quad \dim W = k, \quad \phi \in K(W)^*,$$

where $\tau : \widetilde{W} \to W \subset X$ is the normalisation of a closed $(k+1)$-dimensional subvariety of X.

We write $\mathrm{CH}_k(X) = \mathcal{Z}_k(X)/\mathcal{Z}_k(X)_{\mathrm{rat}}$. For X a smooth n-dimensional variety, we shall also use the notation $\mathrm{CH}^k(X) = \mathrm{CH}_{n-k}(X)$.

If X is n-dimensional and smooth, or more generally, locally UFD, which means that the local rings $\mathcal{O}_{X,x}$ are unique factorisation domains (UFDs), then $\mathcal{Z}_{n-1}(X)$ is the group of Cartier divisors, and the group $\mathrm{CH}_{n-1}(X)$ can be identified with the group Pic X of algebraic line bundles on X modulo isomorphism. Indeed, this isomorphism makes the line bundle $\mathcal{O}_X(D) = \bigotimes_i \mathcal{I}_{D_i}^{\otimes -n_i}$ correspond to a divisor $D = \sum_i n_i D_i$, and D is rationally equivalent to D' if and only if there exists $\phi \in K(X)$ such that $\mathrm{div}(\phi) = D - D'$. Clearly, multiplication by ϕ then induces an isomorphism

$$\phi : \mathcal{O}_X(D) \cong \mathcal{O}_X(D').$$

Conversely, every algebraic line bundle L admits a meromorphic section, by trivialisation on a Zariski dense open set, and is thus isomorphic to a sheaf of the form $\mathcal{O}_X(D)$.

In general, for X reduced and irreducible of dimension n, we have a map

$$s : \mathrm{Pic}\, X \to \mathrm{CH}_{n-1}(X)$$

which to $\mathcal{L}$ associates the class of $\tau_*\mathrm{div}(\sigma)$, where $\tau : \widetilde{X} \to X$ is the normalisation and σ is a non-zero meromorphic section of $\tau^*\mathcal{L}$.

9.1.2 Functoriality: proper morphisms and flat morphisms

If $p : Y \to X$ is a proper morphism of quasi-projective schemes, we have the morphism

$$p_* : \mathcal{Z}_k(Y) \to \mathcal{Z}_k(X)$$

defined above.

Lemma 9.5 *p_* sends $\mathcal{Z}_k(Y)_{\mathrm{rat}}$ to $\mathcal{Z}_k(X)_{\mathrm{rat}}$, and thus induces a morphism*

$$p_* : \mathrm{CH}_k(Y) \to \mathrm{CH}_k(X).$$

Proof Let $\tau : \widetilde{W} \to W \subset Y$ be the normalisation of a closed $(k+1)$-dimensional subvariety of Y, and let $\phi \in K(W)^*$. Assume first that the composition $p \circ \tau : \widetilde{W} \to X$ is generically finite, and let $W' = \mathrm{Im}\, p \circ \tau \subset X$,

$\tau' : \widetilde{W}' \to X$ be its normalisation. We have a factorisation

$$\begin{array}{cccc} \tau : & \widetilde{W} & \longrightarrow & Y \\ & \downarrow{\scriptstyle \tilde{p}} & & \downarrow{\scriptstyle p} \\ \tau' : & \widetilde{W}' & \longrightarrow & X, \end{array}$$

so that $p_* \circ \tau_* = \tau'_* \circ \tilde{p}_* : \mathcal{Z}_k(\widetilde{W}) \to \mathcal{Z}_k(X)$.

The function field $K(\widetilde{W})$ is an algebraic extension of $K(\widetilde{W}')$. Consider the norm morphism

$$N : K(\widetilde{W})^* \to K(\widetilde{W}')^*$$

relative to this field extension.

Lemma 9.6 *For every non-zero rational function on* $\widetilde{W}$, *we have the equality*

$$\tilde{p}_*(\mathrm{div}(\phi)) = \mathrm{div}(N(\phi)).$$

Proof It suffices to show that for every irreducible subvariety $D \subset \widetilde{W}'$ of codimension 1, we have the equality of multiplicities

$$\mathrm{mult}_D(N(\phi)) = \sum_i \deg[K(D_i) : K(D)]\, \mathrm{mult}_{D_i}(\phi),$$

where the D_i are the divisors of $\widetilde{W}$ above D. Localising the sheaf of algebraic functions of $\widetilde{W}'$ along D, we obtain a valuation ring $A' \subset K(\widetilde{W}')$, with valuation v and residue field $K(D)$, and the localisation of $\mathcal{O}_{\widetilde{W}}$ at D is the integral closure $A \subset K(\widetilde{W})$ of A'. The D_i correspond to the primes μ_i of A, and $K(D_i)$ are their residue fields, hence

$$\deg[K(D_i) : K(D)] = l(A/\mu_i),$$

where l is the length as a $K(D)$-module. Thus, we need to prove the formula

$$v(N(\phi)) = \sum_i l(A/\mu_i)\, \mathrm{mult}_{D_i}(\phi)$$

for $\phi \in A$. By definition, we have

$$v(N(\phi)) = v(\det \mu_\phi),$$

where $\mu_\phi : A \to A$ denotes multiplication by $\phi \in A$, seen as a morphism of A'-modules. Thus, $v(N(\phi))$ is also equal to the length of the quotient $A/\phi A$. The formula we want to prove is thus equivalent to

$$l(A/\phi A) = \sum_i l(A/\mu_i)\, \mathrm{mult}_{D_i}(\phi)$$

for $\phi \in A$, and in this form, it can be deduced from the decomposition of ϕ into prime factors. □

This lemma immediately implies that

$$p_*(\tau_*(\operatorname{div}\phi)) = \tau'_*(\operatorname{div}(N(\phi))) \in \mathcal{Z}_k(X)_{\text{rat}}$$

when $p \circ \tau$ is generically finite.

Finally, if $p \circ \tau$ is not finite over its image, as $\dim \widetilde{W} = k+1$, we have either $\dim \operatorname{Im} p \circ \tau < k$, in which case we have

$$p_*(\tau_*(\operatorname{div}\phi)) = 0 \quad \text{in} \quad \mathcal{Z}_k(X)$$

for every $\phi \in K(W)^*$ for reasons of dimension, or $\dim \operatorname{Im} p \circ \tau = k$, in which case the generic fibre of $p \circ \tau$ is a curve (which is smooth since $\widetilde{W}$ is normal). Then, writing $W' = \operatorname{Im}\ p \circ \tau$, it is clear that if ϕ is a rational function on $\widetilde{W}$, we have

$$p_* \circ \tau_*(\operatorname{div}(\phi)) = dW',$$

where d is the sum of the degrees of the maps

$$p \circ \tau : D_i \to W',$$

and the D_i are the components of $\operatorname{div}(\phi)$ which are mapped surjectively onto W', multiplied by the multiplicity of D_i in $\operatorname{div}(\phi)$. But then d is also the degree of the divisor of $\phi_{|F}$, where F is the generic fibre of $p \circ \tau$, so $d = 0$ by the elementary formula $\deg \operatorname{div}(\phi) = 0$ for every rational function on a smooth curve (see volume I, chapter 2, ex. 2). Thus, in this last case we also have

$$p_*(\tau_*(\operatorname{div}\phi)) = 0 \quad \text{in} \quad \mathcal{Z}_k(X).$$

□

Next assume that $p : Y \to X$ is a flat morphism of relative dimension $l = \dim Y - \dim X$. If $Z \subset X$ is a reduced irreducible k-dimensional subscheme, then $p^{-1}(Z)$ is a $(k+1)$-dimensional subscheme of Y, and thus it admits an associated cycle $p^*Z \in \mathcal{Z}_{k+l}(Y)$. Extending this definition by $\mathbb{Z}$-linearity, we thus obtain $p^* : \mathcal{Z}_k(X) \to \mathcal{Z}_{k+l}(Y)$.

Lemma 9.7 *The map p^* defined above sends $\mathcal{Z}_k(X)_{\text{rat}}$ to $\mathcal{Z}_{k+l}(Y)_{\text{rat}}$. Thus, it induces a morphism*

$$p^* : \operatorname{CH}_k(X) \to \operatorname{CH}_{k+l}(Y).$$

Proof The main point is the following compatibility formula (see Fulton 1984, 1.7). Let $j : X' \to X$ be a proper map, and consider the fibred product $Y' = X' \times_X Y$, i.e. the Cartesian square

$$\begin{array}{ccc} j' : & Y' \longrightarrow Y & \\ & p' \downarrow \quad \downarrow p & \\ j : & X' \longrightarrow X. & \end{array} \tag{9.2}$$

Then for $Z \in \mathcal{Z}_k(X')$, assuming the flatness of p, we have

$$p^* \circ j_*(Z) = j'_* \circ p'^*(Z) \in \mathcal{Z}_{k+l}(Y). \tag{9.3}$$

Moreover, if $Z \subset X$ is a k-dimensional subscheme of X, then

$$p^* c(Z) = c((p^{-1}(Z)) \in \mathcal{Z}_{k+l}(Y), \tag{9.4}$$

where $c(Z)$ is the cycle associated to the scheme Z (see (9.1)).

Now let $W \subset X$ be an irreducible $(k+1)$-dimensional subvariety, and $\phi \in K(W)^*$. Let $p' : W' := p^{-1}(W) \to W$, and let $n : \widetilde{W} \to W$, $n' : \widetilde{W}' \to W'$ be the corresponding normalisations. It follows immediately from the formulae (9.3) and (9.4) that writing $\tau : \widetilde{W} \to X$ and $\tau' : \widetilde{W}' \to Y$ for the natural maps, we have

$$\tau'_*(\mathrm{div}\, p'^*\phi) = p^*(\tau_* \mathrm{div}\, \phi),$$

so $p^*(\tau_* \mathrm{div}\, \phi) \in \mathcal{Z}_{k+l}(Y)_{\mathrm{rat}}$. □

By formula (9.3), when $p : Y \to X$ is a flat map and $j : X' \to X$ is a proper map, the morphisms p^* and j_* defined on the level of cycles satisfy the compatibility given in the following proposition, which uses the same notation as in the Cartesian square (9.2).

Proposition 9.8 *For $Z \in \mathrm{CH}_k(X')$, we have the equality*

$$p^* j_*(Z) = j'_* p'^*(Z) \in \mathrm{CH}_{k+r}(Y), \quad r = \dim Y - \dim X.$$

In section 9.2, we will see that the morphism p^* can be defined under much more general hypotheses. One very useful first generalisation is the following. Assume that $i : Y \subset X$ is the (closed) immersion of a Cartier divisor. This means that $Y \subset X$ is locally defined by an equation $f \in \mathcal{O}_X$, and that two such equations f and g defining Y satisfy $f/g \in \mathcal{O}_X^*$ in the intersection of the open sets where they are defined.

We then have $\dim Y - \dim X = -1$, and we can define a morphism

$$i^* : \mathrm{CH}_k(X) \to \mathrm{CH}_{k-1}(Y)$$

as follows.

If Z is an irreducible reduced k-dimensional subvariety of X not contained in Y, then $\dim Z \cap Y = k - 1$ (or $Z \cap Y = \emptyset$), and we set

$$i^*Z := \text{the cycle associated to the scheme } Z \cap Y = c(Z \cap Y).$$

Assume moreover that $Z \subset Y$. Let j be the inclusion of Z into Y. The Cartier divisor $Y \subset X$ determines a line bundle $\mathcal{L}$, whose restriction to Z gives an element $\mathcal{L}_{|Z}$ of $\mathrm{Pic}\, Z$.

Now, we saw that we have a natural map s from $\mathrm{Pic}\, Z$ to $\mathrm{CH}_{k-1}(Z)$, so that we can define

$$i^*Z := j_*(s(\mathcal{L}_{|Z})) \in \mathrm{CH}_{k-1}(Y).$$

Lemma 9.9 *The morphism*

$$i^* : \mathcal{Z}_k(X) \to \mathrm{CH}_{k-1}(Y)$$

defined above factors through $\mathcal{Z}_k(X)_{\mathrm{rat}}$, so it gives a morphism

$$i^* : \mathrm{CH}_k(X) \to \mathrm{CH}_{k-1}(Y).$$

Proof Let W be a $(k+1)$-dimensional subvariety of X, and $\tau : \widetilde{W} \to X$ its normalisation. Let $\phi \in K(W)^*$. We want to show that $i^*(\tau_*\mathrm{div}(\phi)) = 0$ in $\mathrm{CH}_{k-1}(Y)$. If W is contained in Y, we can choose a rational section σ of the line bundle $\mathcal{L}_{|W}$ which does not admit any of the components D_i of $\mathrm{div}(\phi)$ as a pole or a zero. Then, by the definition of i^*, if $\mathrm{div}(\phi) = \sum_i n_i D_i$, we have

$$i^*(\tau_*\mathrm{div}(\phi)) = \sum_i n_i \mathrm{div}(\sigma_{|D_i}).$$

We can now conclude, by noting that if $\mathrm{div}(\sigma) = \sum_i m_j D'_j$, then

$$\sum_i n_i \mathrm{div}(\sigma_{|D_i}) = \sum_j m_j \mathrm{div}\left(\phi_{|D'_j}\right), \tag{9.5}$$

which can be checked locally (see Fulton 1984, 2.4).

If, however, W is not contained in Y and no component of $\tau_*\mathrm{div}(\phi)$ is contained in Y, we set $W_Y = W \cap Y$, $\widetilde{W}_Y = \tau^{-1}(W_Y)$, and we note that

$$\tau'_*\mathrm{div}(\phi_{|W_Y}) = i^* \circ \tau_*(\mathrm{div}(\phi)),$$

where $\tau' : \widetilde{\widetilde{W}}_Y \to Y$ is the normalisation of $\widetilde{W}_Y$. This last formula can be proved by replacing X by $\widetilde{W}$, Y by $\widetilde{W}_Y$, and noting that $\widetilde{W}_Y \subset \widetilde{W}$ is a Cartier divisor locally defined by an equation $f = 0$. We are then reduced to the equality (9.5) for a rational function ϕ on $\widetilde{W}$, having no common zero or pole divisor with f:

$$\mathrm{div}\,(\phi_{|\mathrm{div}\, f}) = \mathrm{div}\,(f_{|\,\mathrm{div}\,\phi}).$$

The mixed case, where W is not contained in Y but one of the components of $\tau_*\mathrm{div}(\phi)$ is, works similarly. □

Corollary 9.10 *The group* Pic X *acts on the Chow groups of* X *; for* $D \in \mathrm{Pic}\, X$, *we have*

$$D : \mathrm{CH}_k(X) \to \mathrm{CH}_{k-1}(X), \quad Z \mapsto D \cdot Z.$$

Proof If $i : D \to X$ is the inclusion of an effective Cartier divisor, we set $D = i_* \circ i^* : \mathrm{CH}_k(X) \to \mathrm{CH}_{k-1}(X)$. By definition, $D \cdot Z$ is also equal to $j_*s(\mathcal{O}_Z(D))$, where $j : Z \to X$ is the inclusion of a k-dimensional subvariety, so it is clear that $D \cdot Z$ depends only on the rational equivalence class of D. This action extends by linearity or using the last formula to the whole Picard group. □

The restriction map i^* also enables us to prove the following algebraic analogue of the homotopy invariance property.

Proposition 9.11 *Let* $A^1_K = \mathrm{Spec}\, K[T]$ *denote the affine line. Then the map*

$$\mathrm{pr}_2^* : \mathrm{CH}_k(X) \to \mathrm{CH}_{k+1}\left(A^1_K \times X\right)$$

is an isomorphism, whose inverse is given by

$$i_0^* : \mathrm{CH}_{k+1}\left(A^1_K \times X\right) \to \mathrm{CH}_k(X),$$

where $i_0 : X \to A^1_K \times X$ *is the inclusion of* $0 \times X$.

Proof Note that i_0 is the inclusion of a Cartier divisor, and that pr_2 is open and thus flat. Therefore, the two maps are well-defined. It is obvious that

$$i_0^* \circ \mathrm{pr}_2^* = \mathrm{Id},$$

since this equality is already satisfied on the level of the groups of cycles (the image of

$$\mathrm{pr}_2^* : \mathcal{Z}_k(X) \to \mathcal{Z}_k\left(A^1_K \times X\right)$$

consists of cycles which are not contained in $0 \times X$).

Thus it remains only to prove that

$$\mathrm{pr}_2^* : \mathrm{CH}_k(X) \to \mathrm{CH}_{k+1}\big(A_K^1 \times X\big)$$

is surjective. Let $Z \in \mathcal{Z}_{k+1}(A_K^1 \times X)$ be the cycle associated to an irreducible $(k+1)$-dimensional subvariety of $A_K^1 \times X$. If $Z' = pr_2(Z)$ is k-dimensional, then $Z = \mathrm{pr}_2^{-1}(\mathrm{pr}_2(Z))$, so $Z \in \mathrm{Im}\,\mathrm{pr}_2^*$. Otherwise, $\dim Z' = k+1$, and $Z \subset \mathrm{pr}_2^{-1}(Z')$ is a divisor. Let U be the smooth locus of Z'. Then $Z \cap \mathrm{pr}_2^{-1}(U)$ is a Cartier divisor in $\mathrm{pr}_2^{-1}(U) = A_K^1 \times U$. We have

$$\mathrm{Pic}\, A_K^1 \times U = \mathrm{pr}_2^*(\mathrm{Pic}\, U),$$

so there exists a rational function $\phi \in K(A_K^1 \times U)^*$ such that

$$Z \cap \mathrm{pr}_2^{-1}(U) - \mathrm{div}(\phi) = \mathrm{pr}_2^{-1}(Z''),$$

where Z'' is a Cartier divisor of U. But then ϕ is a rational function on $A_K^1 \times \widetilde{Z}'$, where $\tau : \widetilde{Z}' \to X$ is the normalisation of Z', and we have

$$Z - (\mathrm{Id}, \tau)_* \mathrm{div}(\phi) \subset A_K^1 \times (Z' - U).$$

But as $A_K^1 \times (Z' - U)$ is $k+1$-dimensional and all of its irreducible components are of the form $A_K^1 \times D$, where D is an irreducible component of $Z' - U$, we have

$$Z - (\mathrm{Id}, \tau)_* \mathrm{div}(\phi) = \mathrm{pr}_2^*(Z''),$$

where Z'' is a k-cycle of X. □

9.1.3 Localisation

Let X be a quasi-projective scheme, and let $F \xrightarrow{l} X$ be the inclusion of a closed subscheme. Let $j : U = X - F \to X$ be the inclusion of the complement. The morphism j is flat since it is open, and the morphism l is proper since it is a closed immersion. Thus, we can define the inverse image morphism j^* and the direct image morphism l_*. Moreover, it is clear that $j^* \circ l_* = 0$, since cycles supported on F do not intersect U.

Lemma 9.12 *The following sequence, known as the localisation sequence, is exact:*

$$\mathrm{CH}_k(F) \xrightarrow{l_*} \mathrm{CH}_k(X) \xrightarrow{j^*} \mathrm{CH}_k(U) \to 0. \tag{9.6}$$

Proof The surjectivity on the right follows from the fact that if $Z \subset U$ is a k-dimensional subvariety, then its Zariski closure $\overline{Z} \subset X$ is a k-dimensional subvariety whose intersection with U is equal to Z.

Furthermore, let $Z \in \mathcal{Z}_k(X)$ be such that $j^*Z \in \mathcal{Z}_k(U)_{\text{rat}}$. Then there exists $W_i \subset U$, $\dim W_i = k+1$, $\phi_i \in K(W_i)^*$, and integers n_i, such that

$$Z \cap U = \sum_i n_i \tau_{i*}(\text{div}(\phi_i)), \tag{9.7}$$

where

$$\tau_i : \widetilde{W_i} \to U$$

is the normalisation of W_i. Then, if $\overline{W_i}$ is the closure of W_i in X and $\overline{\tau}_i : \widetilde{\overline{W_i}} \to X$ is its normalisation, we have $\phi_i \in K(\overline{W_i})^*$, and the equality (9.7) shows that

$$Z - \sum_i n_i \overline{\tau}_{i*}(\text{div}(\phi_i)) \in \mathcal{Z}_k(F).$$

This proves the exactness of the sequence (9.6) at the middle. □

We will use this result to prove the following theorem.

Theorem 9.13 *Let $\pi : E \to X$ be a vector bundle of rank r. Then the map*

$$\pi^* : \text{CH}_k(X) \to \text{CH}_{k+r}(E)$$

is an isomorphism for every k.

Proof We proceed by induction on the dimension of X. Note that π is flat since it is open, so π^* is well-defined. Let $U = X - F$ be a dense Zariski open set of X on which E is trivial. Then $E_U := \pi^{-1}(U)$ is isomorphic to $A^r_K \times U$, so applying proposition 9.11 r times, we obtain an isomorphism

$$\pi_U^* : \text{CH}_k(U) \cong \text{CH}_{k+r}(E_U).$$

Furthermore, the induction hypothesis also implies that we have an isomorphism

$$\pi_F^* : \text{CH}_k(F) \cong \text{CH}_{k+r}(E_F),$$

where $E_F = \pi^{-1}(F)$, $\pi_F = \pi_{|E_F}$. In the following commutative diagram of localisation exact sequences

$$\begin{array}{ccccccc}
\text{CH}_k(F) & \xrightarrow{l^*} & \text{CH}_k(X) & \xrightarrow{j^*} & \text{CH}_k(U) & \longrightarrow & 0 \\
\downarrow \pi_F^* & & \downarrow \pi^* & & \downarrow \pi_U^* & & \\
\text{CH}_{k+r}(E_F) & \xrightarrow{l^*} & \text{CH}_{k+r}(E) & \xrightarrow{j^*} & \text{CH}_{k+r}(E_U) & \longrightarrow & 0,
\end{array}$$

given by proposition 9.8 and lemma 9.12, the arrows π_F^* and π_U^* are surjective, so we can conclude that $\pi^* : \mathrm{CH}_k(X) \to \mathrm{CH}_{k+r}(E)$ is surjective.

The injectivity is proved as follows. In the case where E is a line bundle, we have the inclusion $i : X \to E$ of the section 0, which is the inclusion of a Cartier divisor, and it is obvious that $i^* \circ \pi^* = \mathrm{Id}$. It follows that the result holds, more generally, for a bundle E admitting a filtration $F^i E$, $0 \leq i \leq r$, with $F^i E / F^{i+1} E$ of rank 1 for every $i \leq r - 1$. We then use the splitting principle (corollary 9.26). □

9.2 Intersection and cycle classes

9.2.1 Intersection

Fulton (1984) proposes an intersection theory

$$\mathrm{CH}_k(X) \times \mathrm{CH}_l(X) \to \mathrm{CH}_{k+l-n}(X),$$

or better

$$\mathrm{CH}^k(X) \times \mathrm{CH}^l(X) \to \mathrm{CH}^{k+l}(X),$$

for a smooth n-dimensional variety X, which is particularly well adapted to computing the excess formulae. If Z and Z' are two irreducible reduced subschemes of X, of dimensions k and l respectively, which intersect properly, i.e. such that $\dim Z \cap Z' = k + l - n$, then one classically defines $Z \cdot Z'$ to be the cycle associated to the scheme $Z \cap Z'$. By bilinearity, we can define the intersection $Z \cdot Z'$ in this way for any pair of cycles whose supports intersect properly.

If Z and Z' do not intersect properly, the classical theory replaces Z by a cycle $\widetilde{Z}$ which is rationally equivalent to Z and intersects Z' properly (such a cycle exists by 'Chow's moving lemma'), and defines $Z \cdot Z'$ to be the class of $\widetilde{Z} \cap Z'$ in $\mathrm{CH}_{k+l-n}(X)$.

Let $|Z| = \bigcup_i Z_i$ denote the support of the cycle $Z = \sum_i n_i Z_i$. Fulton's theory avoids the recourse to 'Chow's moving lemma', and gives a refined intersection, i.e. a cycle

$$Z \cdot Z' \in \mathrm{CH}_{k+l-n}(|Z| \cap |Z'|)$$

for every pair of cycles Z and Z', whose image in $\mathrm{CH}_{k+l-n}(X)$ is $Z \cdot Z'$. Thus it provides an exact answer to the problems of excess, i.e. the 'explicit' computation of $Z \cdot Z'$ when Z and Z' do not intersect properly.

The steps are as follows.

1. To define

$$Z \cdot Z' \in \mathrm{CH}_{k+l-n}(Z \cap Z'),$$

where Z and Z' are subschemes of dimension k and l of X, it suffices to consider the case where Z is smooth. Indeed, if $\Delta \subset X \times X$ denotes the diagonal, then $\Delta \cong X$ is smooth and we have

$$Z \cap Z' = (Z \times Z') \cap \Delta \subset \Delta \cong X.$$

We then set

$$Z \cdot Z' = (Z \times Z') \cdot \Delta \in \mathrm{CH}_{k+l-n}(Z \cap Z').$$

2. We use theorem 9.13 to define the morphism

$$i_0^* : \mathrm{CH}_k(E) \to \mathrm{CH}_{k-r}(Y)$$

for every vector bundle $\pi : E \to Y$ of rank r, where i_0 is the section 0. Here i_0^* is defined as the inverse of the isomorphism $\pi^* : \mathrm{CH}_{k-r}(Y) \to \mathrm{CH}_k(E)$.

3. Consider the inclusion of $Z \cap Z'$ into Z'. It admits a cone (which generalises the normal bundle of locally complete intersection schemes)

$$C_{Z\cap Z'/Z'} := \mathrm{Spec} \bigoplus_n \left((\mathcal{I}^{Z'}_{Z\cap Z'})^n / (\mathcal{I}^{Z'}_{Z\cap Z'})^{n+1}\right).$$

We have an obvious inclusion of $C_{Z\cap Z'/Z'}$ into the vector bundle $N_{Z/X\,|Z\cap Z'}$, since

$$N_{Z/X\,|Z\cap Z'} = \mathrm{Spec} \bigoplus_n \left(((\mathcal{I}^X_Z)^n / (\mathcal{I}^X_Z)^{n+1}) \otimes_{\mathcal{O}_Z} \mathcal{O}_{Z\cap Z'}\right),$$

and we have the surjectivity of the restriction map between ideal sheaves

$$\mathcal{I}^X_Z \to \mathcal{I}^{Z'}_{Z\cap Z'}.$$

Note also that $C_{Z\cap Z'/Z'}$ is of dimension equal to $l = \dim Z'$, whereas $N_{Z/X\,|Z\cap Z'}$ is a vector bundle of rank equal to

$$\dim X - \dim Z = n - k$$

on $Z \cap Z'$. Thus, $C_{Z\cap Z'/Z'}$ defines a cycle $[C_{Z\cap Z'/Z'}]$ in $\mathrm{CH}_l(N_{Z/X\,|Z\cap Z'})$, and finally, we obtain the class

$$Z \cdot Z' := i_0^*([C_{Z\cap Z'/Z'}]) \in \mathrm{CH}_{l-n+k}(Z \cap Z').$$

Fulton shows that the image of the cycle $Z \cdot Z'$ in $\mathrm{CH}_{k+l-n}(X)$ depends only on the classes of Z and Z' in $\mathrm{CH}(X)$. Similarly, its image in $\mathrm{CH}_{k+l-n}(Z)$ depends only on the rational equivalence class of Z'. Furthermore, it is quite easy to see that if the intersection is proper, then $Z \cdot Z' \in \mathrm{CH}_{k+l-n}(X)$ is equal to the class of the cycle associated to $Z \cap Z'$ (comparing the multiplicities along each component of the intersection). The intersection product thus constructed on $\mathrm{CH}(X)$ coincides with the classical product.

The preceding construction used only the fact that the normal bundle or cone of Z in X is a vector bundle on Z. Thus, it easily generalises to the case where Z is locally a complete intersection in X (which we no longer assume to be smooth), i.e. that its sheaf of ideals is locally generated by $n - k$ equations. In particular, for every inclusion $j : Z \hookrightarrow X$ of a locally complete intersection subscheme, it gives a morphism

$$j^* : \mathrm{CH}_k(X) \to \mathrm{CH}_{k-r}(Z), \quad r = \operatorname{codim} Z$$

defined by $j^*Z' = l_*(Z \cdot Z')$, where $Z \cdot Z' \in \mathrm{CH}_{k-r}(Z \cap |Z'|)$ is the refined intersection defined above, and $l : Z \cap |Z'| \to Z$ is inclusion.

A very important application of this generalisation concerns the definition of the morphisms

$$f^* : \mathrm{CH}_l(X) \to \mathrm{CH}_k(Y),$$

where $k - l = \dim Y - \dim X$, under the hypothesis that X is smooth, but with no flatness condition on f. We simply note that if X is smooth, then the graph of f,

$$i : \Gamma_f \hookrightarrow Y \times X,$$

is a locally complete intersection in $Y \times X$, and is of course isomorphic to Y. Moreover, the projection

$$\mathrm{pr}_2 : Y \times X \to X$$

is flat. We then define f^* as the composition

$$\mathrm{CH}_l(X) \stackrel{\mathrm{pr}_2^*}{\to} \mathrm{CH}_{l+\dim Y}(Y \times X) \stackrel{i^*}{\to} \mathrm{CH}_k(\Gamma_f) = \mathrm{CH}_k(Y).$$

Proposition 9.8 no longer holds for the map f^* defined above: given a Cartesian diagram

$$\begin{array}{ccc} i' : Z' & \longrightarrow & X \\ {\scriptstyle f'}\downarrow & & \downarrow{\scriptstyle f} \\ i : Z & \longrightarrow & Y, \end{array}$$

where i is proper and Z and Y are smooth, we do not in general have the equality $i'_* \circ f'^* = f^* \circ i_* : \mathrm{CH}(Z) \to \mathrm{CH}(X)$ (in fact, the dimensions do not even coincide in general). This defect is described precisely by Fulton's excess formulae.

However, we have the following compatibility between the intersection product and the morphisms p_*, p^*, where X, Y are smooth and $p : X \to Y$ is a morphism.

Proposition 9.14

(i) (Projection formula; Fulton 1984, 8.1) *If p is proper, then for $Z \in \mathrm{CH}(Y)$ and $Z' \in \mathrm{CH}(X)$, we have*

$$p_*(p^*Z \cdot Z') = Z \cdot p_*Z' \in \mathrm{CH}(Y). \tag{9.8}$$

(ii) With no properness hypothesis on p, for Z, $Z' \in \mathrm{CH}(Y)$, we have

$$p^*(Z \cdot Z') = p^*Z \cdot p^*Z' \in \mathrm{CH}(X).$$

The following corollary is a consequence of the projection formula.

Corollary 9.15 *If $p : X \to Y$ is a proper morphism with* $\dim X = \dim Y$, *then*

$$p_* \circ p^*Z = \deg p\, Z$$

for $Z \in \mathrm{CH}(Y)$, where $\deg p$ *is defined to be equal to* 0 *if p is not dominant, and to be equal to the degree of the field extension $K(Y) \subset K(X)$ otherwise.*

To prove this, we apply formula (9.8) with $Z' = c(X) \in \mathrm{CH}_n(X)$, $n = \dim X$. Indeed, by the definition of p_*, we have $p_*(c(X)) = \deg p\, c(Y) \in \mathrm{CH}(Y)$. □

9.2.2 Correspondences

Definition 9.16 *A correspondence between two smooth varieties X and Y is a cycle $\Gamma \in \mathrm{CH}(X \times Y)$.*

If X is projective, the second projection $X \times Y \to Y$ is proper. A correspondence $\Gamma \in \mathrm{CH}_k(X \times Y)$ then defines a morphism

$$\Gamma_* : \mathrm{CH}_l(X) \to \mathrm{CH}_{l+k-\dim X}(Y)$$

given by

$$\Gamma_*(Z) = \mathrm{pr}_{2*}(\mathrm{pr}_1^*(Z) \cdot \Gamma), \quad Z \in \mathrm{CH}(X).$$

If Y is projective, we can also consider the morphism

$$\Gamma^* : \mathrm{CH}(Y) \to \mathrm{CH}(X)$$

given by

$$\Gamma^*(Z) = \mathrm{pr}_{1*}(\mathrm{pr}_2^*(Z) \cdot \Gamma), \quad Z \in \mathrm{CH}(Y).$$

Now, assume that X, Y and W are smooth varieties, with X and Y projective, and let $\Gamma \in \mathrm{CH}^k(X \times Y)$, $\Gamma' \in \mathrm{CH}^{k'}(Y \times W)$ be correspondences. We can then define the composition $\Gamma' \circ \Gamma \in \mathrm{CH}^{k''}(X \times W)$, where $k'' = k + k' - \dim Y$, by the formula

$$\Gamma' \circ \Gamma = p_{13*}(p_{12}^*\Gamma \cdot p_{23}^*\Gamma'), \tag{9.9}$$

where the p_{ij}, $i = 1, 2, 3$ are the projections of $X \times Y \times W$ onto the product of its ith and jth factors.

We can show that the composition of correspondences is associative. In particular, it equips the group $\mathrm{CH}(X \times X)$ with the structure of a (non-commutative) ring.

Finally, we have the following essential formula.

Proposition 9.17 *Let*

$$\Gamma_* : \mathrm{CH}(X) \to \mathrm{CH}(Y), \quad \Gamma'_* : \mathrm{CH}(Y) \to \mathrm{CH}(W)$$

be the morphisms associated to the correspondences Γ, Γ'. *Then*

$$(\Gamma' \circ \Gamma)_* = \Gamma'_* \circ \Gamma_*.$$

Proof This is a consequence of the projection formula (9.8) and of proposition 9.8. Let $Z \in \mathrm{CH}(X)$. Then

$$(\Gamma' \circ \Gamma)_* Z = p_{W*}(p_X^* Z \cdot \Gamma' \circ \Gamma),$$

where p_X, p_W are the projections of $X \times W$ onto X and W respectively. Replacing $\Gamma' \circ \Gamma$ by its expression (9.9), and applying the projection formula (9.8) to the morphism p_{13}, we obtain

$$\begin{aligned}(\Gamma' \circ \Gamma)_* Z &= p_{W*}(p_{13*}(p_1^* Z \cdot p_{12}^*\Gamma \cdot p_{23}^*\Gamma')) \\ &= p_{3*}(p_1^* Z \cdot p_{12}^*\Gamma \cdot p_{23}^*\Gamma') \\ &= p'_{W*}(p_{23*}(p_1^* Z \cdot p_{12}^*\Gamma \cdot p_{23}^*\Gamma')),\end{aligned}$$

where p_i is the projection of $X \times Y \times W$ onto its ith factor, and in the last expression, p'_W is the projection of $Y \times W$ onto W.

We can now apply the projection formula to the morphism p_{23}, which gives

$$(\Gamma' \circ \Gamma)_* Z = p'_{W*}(\Gamma' \cdot (p_{23*}(p_1^* Z \cdot p_{12}^* \Gamma))).$$

But thanks to proposition 9.8, we have

$$p_{23*}(p_1^* Z \cdot p_{12}^* \Gamma) = p_Y^*(\Gamma_* Z) \in \mathrm{CH}(Y \times W),$$

where p_Y is the projection of $Y \times W$ onto Y, so

$$(\Gamma' \circ \Gamma)_* Z = \Gamma'_* \circ \Gamma_*(Z) \in \mathrm{CH}(W).$$

□

9.2.3 Cycle classes

Assume that X is a smooth complex quasi-projective variety. In vI.11.1.2, we defined the class $[Z] \in H^{2n-2k}(X, \mathbb{Z})$ of a cycle $Z \in \mathcal{Z}_k(X)$, $n = \dim X$.

Lemma 9.18 *If Z is rationally equivalent to 0, then $[Z] = 0$. The map $Z \mapsto [Z]$ thus gives the 'class' map*

$$\mathrm{cl} : \mathrm{CH}_l(X) \to H^{2n-2l}(X, \mathbb{Z}).$$

Its kernel will be denoted by $\mathrm{CH}_e(X)_{\mathrm{hom}}$.

Proof of lemma 9.18 We can easily reduce to the case where X is projective. Indeed, X is a Zariski open set in a smooth projective variety $\overline{X}$. If $Z \in \mathcal{Z}_k(X)$ is a cycle rationally equivalent to 0, then its closure $\overline{Z}$ in $\overline{X}$ satisfies $\overline{Z} = Z_1 + Z_2$, where Z_1 is rationally equivalent to 0 and Z_2 is supported in $\overline{X} - X$. But then $[Z_1] = 0$ if the lemma holds for $\overline{X}$, and $[Z_2]$ vanishes in $H^{2n-2k}(X, \mathbb{Z})$, since by construction, the class $[Z]$ of a cycle Z of $\overline{X}$ vanishes on $\overline{X} - \operatorname{Supp} Z$. Thus, if the lemma holds for $\overline{X}$, then

$$[\overline{Z}]_{|X} = [Z] = 0.$$

Assume, then, that X is projective, and let $W \subset X$ be a projective subvariety of dimension $k+1$, and $\phi \in \mathbb{C}(W)^*$. We must show that $[\tau_* \mathrm{div}(\phi)] = 0$ in $H^{2n-2k}(X, \mathbb{Z})$, where $\tau : \widetilde{W} \to X$ is the normalisation of W. Here, we can replace $\widetilde{W}$ by a desingularisation $\widetilde{\widetilde{W}}$ of W. It is then easy to see that

$$[\tau_* \mathrm{div}(\phi)] = \tau_*([\mathrm{div}(\phi)]),$$

where $\tau_* : H^2\left(\widetilde{\widetilde{W}}, \mathbb{Z}\right) \to H^{2n-2k}(X, \mathbb{Z})$ is the Gysin morphism. Indeed (see vI.11.1.2), a possible definition of $[Z]$ for $Z = \sum_i n_i Z_i$ is given by

$$[Z] = \sum_i n_i [Z_i], \quad [Z_i] = \mathrm{PD}(j_{i*}([\widetilde{Z}_i]_{\mathrm{fund}})),$$

where $j_i : \widetilde{Z}_i \to X$ is a desingularisation of Z_i, $[\widetilde{Z}_i]_{\mathrm{fund}}$ is the fundamental homology class of the smooth oriented manifold $\widetilde{Z}_i$, and PD denotes Poincaré duality.

It thus suffices to show that $[\mathrm{div}(\phi)] = 0$ in $H^2\left(\widetilde{\widetilde{W}}, \mathbb{Z}\right)$. Now, the line bundle $\mathcal{O}_{\widetilde{\widetilde{W}}}(\mathrm{div}(\phi))$ is trivial, and in vI.11.3.1, we proved the equality

$$[D] = c_1(\mathcal{O}_Y(D)) \in H^2(Y, \mathbb{Z})$$

(Lelong's formula) for every divisor D in a smooth variety Y. □

Next, if $Z \in \mathcal{Z}_k(X)_{\mathrm{hom}}$, then we have the Abel–Jacobi invariant $\Phi_X^{n-k}(Z) \in J^{2n-2k-1}(X)$ (see section 7.1).

Lemma 9.19 *If $Z \in \mathcal{Z}_k(X)_{\mathrm{rat}}$, then $\Phi_X^{n-k}(Z) = 0$ in $J^{2n-2k-1}(X)$.*

Proof In vI.12.1.2, we proved that if

$$\mathcal{Z} \subset C \times X$$

is a relative cycle of dimension $k+1$, where C is a smooth connected curve, and $0 \in C$, then the map

$$f : C \to J^{2n-2k-1}(X)$$
$$t \mapsto \Phi_X^{n-k}(Z_t - Z_0) \in J^{2n-2k-1}(X)$$

is holomorphic. (This is actually a special case of theorem 7.9.)

Now, as Z is rationally equivalent to 0, there exists a variety $W \subset X$ and a rational function $\phi \in \mathbb{C}(W)$ such that $Z = \tau_*\mathrm{div}(\phi)$, where $\tau : \widetilde{W} \to X$ is a desingularisation of W. We can write $\phi = \sigma_0/\sigma_\infty$, where σ_0, σ_∞ are holomorphic sections of a line bundle L on $\widetilde{W}$. Consider the family $(Z_t)_{t\in\mathbb{P}^1}$ of cycles of X defined by $Z_t = \tau_*(\mathrm{div}(\sigma_t) - \mathrm{div}(\sigma_\infty))$, where $\sigma_t = \sigma_0 + t\sigma_\infty$. Then

$$Z_0 = Z, \quad Z_\infty = 0.$$

But there exists no non-constant holomorphic map f sending $\mathbb{P}^1$ to a complex torus, since $\mathbb{P}^1$ has no non-zero holomorphic forms of degree 1, so that the pullbacks by f of the holomorphic 1-forms on the torus must vanish, which is equivalent to $df = 0$. Thus, we have

$$\Phi_X^{n-k}(Z) = f(\infty) = f(0) = \Phi_X^{n-k}(0) = 0.$$

□

9.2.4 Compatibilities

Recall that for a smooth complex n-dimensional projective variety X, we have the cycle class map $\mathrm{cl} : \mathrm{CH}^k(X) \to H^{2k}(X, \mathbb{Z})$, the intersection product

$$\cdot : \mathrm{CH}^l(X) \times \mathrm{CH}^k(X) \to \mathrm{CH}^{k+l}(X),$$

and the cup-product

$$\cup : H^{2l}(X, \mathbb{Z}) \times H^{2k}(X, \mathbb{Z}) \to H^{2k+2l}(X, \mathbb{Z}).$$

Proposition 9.20 *For $Z \in \mathrm{CH}^l(X)$, $Z' \in \mathrm{CH}^k(X)$, we have*

$$\mathrm{cl}(Z \cdot Z') = \mathrm{cl}(Z) \cup \mathrm{cl}(Z') \in H^{2k+2l}(X, \mathbb{Z}).$$

Proof To begin with, we have

$$[Z \times Z'] = \mathrm{pr}_1^*[Z] \cup \mathrm{pr}_2^*[Z'] \in H^{2k+2l}(X \times X, \mathbb{Z}),$$

where $Z \times Z' := \sum_{i,j} n_i m_j Z_i \times Z'_j$ for $Z = \sum_i n_i Z_i$, $Z' = \sum_j m_j Z'_j$. Indeed, we may assume that X is projective, in which case this follows from the definition

$$[Z] = \mathrm{PD}(\tau_*[\widetilde{Z}]_{\mathrm{fund}}),$$

for a subvariety $Z \hookrightarrow X$ of codimension k of X and a desingularisation $\tau : \widetilde{Z} \to Z \to X$ of Z.

If on the other hand $i : X = \Delta_X \subset X \times X$ is the inclusion of the diagonal, we know that

$$Z \cdot Z' = i^*(Z \times Z') \in \mathrm{CH}^{k+l}(\Delta_X), \quad [Z] \cup [Z'] = i^*(\mathrm{pr}_1^*[Z] \cup \mathrm{pr}_2^*[Z]).$$

The equality $[Z \times Z'] = [Z] \cup [Z']$ then follows from the equality

$$\mathrm{cl}(i^*(Z \times Z')) = i^*(\mathrm{cl}(Z \times Z')),$$

which is proved in proposition 9.21. □

The cycle class map possesses the following functoriality properties.

Proposition 9.21 *Let $i : Y \to X$ be a morphism between smooth varieties.*

(i) If $Z \in \mathrm{CH}^k(X)$, then

$$i^*\mathrm{cl}(Z) = \mathrm{cl}(i^*Z) \in H^{2k}(Y, \mathbb{Z}).$$

(ii) If i is proper and $Z \in \mathrm{CH}^k(Y)$, then

$$\mathrm{cl}(i_*Z) = i_*\mathrm{cl}(Z) \in H^{2k-2\dim Y+2\dim X}(X, \mathbb{Z}).$$

It follows that the cycle class map is compatible with correspondences. If X and Y are proper and smooth, and $\Gamma \in \mathrm{CH}^r(X \times Y)$, then for every $Z \in \mathrm{CH}^k(X)$, we have

$$\mathrm{cl}(\Gamma_*(Z)) = [\Gamma]_*(\mathrm{cl}(Z)),$$

where $[\Gamma]_* : H^{2k}(X, \mathbb{Z}) \to H^{2l}(Y, \mathbb{Z}), l = r + k - \dim X$ *is defined by*

$$[\Gamma]_*(\alpha) = \mathrm{pr}_{2*}(\mathrm{pr}_1^*\alpha \cup [\Gamma])$$

(see vI.11.3.3).

Proof Property (ii) was proved in the proof of lemma 9.18. As for property (i), it actually suffices to prove it for immersions. Indeed, the general case can be deduced from this one by introducing the graph of i and using the fact noted above, that $\mathrm{cl}(Y \times Z) = p_2^*\mathrm{cl}(Z) \in H^{2k}(Y \times X, \mathbb{Z})$ for $Z \in \mathrm{CH}^k(X)$, where $p_2 : Y \times X \to X$ is the second projection.

By the invariance of the map cl under rational equivalence, we may assume by lemma 9.22 below (the Chow moving lemma) that the intersection $\mathrm{Supp}\, Z \cap Y$ is proper, that the components Z_i of Z are generically smooth along $Z_i \cap Y$, and that the intersection $Z_i \cap Y$ is generically transverse along $Z_i \cap Y$. It suffices to prove the equality for each of the components Z_i of Z. Let $W \subset Y \cap Z_i$ be the set of points where either Z_i is not smooth, or the intersection $Z_i \cap Y$ is not transverse. Then, because $\mathrm{codim}\, W > k$, restriction shows that (see vI.11.1.2)

$$H^{2k}(Y, \mathbb{Z}) \cong H^{2k}(Y - W, \mathbb{Z}).$$

It thus suffices to prove the result after replacing X by $X' = X - W$, Y by $Y' = Y - W$ and Z_i by $Z_i' = Z_i - Z_i \cap W$. Then the intersection $Z_i' \cap Y'$ is transverse, and the result follows immediately from the definition of the cohomology classes $[Z_i']$, $[Z_i' \cap Y]$, using the fact that the normal bundles of Z_i' in X' and of $Z_i' \cap Y'$ in Y' can be identified along $Z_i' \cap Y$ via the natural map

$$N_{Z_i' \cap Y'/Y'} \to N_{Z_i'/X'}|_{Z_i' \cap Y'}$$

induced by the inclusion

$$T_{Y'}|_{Z_i' \cap Y'} \hookrightarrow T_{X'}|_{Z_i' \cap Y'}.$$

□

Lemma 9.22 *Let X be a smooth algebraic variety, and $Y \subset X$ a smooth closed subvariety. Let $Z \in \mathrm{CH}^k(X)$. Then there exists a cycle $T = \sum_i n_i T_i$ of X of codimension k such that T is rationally equivalent to Z, and the components T_i*

of T intersect Y properly. Moreover, we may also assume that T_i is generically smooth along $Y \cap T_i$, and the intersection $T_i \cap Y$ is generically transverse.

Proof We may assume that Z is represented by an irreducible subvariety of X of codimension k. Let L be a very ample linear system on X, giving an embedding $X \hookrightarrow \mathbb{P}^N$. If $\dim X = n$, consider a generic linear projection

$$\mathbb{P}^N \to \mathbb{P}^n,$$

which restricted to X gives a morphism

$$\phi : X \to \mathbb{P}^n.$$

For a suitable choice of such a projection, $\phi_{|Z}$ is of degree 1 on its image, and ϕ is a local isomorphism at a generic point of Z. Set $Z' = \phi^{-1}(\phi(Z)) - Z$. As the cycle $\phi^{-1}(\phi(Z))$ is rationally equivalent to $\deg(\phi(Z))H_1 \cdot \cdots \cdot H_k$, where the H_i are arbitrary elements of L, we can write

$$Z \equiv T - Z',$$

where T is a smooth complete intersection of codimension k which can be chosen so that $T \cap Y$ is smooth.

One sees easily that if $\operatorname{codim} Z \cap Y \subset Y = k - l$, $l \geq 1$, for a generic choice of projection, then $\operatorname{codim} Z' \cap Y \subset Y \geq k - l + 1$. Thus, iterating this construction, we obtain a cycle rationally equivalent to Z which intersects Y properly. The assertion concerning the transversality is easy. □

We next consider the compatibility of the intersection of cycles with the Abel–Jacobi map. Let X be a smooth projective variety. If $Z \in \mathrm{CH}^k(X)$, the class $\mathrm{cl}(Z)$ is a Hodge class of degree $2k$, so the cup-product $\mathrm{cl}(Z)\cup$ gives a morphism of Hodge structures

$$\mathrm{cl}(Z)\cup : H^{2l-1}(X, \mathbb{Z}) \to H^{2l+2k-1}(X, \mathbb{Z}),$$

so it induces a morphism of complex tori

$$\mathrm{cl}(Z)\cup : J^{2l-1}(X) \to J^{2l+2k-1}(X)$$

between the corresponding intermediate Jacobians. Now let $Z' \in \mathrm{CH}^l(X)_{\mathrm{hom}}$; by proposition 9.20, $Z \cdot Z' \in \mathrm{CH}^{k+l}(X)_{\mathrm{hom}}$, and we have the following result.

Proposition 9.23 *The Abel–Jacobi invariant of $Z \cdot Z'$ is given by the formula*

$$\Phi_X^{k+l}(Z \cdot Z') = \mathrm{cl}(Z) \cup \Phi_X^l(Z').$$

Proof We restrict ourselves to giving the proof in the case where Z' is algebraically equivalent to 0 (see subsection 8.2.1), since it is then a consequence of proposition 9.20. The general case is included in theorem 9.24 and will not be proved here.

By hypothesis, there exist a curve C and a relative cycle $\mathcal{Z}'$ of codimension l in $C \times X$ such that $Z' = \sum_i n_i \mathcal{Z}'_{t_i}$, $\sum_i n_i = 0$. We then know (see vI.12.2.1) that we have a morphism of complex tori

$$\alpha : J(C) = \mathrm{CH}_0(C)_{\mathrm{hom}} \to J^{2l-1}(X)$$

such that

$$\Phi^l_X(\mathcal{Z}'_c) = \alpha(\mathrm{alb}(c))$$

for every $c = \sum_j m_j t_j \in \mathrm{CH}_0(C)_{\mathrm{hom}}$. Moreover, the morphism α is determined by the underlying morphism of Hodge structures

$$\alpha_{\mathbb{Z}} : H_1(C, \mathbb{Z}) \to H^{2l-1}(X, \mathbb{Z});$$

furthermore, by the Künneth decomposition and the Poincaré duality on C, $\alpha_{\mathbb{Z}}$ can be identified with the Künneth component of type $(1, 2l-1)$ of the class $\mathrm{cl}(\mathcal{Z}') \in H^{2l}(C \times X, \mathbb{Z})$. Now, the cycle $Z \cdot Z'$ is equal to $\sum_i n_i \mathcal{Z}''_{t_i}$, where $\mathcal{Z}'' := (C \times Z) \cdot \mathcal{Z}' \subset C \times X$. By proposition 9.20, we have

$$\mathrm{cl}(\mathcal{Z}'') = \mathrm{cl}(C \times Z) \cup \mathrm{cl}(\mathcal{Z}') = \mathrm{pr}_2^* \mathrm{cl}(Z) \cup \mathrm{cl}(\mathcal{Z}') \in H^{2k+2l}(C \times X, \mathbb{Z}).$$

It follows immediately that the morphism of Hodge structures

$$\beta_{\mathbb{Z}} : H^1(C, \mathbb{Z}) \to H^{2k+2l-1}(X, \mathbb{Z})$$

associated to the class of $\mathcal{Z}''$ is equal to $(\mathrm{cl}(Z)\cup) \circ \alpha_{\mathbb{Z}}$. Thus, for the same reason as above, the Abel–Jacobi map

$$\beta : J(C) \to J^{2k+2l-1}(X)$$

associated to the cycle $\mathcal{Z}''$ is equal to $(\mathrm{cl}(Z)\cup) \circ \alpha$. Therefore, we have

$$\begin{aligned} \Phi^{k+l}_X(Z \cdot Z') = \beta\left(\mathrm{alb}\left(\sum_i n_i t_i\right)\right) &= \mathrm{cl}(Z) \cup \alpha\left(\mathrm{alb}\left(\sum_i n_i t_i\right)\right) \\ &= \mathrm{cl}(Z) \cup \Phi^l_X(Z'). \end{aligned}$$

□

Propositions 9.20 and 9.23 are in fact special cases of the following general compatibility theorem. Recall that the Deligne cohomology $H^k_{\mathrm{D}}(X, \mathbb{Z}(p))$ is defined to be the hypercohomology of the Deligne complex

$$\mathbb{Z}_{\mathrm{D}}(p) = 0 \to \mathbb{Z} \to \mathcal{O}_X \xrightarrow{d} \cdots \to \Omega_X^{p-1} \to 0,$$

where $\mathbb{Z}$ is placed in degree 0.

The Deligne cohomology group

$$H_{\mathrm{D}}^{2p}(X, \mathbb{Z}(p)) := \mathbb{H}^{2p}(X, \mathbb{Z}_D(p))$$

is an extension

$$0 \to J^{2p-1}(X) \to H_{\mathrm{D}}^{2p}(X, \mathbb{Z}(p) \xrightarrow{c} \mathrm{Hdg}^{2p}(X, \mathbb{Z}) \to 0,$$

where $\mathrm{Hdg}^{2p}(X, \mathbb{Z}) = H^{2p}(X, \mathbb{Z}) \cap H^{p,p}(X)$ is the set of integral Hodge classes of X. In vI.12.3.3, we proved the existence of a 'Deligne cycle class map'

$$\mathrm{cl}_{\mathrm{D}} : \mathrm{CH}_k(X) \to H_{\mathrm{D}}^{2n-2k}(X, \mathbb{Z}(n-k))$$

satisfying $c \circ \mathrm{cl}_{\mathrm{D}} = \mathrm{cl}$ and also $\mathrm{cl}_{\mathrm{D}|\mathrm{CH}_k(X)_{\mathrm{hom}}} = \Phi_X^{n-k}$.

Moreover, there exists a product

$$H_{\mathrm{D}}^k(X, \mathbb{Z}(p)) \otimes H_{\mathrm{D}}^l(X, \mathbb{Z}(q)) \to H_{\mathrm{D}}^{k+l}(X, \mathbb{Z}(p+q)),$$

induced by a product on the Deligne complexes, i.e. a morphism of complexes

$$\mathbb{Z}_{\mathrm{D}}(p) \otimes \mathbb{Z}_{\mathrm{D}}(q) \to \mathbb{Z}_{\mathrm{D}}(p+q)$$

(see Esnault & Viehweg 1988). This product can be computed as follows.

In vI.12.3.2, we represented the Deligne cohomology $H_{\mathrm{D}}^{2p}(X, \mathbb{Z}(p))$ as a quotient of the subgroup $\Xi_{\mathrm{diff}}^{p,p}(X)$ of the group of differential characters

$$\Xi_{\mathrm{diff}}^{2p-1}(X) = \Big\{\phi : Z_{2p-1}^{\mathrm{diff}}(X) \to \mathbb{R}/\mathbb{Z} \Big|$$
$$\exists\, \omega_\phi \in A^{2p}(X) \text{ such that } \phi(\partial\gamma) = \int_\gamma \omega_\phi \bmod \mathbb{Z}, \quad \forall \gamma \in C_{2p}^{\mathrm{diff}}(X)\Big\}.$$

Here, the groups $C_k^{\mathrm{diff}}(X)$ (resp. $Z_k^{\mathrm{diff}}(X)$) are the groups of differentiable (resp. closed differentiable) k-chains of X. The subgroup $\Xi_{\mathrm{diff}}^{p,p}(X)$ is defined by the condition that the form ω is of type (p, p).

Using this representation of the Deligne cohomology, we can compute the product

$$\cdot_{\mathrm{D}} : H_{\mathrm{D}}^{2p}(X, \mathbb{Z}(p)) \otimes H_{\mathrm{D}}^{2q}(X, \mathbb{Z}(q)) \to H_{\mathrm{D}}^{2(p+q)}(X, \mathbb{Z}(p+q)),$$

since it is induced (for $k = 2p,\ l = 2q$) by a product

$$\Xi^{k-1}_{\mathrm{diff}}(X) \times \Xi^{l-1}_{\mathrm{diff}}(X) \to \Xi^{k+l-1}_{\mathrm{diff}}(X),$$
$$(\phi, \psi) \mapsto \phi \cdot \psi$$

on the groups of differential characters. Here, the differential character $\phi \cdot \psi$ is constructed as follows. For $Z \in Z^{\mathrm{diff}}_{k+l-1}(X)$, we must define $\phi \cdot \psi(Z) \in \mathbb{R}/\mathbb{Z}$. Consider the inclusion of the diagonal

$$\mathrm{diag} : X \hookrightarrow X \times X.$$

By the Künneth formula, there exist cycles $Z_i, \in Z^{\mathrm{diff}}_{k_i}(X)$, $Z'_i \in Z^{\mathrm{diff}}_{l_i}(X)$, with $k_i + l_i = k + l - 1$, and a differentiable chain Γ of dimension $k + l$ on $X \times X$ such that

$$\mathrm{diag}(Z) = \sum_i Z_i \times Z'_i + \partial\Gamma.$$

We then set

$$\phi \cdot \psi(Z) = \sum_{i, k_i = k-1} \phi(Z_i) \int_{Z'_i} \omega_\psi + \sum_{i, k_i = k} \psi(Z'_i) \int_{Z_i} \omega_\phi$$
$$+ \int_\Gamma p_1^* \omega_\phi \wedge p_2^* \omega_\psi \in \mathbb{R}/\mathbb{Z}.$$

Note that the form ω_ψ is closed with integral class, so that the integral $\int_{Z'_i} \omega_\psi$ is an integer, and the product $\phi(Z_i) \int_{Z'_i} \omega_\psi$ makes sense in $\mathbb{R}/\mathbb{Z}$. We easily check that $\phi \cdot \psi(Z)$ is well-defined, independently of the choice of the Z_i, Z'_i, and Γ, and that we have

$$\phi \cdot \psi(\partial\Gamma) = \int_\Gamma \omega_\phi \wedge \omega_\psi$$

for a differentiable chain Γ of dimension $k + l$. Thus, $\phi \cdot \psi$ is a differential character, whose associated form is $\omega_\phi \wedge \omega_\psi$.

Having constructed this product, we obtain the following result, which is easy to prove using the definition of the product given above, and the definition given in vI.12.3.3 of the Deligne cycle class map.

Theorem 9.24 *For $Z \in \mathrm{CH}_k(X)$ and $Z' \in \mathrm{CH}_l(X)$, we have*

$$\mathrm{cl}_{\mathrm{D}}(Z \cdot Z') = \mathrm{cl}_{\mathrm{D}}(Z) \cdot_{\mathrm{D}} \mathrm{cl}_{\mathrm{D}}(Z').$$

9.3 Examples

9.3.1 Chow groups of curves

If X is a smooth projective curve, one has $\mathrm{CH}_1(X) \simeq \mathbb{Z}^N$, where N is the number of connected components of X. Next $\mathrm{CH}_0(X) \simeq \mathrm{Pic}X$, and $\mathrm{CH}_0(X)_{\mathrm{hom}} \simeq \mathrm{Pic}^0 X$. By the results of vI.12.1.3, it follows that

$$\mathrm{alb}_X : \mathrm{CH}_0(X)_{\mathrm{hom}} \to \mathrm{Alb}X \simeq J^1(X)$$

is an isomorphism.

9.3.2 Chow groups of projective bundles

Let $E \to X$ be a vector bundle of rank r, and let $\pi : \mathbb{P}(E) \to X$ be the corresponding projective bundle

$$\mathbb{P}(E) = \mathrm{Proj} \oplus_k \mathrm{Sym}^k \mathcal{E}^*,$$

where $\mathcal{E}$ is the sheaf of sections of E. $\mathbb{P}(E)$ is equipped with the invertible bundle $\mathcal{O}_{\mathbb{P}(E)}(1)$, which is the dual of the tautological subbundle $\mathcal{S} \subset \pi^*\mathcal{E}$; we write $h \in \mathrm{Pic}(\mathbb{P}(E))$ for its class. By corollary 9.10, the Picard group acts on the Chow groups, so we have

$$h : \mathrm{CH}_l(\mathbb{P}(E)) \to \mathrm{CH}_{l-1}(\mathbb{P}(E)).$$

Theorem 9.25 *We have an isomorphism*

$$I : \bigoplus\nolimits_{0\le k\le r-1} \mathrm{CH}_{l-r-1+k}(X) \cong \mathrm{CH}_l(\mathbb{P}(E)),$$
$$(Z_0, \ldots, Z_{r-1}) \mapsto \sum_k h^k \pi^* Z_k.$$

Proof Note first that the composition

$$\pi_* \circ h^i \circ \pi^* : \mathrm{CH}_l(X) \to \mathrm{CH}_{l+r-1-i}(X) \tag{9.10}$$

is equal to the identity for $i = r-1$, and is zero for $i < r-1$.

Indeed, if $j : Z \hookrightarrow X$ is the inclusion of a subvariety, and $\pi_Z : \mathbb{P}(E/Z) \to Z$ is the restricted projective bundle, then we first have the equality

$$j_* \circ \pi_{Z*} \circ h^i \circ \pi_Z^* = \pi_* \circ h^i \circ \pi^* \circ j_* : \mathrm{CH}_l(Z) \to \mathrm{CH}_{l+r-1-i}(X),$$

which follows from proposition 9.8. As $\mathrm{CH}_l(X)$ is generated by the $j_*c(Z)$, where $j : Z \hookrightarrow X$ is the inclusion of an irreducible subvariety of dimension l, it suffices to prove (9.10) for $l = n = \dim X$, with X reduced and irreducible. The second equality is then obvious for reasons of dimension. As for the first, by the localisation exact sequence, we can restrict ourselves to proving it after replacing X by a Zariski open set U on which E is trivial. It then follows immediately from the fact that $h^{r-1} \in \mathrm{CH}^{r-1}(U \times \mathbb{P}^{r-1})$ is the class of the subvariety $U \times p$, where p is a point of $\mathbb{P}^{r-1}$.

This immediately implies the injectivity. Indeed, if $I(Z_0, \ldots, Z_{r-1}) = 0$, then

$$\pi_*(I(Z_0, \ldots, Z_{r-1})) = 0 = Z_{r-1}.$$

More generally, if $I(Z_0, \ldots, Z_k, 0, \ldots, 0) = 0$ for $k \leq r-1$, then

$$\pi_*(h^{r-1-k} \cdot I(Z_0, \ldots, Z_k, 0, \ldots, 0)) = 0 = Z_k.$$

Introducing the smallest integer k such that $Z_k \neq 0$, we conclude that $I(Z_0, \ldots, Z_{r-1}) = 0$ implies $(Z_0, \ldots, Z_{r-1}) = 0$.

It remains to prove the surjectivity. As in the proof of theorem 9.13, using a local trivialisation of E, localisation, and induction on the dimension of X, we reduce to the case where E is the trivial bundle, i.e. $\mathbb{P}(E) = X \times \mathbb{P}^{r-1}$. We then use induction on r. Let H be a hyperplane of $\mathbb{P}^{r-1}$, and consider the inclusion

$$j : X \times H \subset X \times \mathbb{P}^{r-1}.$$

Lemma 9.12 gives the exact sequence

$$\mathrm{CH}_l(X \times H) \to \mathrm{CH}_l(X \times \mathbb{P}^{r-1}) \to \mathrm{CH}_l\left(X \times A_K^{r-1}\right) \to 0,$$

and proposition 9.11 shows that

$$\pi^* : \mathrm{CH}_{l-r-1}(X) \to \mathrm{CH}_l\left(X \times A_K^{r-1}\right)$$

is surjective. It follows that

$$\pi^* + j_* : \mathrm{CH}_{l-r-1}(X) \oplus \mathrm{CH}_l(X \times H) \to \mathrm{CH}_l(X \times \mathbb{P}^{r-1})$$

is surjective.

Now, writing $\pi_H = \pi_{|X \times H}$, we clearly have $j_* \circ \pi_H^* = h \circ \pi^*$. Writing $h_H = h_{|X \times H}$, by the induction hypothesis, we have

$$\mathrm{CH}_l(X \times H) = \bigoplus\nolimits_{0 \leq k \leq r-2} h_H^k \pi_H^* \mathrm{CH}_{l-r+k+2}(X),$$

so

$$j_* \mathrm{CH}_l(X \times H) = \sum\nolimits_{0 \leq k \leq r-2} h^{k+1} \pi^* \mathrm{CH}_{l-r+k+2}(X).$$

Thus, we have

$$\mathrm{CH}_l(X \times \mathbb{P}^{r-1}) = \sum\nolimits_{1 \leq k \leq r-1} h^k \pi^* \mathrm{CH}_{l-r+k+1}(X) + \pi^* \mathrm{CH}_{l-r+1}(X),$$

and I is surjective. □

As a corollary of the injectivity of the map $\pi^* : \mathrm{CH}_k(X) \to \mathrm{CH}_{k+r-1}(\mathbb{P}(E))$, we obtain the following result.

Corollary 9.26 *Given a vector bundle $E \to X$ of rank r, there exists a variety Y and a flat morphism $\pi : Y \to X$ such that the bundle $\pi^* E$ admits a filtration $F^i E$, $0 \le i \le r$, with $F^i E / F^{i+1} E$ free of rank 1 for $i \le r-1$, and such that the maps*

$$\pi^* : \mathrm{CH}(X) \to \mathrm{CH}(Y), \quad \pi_E^* : \mathrm{CH}(E) \to \mathrm{CH}(E_Y)$$

are injective, where $\pi_E : \pi^* E \to E$ is the fibred product of E and Y over X.

Proof We know that for a projective bundle $\pi : Y \to X$, the map π^* is injective on Chow groups. If $E \to X$ is a vector bundle, consider $\pi : Y = \mathbb{P}(E) \to X$. Then the map $\pi_E : \pi^* E \to E$ is also a projective bundle, so the maps π_E^* are also injective on Chow groups. Moreover, we have the tautological subbundle $\mathcal{O}_{\mathbb{P}(E)}(-1) \subset \pi^* E$, and the quotient Q is of rank $r-1$. We can thus use induction on the rank of E to conclude the proof. □

9.3.3 Chow groups of blowups

Let X be a smooth variety, and $Y \subset X$ a smooth subvariety of codimension r. Let $\tau : \widetilde{X}_Y \to X$ be the blowup of X along Y, and

$$D = \tau^{-1}(Y) \xrightarrow{\tilde{j}} \widetilde{X}_Y$$

the exceptional divisor of the blowup. The map $\tau_D : D \to Y$ makes D into a projective bundle of rank $r-1$, given precisely by $D = \mathbb{P}(N_{Y/X})$, and we have

$$\mathcal{O}_{\widetilde{X}_Y}(D)_{|D} = \mathcal{O}_{\mathbb{P}(N_{Y/X})}(-1)$$

(see vI.3.3.3). Let us write h for the class of $\mathcal{O}_{\mathbb{P}(N_{Y/X})}(1)$ in Pic D; then $h = -D_{|D}$. The Chow groups of $\widetilde{X}_Y$ are determined by the following formula (see Manin 1968).

Theorem 9.27 *For every l, the map*

$$\tilde{I} : \bigoplus\nolimits_{0 \le k \le r-2} \mathrm{CH}_{l-r+1+k}(Y) \oplus \mathrm{CH}_l(X) \to \mathrm{CH}_l(\widetilde{X}_Y),$$
$$(Z_0, \ldots, Z_{r-2}, Z) \mapsto \sum_k \tilde{j}_* h^k \tau_D^* Z_k + \tau^* Z$$

is an isomorphism.

Proof Let $U = \widetilde{X}_Y - D = X - Y$. It follows immediately from the localisation exact sequences

$$\mathrm{CH}(Y) \to \mathrm{CH}(X) \to \mathrm{CH}(U) \to 0,$$
$$\mathrm{CH}(D) \to \mathrm{CH}(\widetilde{X}_Y) \to \mathrm{CH}(U) \to 0$$

that

$$\mathrm{CH}_l(\widetilde{X}_Y) = \tau^*\mathrm{CH}_l(X) + \tilde{j}_*\mathrm{CH}_l(D).$$

Now, by theorem 9.25, we have

$$\mathrm{CH}_l(D) = \bigoplus\nolimits_{0\leq k\leq r-1} h^k\tau_D^*\mathrm{CH}_{l-r+1+k}(Y).$$

Thus, we see that

$$\mathrm{CH}_l(\widetilde{X}_Y) = \tau^*\mathrm{CH}_l(X) + \sum_{k\leq r-1} \tilde{j}_*h^k\tau_D^*\mathrm{CH}_{l-r+1+k}(Y). \tag{9.11}$$

But clearly, if $Z \in \mathrm{CH}_l(Y)$, then

$$\tau_*(\tilde{j}_*h^{r-1}\tau_D^*Z) = j_*Z \in \mathrm{CH}_l(X),$$

where $j : Y \to X$ is inclusion. Also, since τ is birational, it is of degree 1, so $\tau_* \circ \tau^* = \mathrm{Id}$ by corollary 9.15, and we have

$$\tau_*(\tau^* j_*Z) = j_*Z, \quad Z \in \mathrm{CH}_l(Y).$$

Thus we obtain the relations

$$\tilde{j}_*h^{r-1}\tau_D^*Z - \tau^* j_*Z \in \mathrm{Ker}\,\tau_*, \quad Z \in \mathrm{CH}_l(Y).$$

Using the excess formulae (see Fulton 1984, 6.7), we can show that more precisely, we actually have

$$\tilde{j}_*h^{r-1}\tau_D^*Z - \tau^* j_*Z \in \sum_{k\leq r-2} \tilde{j}_*h^k\tau_D^*\mathrm{CH}_{l-r+1+k}(Y). \tag{9.12}$$

The equality (9.11) and the relations (9.12) together imply immediately that the map $\tilde{I}$ is surjective.

To prove injectivity, consider $(Z_0, \ldots, Z_{r-2}, Z) \in \mathrm{Ker}\,\tilde{I}$. Applying τ_*, we immediately find that $Z = 0$. Moreover, note that by the definition of the morphism $\tilde{j}^*$, we have

$$\tilde{j}^* \circ \tilde{j}_* = [D] = -h : \mathrm{CH}_k(D) \to \mathrm{CH}_{k-1}(D).$$

Applying $\tilde{j}^*$ to the equality

$$\sum_{k\leq r-2} \tilde{j}_*h^k\tau_D^*Z_k = 0,$$

we thus obtain

$$\sum_{k\leq r-2} h^{k+1}\tau_D^* Z_k = 0 \quad \text{in} \quad \mathrm{CH}_{l-1}(D)$$

by the fact that $D_{|D} = -h \in \operatorname{Pic} D$. But then the uniqueness in theorem 9.25 shows that $Z_k = 0$ for all k.

9.3.4 Chow groups of hypersurfaces of small degree

We will use theorem 9.25 to prove the following result.

Theorem 9.28 (Esnault *et al.* 1997) *Let X be a smooth hypersurface of $\mathbb{P}^n$, covered by projective spaces $\mathbb{P}^r$. Then the groups $\mathrm{CH}_l(X)_{\mathrm{hom}}$ are torsion for $l \leq r-1$.*

Note that if $d \geq 2$, the hypothesis implies that $r \leq (n-1)/2$, so that for $l < r$, the Lefschetz theorem yields $H_{2l}(X,\mathbb{Z}) = H_{2l}(\mathbb{P}^n,\mathbb{Z}) = \mathbb{Z}$. The theorem thus says that $\mathrm{CH}_l(X)_{\mathbb{Q}} := \mathrm{CH}_l(X)\otimes\mathbb{Q} = \mathbb{Q}$ for $l < r$.

The proof first uses the following observation.

Lemma 9.29 *Let $h = c_1(\mathcal{O}_X(1))$. Then the map*

$$dh : \mathrm{CH}_l(X)_{\mathrm{hom}} \to \mathrm{CH}_{l-1}(X)_{\mathrm{hom}}$$

is equal to 0.

Proof By the definition of the map j^*, when j is the inclusion of a Cartier divisor, and by the fact that the class of Y in $\operatorname{Pic}\mathbb{P}^n$ is equal to dh, we find that

$$j^*\circ j_* = dh : \mathrm{CH}_l(X)_{\mathrm{hom}} \to \mathrm{CH}_{l-1}(X)_{\mathrm{hom}}.$$

But as $\mathrm{CH}_*(\mathbb{P}^n)_{\mathrm{hom}} = 0$, the map

$$j_* : \mathrm{CH}_l(X)_{\mathrm{hom}} \to \mathrm{CH}_l(\mathbb{P}^n)_{\mathrm{hom}}$$

is zero. □

Proof of theorem 9.28 Let F be the variety parametrising the r-planes contained in X. Up to replacing F by a desingularisation, we may assume that F is smooth.

Let $Q = \{(x, P) \in X \times F \mid x \in P\}$ be the incidence variety, and p, q the projections

$$\begin{array}{ccc} Q & \xrightarrow{q} & X \\ {\scriptstyle p}\downarrow & & \\ F. & & \end{array}$$

By hypothesis, q is surjective, and up to replacing F by a subvariety, we may assume that q is generically finite of degree $N > 0$. Then we know that the image of the map

$$q_* : \mathrm{CH}_l(Q)_{\mathrm{hom}} \to \mathrm{CH}_l(X)_{\mathrm{hom}}$$

contains $N\mathrm{CH}_l(X)_{\mathrm{hom}}$, since $q_* \circ q^* = N \cdot \mathrm{Id}$ on $\mathrm{CH}_l(X)_{\mathrm{hom}}$.

Moreover, the projection p makes Q into a projective bundle of rank r on F, and $q^*(\mathcal{O}_X(1))$ is a divisor $\mathcal{O}_Q(1)$ on the projective bundle $Q \to F$. Writing $H = q^*c_1(\mathcal{O}_X(1))$, we can thus apply theorem 9.25, which yields

$$\mathrm{CH}_l(Q) = \bigoplus\nolimits_{0 \leq k \leq r} H^k p^* \mathrm{CH}_{l-r+k}(F).$$

By the fact that we also have the corresponding decomposition on the cohomology

$$H^m(Q, \mathbb{Z}) = \bigoplus\nolimits_{0 \leq k \leq r} [H]^k p^* H^{m-2k}(F, \mathbb{Z}), \quad [H] \in H^2(Q, \mathbb{Z})$$

for every m (see vI.7.3.3), we immediately deduce the decomposition

$$\mathrm{CH}_l(Q)_{\mathrm{hom}} = \bigoplus\nolimits_{0 \leq k \leq r} H^k p^* \mathrm{CH}_{l-r+k}(F)_{\mathrm{hom}}.$$

In particular, for $l < r$, the map

$$H : \mathrm{CH}_{l+1}(Q)_{\mathrm{hom}} \to \mathrm{CH}_l(Q)_{\mathrm{hom}}$$

is surjective. Thus, for $Z \in \mathrm{CH}_l(X)_{\mathrm{hom}}$, there exists $Z' \in \mathrm{CH}_l(Q)_{\mathrm{hom}}$, $Z'' \in \mathrm{CH}_{l+1}(Q)_{\mathrm{hom}}$ such that

$$NZ = q_*Z' \in \mathrm{CH}_l(X), \quad Z' = HZ'' \in \mathrm{CH}_l(Q).$$

Now, for $Z'' \in \mathrm{CH}_{l+1}(Q)$, we have the projection formula

$$q_*(HZ'') = q_*(q^*hZ'') = hq_*Z'' \text{ in } \mathrm{CH}_l(X).$$

Thus, we obtain finally $dNZ = dhq_*Z'' = 0$ by lemma 9.29. □

It is easy to see that the hypothesis is satisfied whenever n is sufficiently large with respect to d. Indeed, the variety $F_r(X)$ of r-dimensional projective spaces contained in X is the locus of zeros of the section σ_X of the holomorphic vector bundle E_d on Grass(r, n), with fibre $E_d = H^0(\mathcal{O}_P(d))$ at the point P, given by the evaluation $\sigma_X(P) = F_{|P}$.

For sufficiently large n, it is then clear that the dimension of the incidence variety parametrised by W,

$$Q \subset W \times X, \quad Q = \{(P, x) \mid x \in P,\ P \subset X\},$$

is greater than or equal to $n - 1$. Moreover, we can prove the surjectivity of the second projection $Q \to X$ by computing the intersection numbers on the Grassmannian.

The following result, which refines a theorem due to Paranjape (1994), can thus be deduced from theorem 9.28.

Corollary 9.30 *For sufficiently large n, with d and r fixed, $\mathrm{CH}_l(X)_{\mathrm{hom}}$ is a torsion group for $l \leq r - 1$. More precisely, for $d > 2$, this is satisfied for n such that*

$$(r + 1)(n - r) - C_d^{r+d} + r \geq n - 1,$$

i.e. $\dim Q = \dim F_r(X) + r \geq \dim X$.

Exercises

1. *Chow groups of conic bundles.* Let X be a smooth projective $(n + 1)$-dimensional variety, and let $f : X \to \mathbb{P}^n$ be a morphism satisfying the following properties:
   - There exists a smooth hypersurface $Y \subset \mathbb{P}^n$ such that for $x \in U := \mathbb{P}^n - Y$, the fibre $f^{-1}(x)$ is a smooth curve isomorphic to $\mathbb{P}^1$.
   - For $x \in Y$, the fibre $f^{-1}(x)$ is the union of two smooth curves isomorphic to $\mathbb{P}^1$, which intersect transversally in a point.

   (Such a variety is called a conic bundle over $\mathbb{P}^n$ (see Beauville 1977)).

   (a) Show that $\mathrm{CH}_0(X) = \mathbb{Z}$.

   (b) Let $X_U := f^{-1}(U)$. Show that there exists a smooth variety which is a double cover $r : S \to U$, such that $X_S := X \times_U S$ is a projective bundle on S. (Consider the hypersurfaces in $|K_{X/\mathbb{P}^n}^{-1} \otimes f^*\mathcal{O}(k)|$, for large k.)

   (c) Deduce from theorem 9.25 applied to X_S, and from the facts that $\mathrm{CH}_0(U) = 0$ and $\mathrm{CH}_1(U)$ is torsion, that $\mathrm{CH}^1(X_U)$ is torsion.

   Assume from now on that the variety $\widetilde{Y}$ which parametrises a point of Y and a component of the fibre $f^{-1}(y)$ is irreducible. ($\widetilde{Y}$ is smooth, as its schematic structure is given by the Hilbert scheme of the components of the fibres.)

(d) Show that the group $H_2(X, \mathbb{Z})$ is of rank 2, and can be inserted into an exact sequence

$$0 \to [l]\mathbb{Z} \to H_2(X, \mathbb{Z}) \stackrel{f_*}{\to} H_2(\mathbb{P}^n, \mathbb{Z}) \to T \to 0,$$

where T is torsion and $[l]$ is the class of a component of a fibre.

(e) Deduce from the localisation exact sequence together with theorem 9.25 applied to the projective fibration $\widetilde{X}_Y \to \widetilde{Y}$, where $\widetilde{X}_Y$ is the normalisation of $X_Y := f^{-1}(Y)$, that $\mathrm{CH}_1(X)_{\mathrm{hom}}$ is generated up to torsion by the cycles $l_c - l'_c$, where $c \in Y$ and l_c, l'_c are the two components of $f^{-1}(c)$.

2. *Decomposition of the diagonal of a surface.* Let S be a smooth projective complex surface. In this exercise, we propose to prove the following theorem.

Theorem 9.31 (Murre 1990) *There exist cycles $\Delta_i \in \mathrm{CH}^2(S \times S)_{\mathbb{Q}}$, $i = 0, \ldots, 4$, satisfying*

- *The class $[\Delta_i] \in H^4(S \times S)$ is the ith Künneth component of the class $[\Delta]$ of the diagonal. (Equivalently, $[\Delta_i]^* : H^*(S, \mathbb{Q}) \to H^*(S, \mathbb{Q})$ is the identity on $H^i(S, \mathbb{Q})$ and 0 on $H^j(S, \mathbb{Q})$, $j \neq i$; see vI.11.3.3.)*
- $\sum_i \Delta_i = \Delta \in \mathrm{CH}^2(S \times S)_{\mathbb{Q}}$.
- $\Delta_i \circ \Delta_i = \Delta_i$ *and* $\Delta_i \circ \Delta_j = 0$ *for* $i \neq j$.

(a) Let $x_0 \in S$. Show that $\Delta_0 = S \times x_0$ and $\Delta_4 = x_0 \times S$ satisfy:
- The classes $[\Delta_0]$ and $[\Delta_4]$ are respectively the 0th and the 4th Künneth components of Δ.
- $\Delta_i \circ \Delta_i = \Delta_i$, $i = 0, 4$, and $\Delta_0 \circ \Delta_4 = \Delta_4 \circ \Delta_0 = 0$.

(b) Let $C \stackrel{j}{\to} S$ denote the inclusion of a smooth ample curve. Using the Lefschetz theorem 1.23 and the fact that the Hodge structures considered here are polarised, show that there exists a decomposition as a direct sum of Hodge structures

$$H^1(C, \mathbb{Q}) = L \oplus H^1(S, \mathbb{Q}), \tag{9.13}$$

where $H^1(S, \mathbb{Q})$ can be identified with $\mathrm{Im}\,(j^* : H^1(S, \mathbb{Q}) \to H^1(C, \mathbb{Q}))$.

(c) Using the Lefschetz theorem on (1, 1)-classes, show that there exists a cycle $D \in \mathrm{CH}^1(S \times C)_{\mathbb{Q}}$ such that the class of D in $H^2(S \times C, \mathbb{Q})$ can be identified, via the Künneth decomposition

$$H^2(S \times C, \mathbb{Q}) \cong \mathrm{Hom}\,(H^*(C, \mathbb{Q}), H^*(S, \mathbb{Q})),$$

with the projection $H^1(C, \mathbb{Q}) \to H^1(S, \mathbb{Q})$ given by the decomposition (9.13).

(d) Let $\Delta_1 = (1, j)_* D \in \mathrm{CH}^2(S \times S)$, and let $\Delta_3 = {}^t\Delta_1$ (i.e. the image of Δ_1 by the map exchanging the factors of $S \times S$). Show that $[\Delta_1]$ and $[\Delta_3]$ are respectively the first and third Künneth components of $[\Delta]$.

(f) Show that the cycles Δ_i, $0 \leq i \leq 4$, $i \neq 2$ satisfy $\Delta_i \circ \Delta_i = \Delta_i$ and $\Delta_i \circ \Delta_j = 0$, $i \neq j$.

(e) Deduce that if $\Delta_2 = \Delta - \Delta_0 - \Delta_1 - \Delta_3 - \Delta_4$, then the Δ_i satisfy the conditions of the theorem.

10
Mumford's Theorem and its Generalisations

The general theme of this chapter is the interaction between the complexity or the size of the Chow groups and the complexity or the level of the Hodge structures of a smooth complex projective variety X. Using the cycle class map and the Abel–Jacobi map, we can relate the Chow groups of codimension k with the Hodge classes of degree $2k$, and also with the intermediate Jacobian $J^{2k-1}(X)$. The first group is a discrete group, and the algebraic part of the intermediate Jacobian is an abelian variety. By theorem 8.15, the image of the Abel–Jacobi map modulo this algebraic part is a countable group. Thus, the cycle class maps only describe a quotient of the Chow group, isomorphic to a countable union of translates of an abelian variety.

Mumford's theorem implies that in general, the kernel of the Abel–Jacobi map in the case of the 0-cycles on a surface does not have the size of an algebraic variety. It also relates this phenomenon with the non-triviality of the transcendental part of the Hodge structure on $H^2(S)$, whereas the cycle class map has values in $H^4(S, \mathbb{Z})$, and the Abel–Jacobi map has values in $J^3(S) = \mathrm{Alb}\, S$.

Theorem 10.1 *(Mumford)* *Let S be a smooth projective surface such that $H^{2,0}(S) \neq 0$. Then the group $\mathrm{CH}_0(S)$ is infinite-dimensional.*

This notion of infinite-dimensionality works as follows. The symmetric product $S^{(k)}$ can be viewed as parametrising the effective 0-cycles of degree k of S. Thus, assuming that S is connected, we have a natural map

$$\begin{aligned}\sigma_k : S^{(k)} \times S^{(k)} &\to \mathrm{CH}_0(S)_{\mathrm{hom}},\\ (Z_1, Z_2) &\mapsto Z_1 - Z_2 \text{ mod rational equivalence.}\end{aligned}$$

One can show that the fibres of this map are countable unions of algebraic closed subsets. Thus, we can assign a dimension d_k to the general fibre, and set $\dim \operatorname{Im} \sigma = \dim S^{(k)} \times S^{(k)} - d_k = 4k - d_k$. The Chow groups are then infinite-dimensional in the sense that $\lim_{k\to\infty} \dim \operatorname{Im} \sigma_k = \infty$.

We also have the following theorem, due to Roitman.

Theorem 10.2 *Let X be a smooth complex projective variety whose group of 0-cycles $\mathrm{CH}_0(X)$ is finite-dimensional. Then the map*

$$\mathrm{alb}_X : \mathrm{CH}_0(X)_{\mathrm{hom}} \to \mathrm{Alb}(X) = J^{2n-1}(X), \quad n = \dim X$$

is an isomorphism.

This theorem is one of the rare examples of the determination of a Chow group by Hodge theory, and it is only valid under the very restrictive hypothesis of finite-dimensionality. Conjecturally however, as we will see in the following chapter, this type of statement should be generalisable, and yield a complete determination of the Chow groups with rational coefficients of a variety by its Hodge structures.

Mumford's theorem admits the following version.

Theorem 10.3 *Let S be a smooth projective surface such that there exists a curve $j : C \hookrightarrow S$ satisfying the condition that $j_* : \mathrm{CH}_0(C) \to \mathrm{CH}_0(S)$ is surjective. Then $H^{2,0}(S) = 0$.*

In this form, it admits the following generalisation, which can also be attributed to Roitman.

Theorem 10.4 *Let X be a smooth projective variety such that there exists a k-dimensional subvariety $j : W \hookrightarrow X$ satisfying the condition that $j_* : \mathrm{CH}_0(W) \to \mathrm{CH}_0(X)$ is surjective. Then $H^{l,0}(X) = 0$ for all $l > k$.*

The proof that we give here follows the elegant model proposed by Bloch & Srinivas, which, under the hypotheses of theorem 10.4, establishes the existence of a decomposition of the diagonal

$$m\Delta = Z_1 + Z_2 \text{ in } \mathrm{CH}^n(X \times X), \quad n = \dim X$$

with $Z_1 \subset X \times W$, and $Z_2 \subset T \times X$, where $T \subset X$ is a proper algebraic subset.

This decomposition was itself generalised by Paranjape, under hypotheses concerning the Chow groups of small dimension. Using this, one can also obtain the following result, due to Lewis and Schoen independently.

Theorem 10.5 *Let X be a smooth projective n-dimensional variety such that*

$$\mathrm{cl} : \mathrm{CH}_l(X)_{\mathbb{Q}} \to H^{2n-2l}(X, \mathbb{Q})$$

is injective for $l \leq k$. Then $H^{p,q}(X) = 0$ for $p \neq q$, $q \leq k$.

The first section of this chapter is devoted to general results on the notion of finite dimensionality of Chow groups, and to the proof of Roitman's theorem 10.2. In the second section, we establish Bloch–Srinivas decomposition of the diagonal, and deduce Mumford's theorem in its most precise form. We also give some interesting consequences of this decomposition principle, following Bloch & Srinivas (1983).

We end the chapter with the generalised decomposition of the diagonal, and the proof of theorem 10.5.

10.1 Varieties with representable CH_0

10.1.1 Representability

Consider the group $\mathrm{CH}_0(X)$ of 0-cycles of a complex projective variety. Assume that X is connected, so that $H_0(X, \mathbb{Z}) = \mathbb{Z}$, and that the map

$$\mathrm{cl} : \mathrm{CH}_0(X) \to H_0(X, \mathbb{Z})$$

can be identified with the degree map which to $Z = \sum_i n_i p_i$, $p_i \in X$ associates $\deg Z = \sum_i n_i$. A 0-cycle Z can be written as the difference of two effective 0-cycles:

$$Z = Z_1 - Z_2, \quad Z_1 = \sum_{i, n_i > 0} n_i p_i, \quad Z_2 = \sum_{i, n_i < 0} -n_i p_i.$$

Let $d^+ = \sum_{i, n_i > 0} n_i$ and $d^- = \sum_{i, n_i < 0} -n_i$. The 0-cycles Z_1 and Z_2 can be viewed as elements of the symmetric products $X^{(d^+)}$ (resp. $X^{(d^-)}$) which parametrise the unordered sets of d^+ (resp. d^-) points of X. These symmetric products are quotients of the self-products X^{d^+}, X^{d^-} of X by the action of the symmetric groups S_{d^+} (resp. S_{d^-}). Thus, they are singular varieties of dimension nd^+ (resp. nd^-), where $n = \dim X$.

In what follows, we will write $c : X^{(d)} \to \mathrm{CH}_0(X)$ for the map

$$Z \mapsto \mathrm{class}(Z) \text{ mod rational equivalence.}$$

The discussion above shows that for each positive integer d, we have a map

$$\sigma_d : X^{(d)} \times X^{(d)} \to \mathrm{CH}_0(X)_{\mathrm{hom}},$$
$$(Z_1, Z_2) \mapsto c(Z_1) - c(Z_2).$$

Definition 10.6 *We say that* $\mathrm{CH}_0(X)$ *is representable if the map* σ_d *is surjective for sufficiently large* d.

We will show that if $\mathrm{CH}_0(X)$ is not representable, then $\mathrm{CH}_0(X)$ is infinite-dimensional, in a sense to be made precise below.

Lemma 10.7 *The fibres of* σ_d *are countable unions of closed algebraic subsets of* $X^{(d)} \times X^{(d)}$.

Proof Let us take a cycle $Z = Z^+ - Z^-$ of degree 0, with class

$$z = c(Z) \in \mathrm{CH}_0(X)_{\mathrm{hom}},$$

and study the fibre $\sigma_d^{-1}(z)$. It consists of the pairs (Z_1, Z_2) such that $Z_1 - Z_2 - Z$ is rationally equivalent to 0, which means that there exist curves $C_i \subset X$ with normalisation

$$\tau_i : \widetilde{C}_i \to X,$$

and rational functions ϕ_i on C_i, such that in $\mathcal{Z}_0(X)$, we have

$$\sum_i (\tau_i)_* \operatorname{div}(\phi_i) = Z_1 - Z_2 - Z.$$

Then, decomposing into positive and negative parts, we obtain the existence of effective 0-cycles A, B such that

$$\begin{aligned} Z_1 + Z^- + A &= \sum_i (\tau_i)_*\big(\phi_i^{-1}(0)\big) + B, \\ Z_2 + Z^+ + A &= \sum_i (\tau_i)_*\big(\phi_i^{-1}(\infty)\big) + B. \end{aligned} \tag{10.1}$$

Now, the curves C_i are parametrised by a countable union of Hilbert schemes of X, and if we fix these Hilbert schemes and the degrees d_i of the divisors $\phi_i^{-1}(0)$, each function ϕ_i can be viewed as a pencil of degree d_i on C_i. Moreover, if we fix the degrees a and b of A and B, the 0-cycles A and B are parametrised by the symmetric products $X^{(a)}$ and $X^{(b)}$ of X. Thus, on the condition of fixing certain discrete data, the objects $(C_i, \tau_i, \phi_i, A, B)$ are parametrised by a projective algebraic variety K. Moreover, the equations (10.1) define a closed algebraic subset W_K in $X^{(d)} \times X^{(d)} \times K$, and the fibre $\sigma_d^{-1}(z)$ is the union, over the countable set of varieties K, of the projections of W_K to $X^{(d)} \times X^{(d)}$. □

We can thus define the dimension of a fibre of σ_d to be the largest dimension occurring among the dimensions of its algebraic components. Finally, by examining the proof of lemma 10.7, one sees easily that there exists a countable union B of proper algebraic subsets of $X^{(d)} \times X^{(d)}$ such that for $x \in X^{(d)} \times X^{(d)} - B$, the dimension of the fibre $\sigma_d^{-1}(\sigma_d(x))$ is constant and equal to r.

Definition 10.8 *The dimension of* $\mathrm{Im}\,\sigma_d$ *is defined to be equal to* $2nd - r$, *where* r *is defined as above, and* $n = \dim X$ *so* $2nd = \dim X^{(d)} \times X^{(d)}$.

It is easily seen that this dimension increases with d.

Definition 10.9 *We say that* $\mathrm{CH}_0(X)_{\mathrm{hom}}$ *is infinite-dimensional if*

$$\lim_{d\to\infty} \dim \mathrm{Im}\,\sigma_d = \infty,$$

and finite-dimensional otherwise.

Proposition 10.10 *The group* $\mathrm{CH}_0(X)$ *is representable if and only if it is finite-dimensional.*

Proof If $\mathrm{CH}_0(X)$ is representable, there exists an integer D such that σ_D is surjective. For every integer d, consider the set

$$R \subset X^{(d)} \times X^{(d)} \times X^{(D)} \times X^{(D)}$$

consisting of the quadruples (Z_1, Z_2, Z'_1, Z'_2) such that $\sigma_d(Z_1, Z_2) = \sigma_D(Z'_1, Z'_2)$. As σ_D is surjective, the projection

$$p : R \to X^{(d)} \times X^{(d)}$$

is surjective. Now, the proof of lemma 10.7 also shows that R is the countable union of closed algebraic subsets. Thus, by Baire, there exists an algebraic component R_0 of R such that the interior of the image of

$$p : R_0 \to X^{(d)} \times X^{(d)}$$

is non-empty, so that p is surjective. In particular, $\dim R_0 \geq 2nd$. But then, if

$$(Z_1, Z_2, Z'_1, Z'_2) \in R_0 \subset X^{(d)} \times X^{(d)} \times X^{(D)} \times X^{(D)},$$

we have

$$\dim_{(Z_1,Z_2)} R_0 \cap \left(X^{(d)} \times X^{(d)} \times (Z'_1, Z'_2)\right) \geq 2nd - 2nD.$$

Now, this is an algebraic set contained in the fibre $\sigma_d^{-1}(\sigma_D(Z'_1, Z'_2))$. As (Z_1, Z_2) is arbitrary, since the projection

$$p : R_0 \to X^{(d)} \times X^{(d)}$$

is surjective, we conclude that the fibres of σ_d are all of dimension $\geq 2nd - 2nD$, so that we have $\dim \operatorname{Im} \sigma_d \leq 2nD$, and $\mathrm{CH}_0(X)$ is finite-dimensional.

Conversely, assume that $\mathrm{CH}_0(X)$ is finite-dimensional. Let D be such that $\dim \operatorname{Im} \sigma_D = \dim \operatorname{Im} \sigma_{D+1}$. We will show that this implies that $\operatorname{Im} \sigma_D = \operatorname{Im} \sigma_{D+1}$. Indeed, if x is a point of X, we have an inclusion

$$i_x : X^{(D)} \times X^{(D)} \to X^{(D+1)} \times X^{(D+1)},$$
$$(Z_1, Z_2) \mapsto (Z_1 + x, Z_2 + x)$$

such that $\sigma_{D+1} \circ i_x = \sigma_D$. Let F be the fibre of σ_D passing through a general point (Z_1, Z_2) of $X^{(D)} \times X^{(D)}$, and let F' be the fibre of σ_{D+1} passing through a general point of $X^{(D+1)} \times X^{(D+1)}$. We know by assumption that $\dim F' = \dim F + 2n$, where the dimension is defined as the largest dimension of an algebraic component of F passing through the point under consideration. Let F'' be the fibre of σ_{D+1} passing through the point $(Z_1 + x, Z_2 + x)$ of $X^{(D+1)} \times X^{(D+1)}$. By semicontinuity of the dimensions of the fibres of σ_{D+1}, we have $\dim F'' \geq \dim F'$. Let us introduce as before the subset

$$R \subset X^{(D+1)} \times X^{(D+1)} \times X^{(D)} \times X^{(D)},$$
$$R = \{(Z_1, Z_2, Z'_1, Z'_2) \mid \sigma_{D+1}(Z_1, Z_2) = \sigma_D(Z'_1, Z'_2)\}.$$

It is now clear that R has an algebraic component R_0 passing through $(Z_1 + x, Z_2 + x, Z_1, Z_2)$, which dominates $X^{(D)} \times X^{(D)}$, and is such that the fibre of the second projection $q : R_0 :\to X^{(D)} \times X^{(D)}$ is of dimension equal to $\dim F''$, and so of dimension greater than or equal to $\dim F + 2nD$, whereas the fibres of the first projection $p : R_0 \to X^{(D+1)} \times X^{(D+1)}$ are of generic dimension at most equal to $\dim F$. Thus, we have

$$\dim p(R_0) \geq \dim R_0 - \dim F \geq \dim F'' + \dim X^{(D)} \times X^{(D)} - \dim F$$
$$\geq \dim F' - \dim F + 2nD = 2n(D+1) = \dim X^{(D+1)} \times X^{(D+1)}.$$

Therefore p is surjective and this clearly implies that $\operatorname{Im} \sigma_D = \operatorname{Im} \sigma_{D+1}$. It follows immediately that

$$\operatorname{Im}\sigma_k = \operatorname{Im}\sigma_D \quad \text{for all } k \geq D,$$

which shows that σ_D is surjective, since

$$\mathrm{CH}_0(X)_{\mathrm{hom}} = \bigcup_k \operatorname{Im}\sigma_k.$$

□

10.1.2 Roitman's theorem

The term 'representability' can be justified by the following theorem of Roitman, which shows that if $\mathrm{CH}_0(X)_{\mathrm{hom}}$ is finite-dimensional, then it is an algebraic group.

Theorem 10.11 (Roitman 1972) *If $\mathrm{CH}_0(X)$ is representable, then the Albanese map*

$$\mathrm{alb}_X : \mathrm{CH}_0(X)_{\mathrm{hom}} \to \mathrm{Alb}(X)$$

(or the Abel–Jacobi map Φ_X^n (see vI.12.1.3)), which is well-defined by lemma 9.19, is an isomorphism.

The proof of this theorem makes use of the following step.

Proposition 10.12 $\mathrm{CH}_0(X)$ *is representable if and only if for every smooth curve $C = Y_1 \cap \cdots \cap Y_{n-1}$ which is a complete intersection of ample hypersurfaces $Y_i \subset X$, letting $j : C \hookrightarrow X$ be the inclusion of C into X, the map*

$$j_* : \mathrm{CH}_0(C)_{\mathrm{hom}} = J(C) \to \mathrm{CH}_0(X)_{\mathrm{hom}}$$

is surjective.

Proof One direction is obvious, since if $D \subset X$ is a curve satisfying the conclusion of the proposition, then we know that for $g = g(D)$, the map

$$D^{(g)} \to \mathrm{CH}_0(D)_{\mathrm{hom}} = J(D), \quad z \mapsto c(z) - gc(x_0)$$

is surjective (see vI.12.1.3). Then the surjectivity of j_* implies that the map

$$X^{(g)} \to \mathrm{CH}_0(X)_{\mathrm{hom}}, \quad z \mapsto c(z) - gc(x_0)$$

is surjective.

In the other direction, let $0 \in C$ and consider the map

$$\sigma_m^0 : X^{(m)} \to \mathrm{CH}_0(X)_{\mathrm{hom}}, \quad Z \mapsto \sigma_m(Z, m0).$$

where $m0$ is an element of the symmetric product of order m. As $\dim \operatorname{Im} \sigma_m^0$ is bounded by a constant K, the general fibres of σ_m^0 are of dimension greater than or equal to $mn - K$ for sufficiently large m. In fact, we may even assume that they are of dimension equal to $mn - K$ for sufficiently large m. Indeed, the function $\dim \operatorname{Im} \sigma_m^0$ is increasing with m, so if it is bounded, then it is constant for sufficiently large m. It follows that an irreducible component of maximal dimension Z of a general fibre of σ_m^0 cannot be contained in a set of the form $X^{(m-i)} + W$, $i \geq 1$, with $W \subset X^{(i)}$ and $\dim W < i$. Indeed, assume that

$$Z \subset X^{(m-i)} + W, \quad i \geq 1, \quad W \subset X^{(i)}, \quad \dim W < i.$$

As $\dim Z = mn - K$, we must have

$$\dim X^{(m-i)} \geq mn - K - i + 1,$$

so $n(m-i) \geq mn - K - i + 1$, which implies that $i < K/(n-1)$. Now consider the subset $Z' \subset X^{(m-i)} \times W$ given by

$$Z' = \{(z, w) \mid z + w \in Z\}.$$

By hypothesis, this set would dominate Z via the map $(z, w) \mapsto z + w$, so it would be of dimension greater than or equal to $mn - K$. The fibres Z'_w of $\mathrm{pr}_2 : Z' \to W$ would then be of dimension at least equal to $mn - K - i + 1$. Furthermore, as

$$\sigma_m^0(z + w) = \sigma_{m-i}^0(z) + \sigma_i^0(w), \quad z \in X^{(m-i)}, \quad w \in X^{(i)},$$

and σ_m^0 is constant along Z, the map σ_{m-i}^0 would be constant along each Z'_w. Moreover, if Z passes through a general point of $X^{(m)}$, Z'_w should also pass through a general point of $X^{(m-i)}$, for w generic in W so that $\dim Z'_w \leq \dim$ (general fibre of σ_{m-i}^0. As

$$\dim Z'_w \geq mn - K - i + 1,$$

this would contradict the fact that the general fibres of σ_{m-i}^0 are of dimension equal to $(m - i)n - K$, since $m - i > m - K/(n - 1)$ is arbitrarily large.

If $n \geq 2$, we have $mn - K \geq m$ for sufficiently large m, so we can apply the following lemma, whose proof will be given below.

Lemma 10.13 *Let Y_1 be an ample hypersurface of X and let Z be an irreducible subset of $X^{(m)}$ not contained in any subset of the form $X^{(m-i)} + W$ with $i \geq 1$ and $\dim W < i$. Then if $\dim Z \geq m$, Z intersects $Y_1^{(m)}$.*

We thus conclude that a general fibre of σ_m^0 intersects $Y_1^{(m)}$ for sufficiently large m if $n \geq 2$. This means that σ_m^0 and $\sigma^0_{m|Y_1^{(m)}}$ have the same image for sufficiently large m. But the images of the maps $\sigma^0_{m|Y_1^{(m)}}$ are also of bounded dimension, so we can iterate the reasoning to finally conclude that σ_m^0 and $\sigma^0_{m|C^{(m)}}$ have the same image for sufficiently large m. This shows that every 0-cycle of X of the form $Z - m0$ is rationally equivalent to a 0-cycle of the form $Z' - m0$, with $Z' \in C^{(m)}$, so

$$j_* : \mathrm{CH}_0(C) \to \mathrm{CH}_0(X)$$

is surjective. □

Proof of lemma 10.13 Consider the quotient map

$$r : X^m \to X^{(m)},$$

and let $\widetilde{Z} := r^{-1}(Z)$. Let $\widetilde{Z}_0$ be an irreducible component of $\widetilde{Z}$ dominating Z; the hypothesis on Z is equivalent to the fact that for every $i \geq 1$ and every subset I of cardinal i of $\{1, \ldots, m\}$, we have

$$\dim p_I(\widetilde{Z}_0) \geq i,$$

where p_I is the projection on X^i corresponding to the set of indices I. Moreover, as $\widetilde{Z}_0$ dominates Z, it suffices to show that $\widetilde{Z}_0$ intersects Y_1^m. It is easy to see that a complete intersection K of generic ample hypersurfaces in $\widetilde{Z}_0$ satisfying dim $K = m$ also satisfies the hypotheses. Thus, we may assume that $\dim \widetilde{Z}_0 = m$. Consider a desingularisation $\tau : Z' \to \widetilde{Z}_0$ of $\widetilde{Z}_0$. On Z', consider the divisors

$$D_i := (\mathrm{pr}_i \circ \tau)^{-1}(Y_1).$$

These divisors are 'numerically effective', since their cohomology classes $[D_i]$ admit representatives which are semipositive forms of type $(1, 1)$. (This follows from the fact that Y_1 is ample, so that $[Y_1]$ admits a representative which is a form of type $(1, 1)$ that is strictly positive on X.) Moreover, to show that $D_1 \cap \cdots \cap D_m \neq \emptyset$, it suffices to show that

$$D_1 \cdot \cdots \cdot D_m := \int_{Z'} [D_1] \cup \cdots \cup [D_m] > 0. \tag{10.2}$$

The inequality (10.2) follows from the Hodge index theorem applied to divisors on surfaces, which says that the intersection form, restricted to the Néron–Severi group, is non-degenerate of signature $(1, \rho - 1)$.

The proof is by induction on $m \geq 2$. We restrict to giving the proof for $m = 2$. Then Z is a surface, and D_1, D_2 are divisors on Z whose cohomology classes are non-trivial, since they are the pullbacks under the maps $\mathrm{pr}_i, i = 1, 2$, of ample divisors on X, and $\dim \mathrm{Im}\, \mathrm{pr}_i \geq 1$.

Since the D_is are numerically effective, we have $D_i^2 \geq 0$, $D_2^2 \geq 0$. If

$$D_1 \cdot D_2 = 0$$

then the intersection form restricted to $\langle D_1, D_2 \rangle$ is semipositive. By the Hodge index theorem, this is possible only if the cohomology classes $[D_1]$ and $[D_2]$ are proportional. But then $D_1^2 = 0 = D_2^2$, and also $(D_1 + D_2)^2 = 0$. This is not possible since $D_1 + D_2$ is the pullback of an ample divisor on $X \times X$ via a generically finite map. This gives us a contradiction, hence the inequality $D_1 \cdot D_2 > 0$ is proved. □

Proof of theorem 10.11 Proposition 10.12 shows that if $\mathrm{CH}_0(X)$ is representable, then there exist an abelian variety A and a correspondence

$$\Gamma \subset A \times X$$

of dimension $a = \dim A$ satisfying the following properties.

(i) The map

$$\Gamma_* : A \to \mathrm{CH}_0(X),\ a \mapsto \Gamma_*(a)$$

takes values in $\mathrm{CH}_0(X)_{\mathrm{hom}}$ and is a group homomorphism.

(ii) The morphism Γ_* is surjective.

Note that property (i) is by no means automatic: it states that the map

$$\Gamma_* : \mathrm{CH}_0(A) \to \mathrm{CH}_0(X)$$

induced by the correspondence Γ factors through the Albanese map of A, given by

$$\mathrm{alb}_A : \mathrm{CH}_0(A) \to \mathrm{Alb}\, A = A.$$

In fact, when j is the inclusion of a curve C, the morphisms of type

$$j_* : J(C) \overset{\mathrm{alb}_C^{-1}}{\cong} \mathrm{CH}_0(C)_{\mathrm{hom}} \to \mathrm{CH}_0(X)$$

are induced by such a correspondence. To see this, consider the cycle $\Gamma_C \subset J(C) \times C$ constructed as follows. Introduce

$$\phi : C^{(g)} \to J(C), \quad z_1 + \cdots + z_g \mapsto \mathrm{alb}_C\left(\sum_i z_i - gc_0\right),$$

where g is the genus of C and $c_0 \in C$ is a fixed point. We first set

$$\Gamma_C^1 = \{(a, c) \in J(C) \times C \mid \exists z \in C^{(g)}, c \in z, \phi(z) = a\} \subset J(C) \times C,$$
$$\Gamma_C^2 = g(J(C) \times c_0) \subset J(C) \times C,$$

and $\Gamma_C = \Gamma_C^1 - \Gamma_C^2$. It is easy to check that the morphism Γ_{C*} annihilates the cycle 0, so it takes values in $\mathrm{CH}_0(C)_{\mathrm{hom}}$, and is such that $\Gamma_{C*} : \mathrm{CH}_0(J(C)) \to \mathrm{CH}_0(C)_{\mathrm{hom}}$ is described by

$$\Gamma_{C*}(\{a\}) = \mathrm{alb}_C^{-1}(a)$$

for every $a \in J(C)$. Thus, this morphism clearly also satisfies property (i).

It follows immediately that the cycle $\Gamma = (\mathrm{Id}, j)_*\Gamma_C \subset J(C) \times X$ satisfies

$$\Gamma_*(a) = j_*(\mathrm{alb}_C^{-1})(a),$$

so it also satisfies property (i). □

Now let A and Γ satisfy properties (i) and (ii). Lemma 10.7 then shows that the kernel of Γ_* is a countable union of algebraic subsets of A. As it is a subgroup of A, it is a countable union of translates of an abelian subvariety $A_0 \subset \mathrm{Ker}\,\Gamma_*$. Then there exists an abelian subvariety B of A, which is supplementary to A_0 up to isogeny, i.e. such that the morphism

$$A_0 \times B \to A, \quad (a, b) \mapsto a + b$$

is surjective with finite kernel.

Replacing A by B and Γ by $\Gamma_{|B \times X}$, we may thus assume that the kernel of the morphism Γ_* is countable.

Let $x_0 \in X$, and let $R \subset X \times A$ be the subset defined by

$$R = \{(x, a) \mid \Gamma_*(a) = c(x) - c(x_0) \text{ in } \mathrm{CH}_0(X)\}.$$

R is a countable union of algebraic sets whose projection onto X is surjective, since Γ_* is surjective. Thus, R has an algebraic component R_0 which dominates X. Furthermore, as $\mathrm{Ker}\,\Gamma$ is countable, the projection $R_0 \to X$ is finite of degree r. R_0 is thus a correspondence of dimension equal to $\dim X$ between X and A, and provides a morphism

$$\alpha : X \to A,$$

given by $\alpha(x) = \mathrm{alb}_A(R_{0*}(x - x_0))$. By the definition of R_0 and property (i), it is clear that

$$\Gamma_* \circ \alpha(x) = r(x - x_0) \text{ in } \mathrm{CH}_0(X). \tag{10.3}$$

Moreover, by the universal property of the Albanese map (see vI.12.1.3), there exists a morphism of groups

$$\beta : \mathrm{Alb}(X) \to A$$

such that $\alpha = \beta \circ \mathrm{alb}_X$, where the Albanese map is constructed using the point x_0. We thus have morphisms of groups

$$A \stackrel{\Gamma_*}{\to} \mathrm{CH}_0(X)_{\mathrm{hom}} \stackrel{\mathrm{alb}_X}{\to} \mathrm{Alb}(X) \stackrel{\beta}{\to} A,$$

which, by (10.3), satisfy the equality

$$\Gamma_* \circ \beta \circ \mathrm{alb}_X = r\, \mathrm{Id}_{\mathrm{CH}_0(X)_{\mathrm{hom}}}.$$

It follows immediately that $\mathrm{Ker}\,\mathrm{alb}_X$ is torsion, so that

$$\mathrm{alb}_X : \mathrm{CH}_0(X)_{\mathrm{hom}} \to \mathrm{Alb}(X)$$

is an isomorphism up to torsion, since alb_X is always surjective (see vI.12.1.3).
The proof of the theorem can now be concluded by applying the following theorem (see Roitman 1980; Bloch 1979).

Theorem 10.14 *The Albanese map*

$$\mathrm{alb}_X : \mathrm{CH}_0(X)_{\mathrm{hom}} \to \mathrm{Alb}(X)$$

induces an isomorphism on torsion points for every smooth projective variety X. □

10.1.3 Statement of Mumford's theorem

Mumford's theorem proves the existence of smooth projective varieties with non-representable CH_0. Its most precise version in the case of surfaces is as follows.

Theorem 10.15 (Mumford 1968) *Let S be a smooth projective surface such that*

$$H^0(S, K_S) \neq 0.$$

Then for every integer m, the map

$$\sigma_m : S^{(m)} \times S^{(m)} \to \mathrm{CH}_0(S)_{\mathrm{hom}}, \quad (Z_1, Z_2) \mapsto c(Z_1) - c(Z_2)$$

has countable general fibre. In particular, $\mathrm{CH}_0(S)$ is not representable.

Mumford's theorem was generalised by Roitman (1972). Consider a smooth projective n-dimensional complex variety X, and the map

$$\sigma_m : X^{(m)} \times X^{(m)} \to \mathrm{CH}_0(X)_{\mathrm{hom}}.$$

Let d_m denote the dimension of the general fibre of σ_m.

Theorem 10.16 *If $H^0(X, \Omega_X^k) \neq 0$, then for sufficiently large m, we have*

$$d_m < 2(n-k+1)m.$$

When $n = k = 2$, Mumford's theorem 10.15 is stronger. This is due to the 'symplectic' argument, which we will explain below, and which is only applicable to 2-forms.

Roitman's theorem and Mumford's theorem 10.15 are in fact both consequences of the following result, which we will prove in the next section.

Theorem 10.17 *If there exists a subvariety $j : X' \hookrightarrow X$ such that $\dim X' < k$ and the map*

$$j_* : \mathrm{CH}_0(X') \to \mathrm{CH}_0(X)$$

is surjective, then $H^0(X, \Omega_X^k) = 0$.

The case $k = 2$, $n = 2$ of this theorem can be considered as a weak version of Mumford's theorem, thanks to proposition 10.12. This theorem admits the following corollary, which will be generalised below (see theorem 10.31).

Corollary 10.18 *Let X be a smooth projective variety. If the group $\mathrm{CH}_0(X)_{\mathrm{hom}}$ of 0-cycles homologous to 0 on X is zero, then*

$$H^0\big(X, \Omega_X^k\big) = 0 \ \textit{for all}\ k > 0.$$

Proof Under the hypotheses of the corollary, it suffices to take for X' a finite set of points of X, one for each connected component of X, and this set is then 0-dimensional. □

Proof of theorem 10.16 We first show that if $d_m \geq 2(n-k+1)m$, $m \to \infty$, then there exists $X' \subset X$, of dimension $\leq k-1$, such that

$$j_* : \mathrm{CH}_0(X') \to \mathrm{CH}_0(X)$$

is surjective. We can then conclude by applying theorem 10.17.

The argument is the same as in the proof of proposition 10.12. For X', we take the complete intersection of $n-k+1$ ample hypersurfaces, and show that $X'^{(m)} \times X'^{(m)}$ must intersect the fibres of σ_m for sufficiently large m, on the condition that we have $d_m \geq 2(n-k+1)m$, with m arbitrarily large. □

10.2 The Bloch–Srinivas construction

10.2.1 Decomposition of the diagonal

The following result, due to Bloch (1980a), Bloch & Srinivas (1983), gives an elegant proof of theorem 10.17. Let $f : X \to Y$ be a projective fibration, where X and Y are smooth and connected, and let $Z \in \mathrm{CH}^k(X)$ be a cycle of X. Consider the following property:

(*) *There exists a subvariety $X' \subset X$ such that for every $y \in Y$, the cycle $Z_y := Z_{|X_y} := j_y^* Z \in \mathrm{CH}^k(X_y)$ vanishes in $\mathrm{CH}^k(X_y - X'_y)$, where j_y is the inclusion of the fibre $X_y = f^{-1}(y)$ into X.*

Theorem 10.19 *If Z satisfies the property (*), then there exist an integer $m > 0$, and a cycle Z' supported in X', such that we have the equality*

$$mZ = Z' + Z'' \text{ in } \mathrm{CH}^k(X), \tag{10.4}$$

where Z'' is a cycle supported in $f^{-1}(Y')$, for a proper closed algebraic subset $Y' \subset Y$.

In particular, if $X = X_1 \times Y$, f is the second projection, and $X' = X'_1 \times Y$, we obtain the following result.

Corollary 10.20 *Let $\Gamma \in \mathrm{CH}^k(X_1 \times Y)$, and assume that for every $y \in Y$, the cycle $\Gamma^*(y) \in \mathrm{CH}^k(X_1)$ restricts to zero in $\mathrm{CH}^k(X_1 - X'_1)$. Then we have a decomposition*

$$mZ = Z' + Z'' \in \mathrm{CH}^k(X_1 \times Y),$$

where Z' is supported in $X'_1 \times Y$ and Z'' is supported in $X_1 \times Y'$, for a proper closed algebraic subset $Y' \subset Y$.

In particular, we can apply the preceding corollary to the diagonal $Z = \Delta_X \subset X \times X$, assuming that there exists an algebraic closed subset $j : X' \hookrightarrow X$

such that

$$j_* : \mathrm{CH}_0(X') \to \mathrm{CH}_0(X)$$

is surjective. Then $\mathrm{CH}_0(X - X')$ is zero by lemma 9.12, so for every $x \in X$, the cycle $\Delta_{X*}x$ vanishes in $\mathrm{CH}_0(X - X')$. We thus obtain the following decomposition of the diagonal.

Corollary 10.21 *Under this hypothesis, there exists a proper closed algebraic subset $T \subset X$, and a decomposition*

$$m\Delta = Z' + Z'' \in \mathrm{CH}^n(X \times X),$$

where Z' is supported in $T \times X$ and Z'' is supported in $X \times X'$.

Proof of theorem 10.19 The localisation exact sequence shows that the theorem is equivalent to the existence of a cycle Z' supported on X' and of a Zariski dense open set U of Y such that $mZ - Z'$ vanishes in $\mathrm{CH}^k(X_U)$, where $X_U := f^{-1}(U)$. It thus suffices to prove the lemma with Y replaced by a Zariski open set.

Up to replacing Y by a Zariski open set, we may assume that the intersections $Z_i \cap X_y$ and $X' \cap X_y$ are proper for every component of Z. The hypothesis and the localisation exact sequence then say that for every $y \in Y$, there exist subvarieties $W_{y,l}$, $l = 1, \ldots, N_y$ of X_y of codimension $k-1$, rational functions $\phi_{y,l}$ on $W_{y,l}$, and cycles Z'_y contained in X'_y, such that writing $\tau_{y,l} : \widetilde{W}_{y,l} \to X_y$ for the normalisations, we have

$$\sum_l \tau_{y,l_*}(\mathrm{div}(\phi_{y,l})) = Z_y + Z'_y.$$

Now, there exists a countable set of Zariski open sets H_i of relative Hilbert schemes parametrising the tuples consisting of a point y of Y, an integer N_y, N_y irreducible subvarieties W_l of X_y of codimension $k - 1$, Weil divisors D_l of W_l, and a cycle Z'_y of codimension k contained in X'_y. For each i, consider the subset $H'_i \subset H_i$ defined by

$$H'_i = \left\{ (y, (W_l, D_l)_{l=1,\ldots,N_y}) \,\middle|\, \sum_l j_{l*}D_l = Z_y + Z'_y \text{ and } \exists \phi_l \in \mathbb{C}(W_l)^* \text{ such that } D_l = n_{l*}\mathrm{div}(\phi_l) \right\},$$

where $n_l : \widetilde{W}_l \to W_l$ is the normalisation. One can show that H'_i is a countable union of locally closed algebraic subsets of H_i. Now, the hypothesis says that the maps $p_{i|H'_i} \to Y$, where $p_i : H_i \to Y$ are the first projections, satisfy

$$\bigcup_i \operatorname{Im} p_{i|H'_i} = Y.$$

Thus, there exists a closed algebraic subset H of an open set of H_i, contained in H'_i, such that the map $p = p_{i|H}$ is dominant.

Up to replacing H by a subvariety, we may assume that p is generically finite of degree N. Up to restricting Y, we may even assume that p is proper. Finally, by desingularisation, we may assume that H is smooth. Consider the fibred product

$$\begin{array}{lccc} p_X : & X_H & \longrightarrow & X \\ & \downarrow & & \downarrow \\ p : & H & \longrightarrow & Y. \end{array}$$

By the definition of the Hilbert schemes H_i, and by the fact that $H \subset H_i$, there exist universal subschemes

$$\mathcal{W}_l \subset X_H$$

which are of codimension $k-1$ and flat over H, Weil divisors $\mathcal{D}_l$ of $\mathcal{W}_l$, and a relative cycle $\mathcal{Z}'$ of codimension k of X_H supported in $X'_H := p_X^{-1}(X')$, satisfying the properties:

(i) Every fibre $D_{l,h}$ of $\mathcal{D}_l$ over $h \in H$ is a divisor rationally equivalent to 0 in the corresponding fibre $W_{l,h} \subset \mathcal{W}_l$ over h.
(ii) We have the equality

$$\sum_l \mathcal{D}_l = p_X^{-1}(Z) + \mathcal{Z}'$$

of cycles of X_H.

Lemma 10.22 *Under hypothesis (i), for each l, there exists a Zariski open set U_l of H, such that the restriction of the codimension 1 cycle $\mathcal{D}_l$ of $\mathcal{W}_l$ is rationally equivalent to 0 in $p_l^{-1}(U_l)$, where $p_l : \mathcal{W}_l \to H$ is the restriction of f_H to $\mathcal{W}_l$.*

Temporarily admitting this lemma, set $U = \bigcap_l U_l$. Then U contains a Zariski open set of the form $p^{-1}(V)$, where V is a Zariski open set in Y. Lemma 10.22 shows that the cycle $\sum_l \mathcal{D}_l$ of X_H becomes rationally equivalent to 0 in the open set $X_{p^{-1}(V)}$ of X_H. As the map

$$p_X : X_{p^{-1}(V)} \to X_V = f^{-1}(V)$$

is proper of degree N, applying p_{X*} to the equality in (ii), we obtain

$$NZ_{|X_V} + p_{X*}\mathcal{Z}' = 0 \quad \text{in} \quad \mathrm{CH}^k(X_V),$$

and this concludes the proof of theorem 10.19, since $p_{X*}\mathcal{Z}'$ is supported in X'. □

Proof of lemma 10.22 By restricting H, which allows us to consider a simultaneous desingularisation of the family $\mathcal{W}_l \to H$, we easily reduce to the case where $\mathcal{W}_l$ is smooth and p_l is submersive. Then $\mathcal{D}_l$ is the divisor of a rational section of a line bundle $\mathcal{L}$ on $\mathcal{W}_l$, which satisfies the property that the restrictions $\mathcal{L}_{|W_{l,h}}$ are trivial for every $h \in H$. As the fibres $W_{l,h}$ are connected, we have $h^0(\mathcal{D}_{l|W_{l,h}}) = 1$, and the base change theorem then shows that $R^0 p_{l*}\mathcal{L}$ is a line bundle on H. Moreover, the triviality of $\mathcal{L}_{W_{l,h}}$ and the base change theorem (see Hartshorne 1977, III.12) imply that there exists a canonical isomorphism

$$p_l^* R^0 p_{l*}\mathcal{L} \cong \mathcal{L}.$$

It then suffices to take for U_l any Zariski open set on which the line bundle $R^0 p_{l*}\mathcal{L}$ is trivial. □

Remark 10.23 *The proof given here is nearer in spirit to Mumford's arguments. What we have retained from Bloch & Srinivas (1983) is essentially the principle of the decomposition (10.4) of the cycle Z.*

10.2.2 Proof of Mumford's theorem

We first prove theorem 10.17, using corollary 10.21.

Proof of theorem 10.17 By hypothesis, there exists $j : X' \hookrightarrow X$, where X' is of dimension $\leq k-1$, such that $j_*(\mathrm{CH}_0(X')) = \mathrm{CH}_0(X)$. Thus, we can apply corollary 10.21. This gives an equality

$$m\Delta_X = Z' + Z''$$

in $\mathrm{CH}^n(X \times X)$, where $n = \dim X$, Z' is supported in $X \times X'$, and Z'' is supported in $T \times X$, for T a proper algebraic subset of X. By lemma 9.18, we thus have equality of the corresponding cohomology classes:

$$m\,\mathrm{cl}(\Delta_X) = \mathrm{cl}(Z') + \mathrm{cl}(Z'') \text{ in } H^{2n}(X \times X, \mathbb{Z}).$$

We can view these cohomology classes, or rather their Künneth components, as morphisms of Hodge structures for every r

$$m[\Delta_X]^*,\ [Z']^*,\ [Z'']^* : H^r(X,\mathbb{Z}) \to H^r(X,\mathbb{Z})$$

(see vI.11.3.3). Then we have the equality

$$m[\Delta_X]^* = [Z']^* + [Z'']^* \text{ in } \operatorname{Hom}(H^r(X,\mathbb{Z}), H^r(X,\mathbb{Z})). \tag{10.5}$$

Now, the morphisms $[\Delta_X]^*$ are equal to the identity, by the fact that Δ_X is the diagonal. Thus, (10.5) gives

$$m\,\mathrm{Id} = [Z']^* + [Z'']^* \text{ in } \operatorname{Hom}(H^r(X,\mathbb{Z}), H^r(X,\mathbb{Z})).$$

In particular, for $\eta \in H^0(X,\Omega_X^r) \subset H^r(X,\mathbb{C})$, we have

$$m\eta = [Z']^*\eta + [Z'']^*\eta \ in\ H^r(X,\mathbb{C}). \tag{10.6}$$

Now let $l : \widetilde{T} \to X$ be a desingularisation of T. As the cycle Z'' is supported in $T \times X$, it comes from a cycle $\widetilde{Z}''$ of $\widetilde{T} \times X$, $Z'' = (l,\mathrm{Id})_*(\widetilde{Z}'')$, so similarly, we have

$$\mathrm{cl}(Z'') = (l,\mathrm{Id})_*(\mathrm{cl}(\widetilde{Z}'')).$$

Recall that if $\alpha \in H^{2n}(X \times X,\mathbb{Z})$, the corresponding morphism

$$\alpha_r^* : H^r(X,\mathbb{Z}) \to H^r(X,\mathbb{Z})$$

is defined by

$$\alpha_r^*(\beta) = \mathrm{pr}_{1*}(\mathrm{pr}_2^*(\beta) \cup \alpha). \tag{10.7}$$

Formula (10.7) then shows that

$$[Z'']^* = l_* \circ [\widetilde{Z}'']^*. \tag{10.8}$$

Similarly, let $\tilde{j} : \widetilde{X}' \to X$ be a desingularisation of X', and let $\widetilde{Z}'$ be a cycle of $X \times \widetilde{X}'$ such that $(\mathrm{Id},\tilde{j})_*(\widetilde{Z}') = Z'$. Then formula (10.7) shows that

$$[Z']^* = [\widetilde{Z}']^* \circ \tilde{j}^*. \tag{10.9}$$

By equations (10.8) and (10.9), (10.6) now gives

$$m\eta = [\widetilde{Z}']^* \circ \tilde{j}^*\eta + l_* \circ [\widetilde{Z}'']^*\eta.$$

But as $\dim X' < k$, we have for $r \geq k$

$$\tilde{j}^*\eta = 0 \ \text{ in } \ H^0(\widetilde{X}',\Omega_{\widetilde{X}'}^r).$$

Thus, we have $[\widetilde{Z}']^* \circ \tilde{j}^*\eta = 0$ and

$$m\eta = l_* \circ [\widetilde{Z}'']^*\eta. \tag{10.10}$$

Moreover, as $\dim T < \dim X$, the morphism of Hodge structures l_* is of bidegree (s, s) with $s = \operatorname{codim} T > 0$, so the intersection of its image with $H^0(X, \Omega_X^r) = H^{r,0}(X)$ is reduced to 0. Thus, the equality (10.10) implies that $\eta = 0$ for

$$\eta \in H^0\big(X, \Omega_X^r\big), \quad r \geq k.$$

□

More generally, corollary 10.20 and an easy adaptation of the proof of theorem 10.17 yield the following result.

Proposition 10.24 *Let W be a smooth projective variety, and let $Z \subset W \times X$ be a cycle of codimension $n =$ dim X. Assume that there exists a k_0-dimensional subvariety $X' \subset X$ such that for every $w \in W$, the 0-cycle Z_w is rationally equivalent in X to a cycle supported on X'. Then for every $k > k_0$ and every $\eta \in H^0(X, \Omega_X^k)$, we have $[Z]^*\eta = 0$ in $H^0(W, \Omega_W^k)$.*

Let us now prove Mumford's theorem.

Proof of theorem 10.15 For a smooth projective complex surface S and an integer n, consider an algebraic component $R \subset S^n \times S^n$ of the set

$$\left\{(x_1, \ldots, x_{2n}) \in S^{2n} \,\middle|\, \sum_{1 \leq i \leq 2n} n_i c(x_i) = 0 \text{ in } \mathrm{CH}_0(S)\right\},$$

where the integers n_i are non-zero. Assume that R dominates S^n via the first projection.

Lemma 10.25 *If $H^0(S, K_S) \neq 0$, then* dim $R \leq 2n$*, so that the first projection $p_1 : R \to S^n$ is generically finite.*

We first show that this implies that under the same hypothesis, the general fibre of

$$\sigma_n : S^{(n)} \times S^{(n)} \to \mathrm{CH}_0(S)_{\mathrm{hom}}$$

is countable. Indeed, if the general fibre of σ_n were not countable, there would exist an algebraic subvariety

$$R \subset S^{(n)} \times S^{(n)} \times S^{(n)} \times S^{(n)}$$

such that the first projection $R \to S^{(n)} \times S^{(n)}$ is dominant with positive-dimensional fibres, and such that

$$(Z_1, Z_2, Z_3, Z_4) \in R \Rightarrow \sigma_n(Z_1, Z_2) = \sigma_n(Z_3, Z_4).$$

But then, taking the inverse image of R in $S^n \times S^n \times S^n \times S^n$, we find that there exists

$$R' \subset S^n \times S^n \times S^n \times S^n$$

such that the first projection $R' \to S^n \times S^n$ is dominant with positive-dimensional fibres, and such that

$$(x_1, \ldots, x_n, y_1, \ldots, y_n, z_1, \ldots, z_n, t_1, \ldots, t_n) \in R' \Rightarrow \sum_i x_i - y_i - z_i + t_i$$
$$= 0 \in \mathrm{CH}_0(S).$$

But this would contradict lemma 10.25, with n replaced by $2n$ and the n_i equal to ± 1. □

Proof of lemma 10.25 Note that the inclusion $R \subset S^{2n}$ gives a correspondence $Z \subset R \times S$ given by $Z_*(r) = n_1 z_1 + \cdots + n_{2n} z_{2n}$ for $r = (z_1, \ldots, z_{2n}) \in R$. This correspondence is in fact equal to

$$\sum_{1 \le i \le 2n} n_i Z_i,$$

where the correspondence Z_i satisfies

$$Z_{i*}(r) = z_i, \quad r = (z_1, \ldots, z_{2n}).$$

In other words, Z_i is the graph of the restriction to R of the ith projection p_i. We can now apply proposition 10.24, first desingularising R if necessary, to the correspondence Z, and we conclude that for any $\eta \in H^0(S, K_S) = H^0(S, \Omega_S^2)$, we have

$$[Z]^*\eta = 0 \text{ in } H^0(\Omega_R^2).$$

But as $Z = \sum_i n_i Z_i$ with $Z_i = \text{graph}(p_{i|R})$, we find that

$$[Z_i]^*\eta = p_{i|R}^*(\eta) = p_i^*\eta_{|R},$$

$$[Z]^*\eta = \sum_i n_i p_i^*\eta_{|R}.$$

We thus have the relation $\sum_i n_i p_i^*\eta_{|R} = 0$ in $H^0(R, \Omega_R^2)$. Now let $r \in R$ be a generic point. R is smooth at the point r, and the first projection $p : R \to S^n$ is submersive. Moreover, $p(r)$ is a generic point of S^n, so a given non-zero form η is non-zero at every point z_i, $i \leq n$. It follows that the holomorphic 2-form

$$\sum_{i \leq n} n_i p_i^*\eta_{|R}$$

is of rank $2n$ on R in the neighbourhood of r. The equality

$$\sum_{i \leq n} n_i p_i^*\eta_{|R} = -\sum_{n+1 \leq i \leq 2n} n_i p_i^*\eta_{|R}$$

then shows that the form

$$\sum_{n+1 \leq i \leq 2n} n_i p_i^*\eta_{|R}$$

is also of rank $2n$, which implies that η is also non-zero at the points z_j for $j > n$. But then the form

$$\omega = \sum_{1 \leq i \leq 2n} n_i p_i^*\eta$$

is a holomorphic form of degree 2 on S^{2n}, which is non-degenerate in the neighbourhood of r. Now, this form vanishes by restriction to R. Thus, dim $R \leq 2n$, since its tangent space at r is totally isotropic for the 2-form ω on the tangent space of S^{2n} at r, and thus is of dimension at most equal to $\frac{1}{2}$dim $T_{S^{2n},r} = 2n$. □

10.2.3 Other applications

The decomposition given in corollary 10.21 has other applications, which generalise certain previously known results. For example, we have the following.

Proposition 10.26 (Bloch & Srinivas 1983) *Let X be a smooth complex projective variety such that there exists a subvariety $j : X' \hookrightarrow X$, of dimension ≤ 3, such that the map*

$$j_* : \text{CH}_0(X') \to \text{CH}_0(X)$$

is surjective. Then the Hodge conjecture holds for classes of degree 4 on X.

This result was originally proved by Conte & Murre (1978) in the case where X is a 4-dimensional variety covered by rational curves. Such a variety X satisfies the hypothesis, since every point x is contained in a rational curve C_x whose normalisation is isomorphic to $\mathbb{P}^1$. By the definition of rational equivalence, all the points $y \in C_x$ are rationally equivalent in C_x, so also in X, and if X' is an ample hypersurface of X, then C_x intersects X' and x is rationally equivalent in X to any point of $X' \cap C_x$.

Proof of proposition 10.26 Applying corollary 10.21, we see that there exists a proper subset $T \subset X$, which we may assume to be of codimension 1, and n-dimensional cycles Z' and Z'' supported in $T \times X$ and $X \times X'$ respectively, such that

$$m\Delta_X = Z' + Z'' \text{ in } \mathrm{CH}^n(X \times X) \tag{10.11}$$

for some non-zero integer m. Let $k : \widetilde{T} \to X$ and $\tilde{j} : \widetilde{X}' \to X$ be desingularisations of T and X' respectively. By lemma 9.18, the decomposition (10.11) then gives the equality

$$m\,\mathrm{cl}(\Delta_X) = \mathrm{cl}(Z') + \mathrm{cl}(Z'') \text{ in } H^{2n}(X \times X, \mathbb{Z}),$$

and in particular, it yields the equalities of the morphisms of Hodge structure associated to the Künneth components of type $(2n-4, 4)$ of these three Hodge classes:

$$m[\Delta_X]^* = [Z']^* + [Z'']^* : H^4(X, \mathbb{Z}) \to H^4(X, \mathbb{Z}).$$

Now, let $\widetilde{Z}' \subset \widetilde{T} \times X$ be a cycle of codimension n such that

$$(k, \mathrm{Id})_* \widetilde{Z}' = Z',$$

and let $\widetilde{Z}'' \subset X \times \widetilde{X}'$ be a cycle of codimension n such that

$$(\mathrm{Id}, \tilde{j})_* \widetilde{Z}'' = Z''.$$

We have

$$\mathrm{cl}(Z') = (k, \mathrm{Id})_* \mathrm{cl}(\widetilde{Z}'), \quad \mathrm{cl}(Z'') = (\mathrm{Id}, \tilde{j})_* \mathrm{cl}(\widetilde{Z}''),$$

and it follows that for every $\alpha \in H^4(X, \mathbb{Z})$, we have

$$[Z']^*\alpha = k_*[\widetilde{Z}']^*\alpha, \quad [Z'']^*\alpha = [\widetilde{Z}'']^*(\tilde{j}^*\alpha).$$

But as X' is of dimension ≤ 3, the rational Hodge conjecture holds for $\widetilde{X}'$ in every degree. (Indeed, it holds for classes of degree 2 by the Lefschetz theorem on $(1,1)$-classes, and for classes of degree 4 by the Lefschetz isomorphism

$$L = [H] \cup : H^2(X, \mathbb{Q}) \cong H^4(X, \mathbb{Q}),$$

where H is an ample divisor which maps $\mathrm{Hdg}^2(X)$ to $\mathrm{Hdg}^4(X)$ isomorphically, and $[D]$ to $[H] \cup [D] = [H \cdot D]$.) Moreover, the Hodge conjecture is satisfied for classes of degree ≤ 2 and for every smooth projective variety.

If $\alpha \in H^4(X, \mathbb{Q}) \cap H^{2,2}(X)$ is a rational Hodge class, the classes $\tilde{j}^*\alpha \in H^4(\widetilde{X}', \mathbb{Q})$ and $[Z']^*\alpha \in H^2(\widetilde{T}, \mathbb{Q})$ are classes of algebraic cycles of $\widetilde{X}'$ and $\widetilde{T}$ respectively. The relation

$$m[\Delta_X]^*\alpha = m\alpha = k_*[\widetilde{Z}']^*\alpha + [\widetilde{Z}'']^*(\tilde{j}^*\alpha)$$

and the compatibility of the cycle class map with correspondences then show that α is also the class of an algebraic cycle with rational coefficients. □

Another application concerns the Griffiths groups

$$\mathrm{Griff}^k(X) := \mathcal{Z}^k(X)_{\mathrm{hom}}/\mathcal{Z}^k(X)_{\mathrm{alg}}$$

introduced in section 8.2.1.

Proposition 10.27 (Bloch & Srinivas 1983) *Let X be a smooth complex projective variety such that there exists a subvariety $j : S \hookrightarrow X$, of dimension ≤ 2, such that $j_* : \mathrm{CH}_0(S) \to \mathrm{CH}_0(X)$ is surjective. Then the group* $\mathrm{Griff}^2(X)$ *is a torsion group.*

Proof Clearly, the subgroups $\mathrm{CH}_{\mathrm{alg}} \subset \mathrm{CH}$ consisting of the algebraic cycles equivalent to 0 are stable under the action of correspondences. (For example, this follows from the definition of algebraic equivalence together with proposition 9.17.)

Moreover, with the same notation as above, corollary 10.20 yields a decomposition

$$m\Delta_X = Z' + Z'' \text{ in } \mathrm{CH}^n(X \times X) \tag{10.12}$$

for some non-zero integer m, where $T \subset X$ is a proper closed subset which we may assume to be of codimension 1, and Z' and Z'' are cycles of codimension n supported in $T \times X$ and $X \times X'$ respectively. But then, for every cycle $z \in \mathrm{CH}^l(X)$, we have the relation

$$mz = m\Delta_X^* z = Z'^* z + Z''^* z. \tag{10.13}$$

Let $k : \widetilde{T} \to X$ and $\tilde{j} : \widetilde{S} \to X$ be desingularisations of T and S respectively, and let $\widetilde{Z}' \subset \widetilde{T} \times X$, $\widetilde{Z}'' \subset X \times \widetilde{S}$ be such that

$$(k, \mathrm{Id})_* \widetilde{Z}' = Z', \quad (\mathrm{Id}, \tilde{j})_* \widetilde{Z}'' = Z''. \tag{10.14}$$

If now $z \in \mathrm{CH}^2(X)_{\mathrm{hom}}$, we deduce from (10.13) and (10.14) that

$$mz = k_* \tilde{Z}'^* z + \tilde{Z}''^* (\tilde{j}^* z). \qquad (10.15)$$

But as S is a surface, we have $\tilde{j}^* z \in \mathrm{CH}_0(\widetilde{S})_{\mathrm{hom}}$, so it is algebraically equivalent to 0 (see subsection 8.2.1). Similarly, $\tilde{Z}'^* z \in \mathrm{CH}^1(\widetilde{T})_{\mathrm{hom}}$, so $\tilde{Z}'^* z$ is algebraically equivalent to 0 (see subsection 8.2.1). It then follows from (10.15) that mz is algebraically equivalent to 0. □

10.3 Generalisation

10.3.1 Generalised decomposition of the diagonal

The decomposition of the diagonal given in corollary 10.21 yields the following corollary when the smooth projective connected variety X satisfies the hypothesis that $\mathrm{CH}_0(X) = \mathbb{Z}$ (so that we may assume that X' is a point $x \in X$).

Corollary 10.28 *If $\mathrm{CH}_0(X) = \mathbb{Z}$, there exists a decomposition $m\Delta_X = Z' + Z''$ in $\mathrm{CH}^n(X \times X)$, where Z' is supported in $T \times X$ for a proper closed algebraic subset $T \subset X$, and Z'' is supported in $X \times x$.*

Note that the condition $\mathrm{CH}_0(X) = \mathbb{Z}$ is equivalent to the fact that the map cl is injective on $\mathrm{CH}_0(X)$.

This result was generalised by Paranjape (1994) and Laterveer (1996), who actually studied more general triviality hypotheses on Chow groups of small dimension.

Theorem 10.29 *Assume that for $k \leq k_0$, the maps*

$$\mathrm{cl} : \mathrm{CH}_k(X) \otimes \mathbb{Q} \to H^{2n-2k}(X, \mathbb{Q})$$

are injective. Then there exists a decomposition

$$m\Delta_X = Z_0 + \cdots + Z_{k_0} + Z' \in \mathrm{CH}^n(X \times X), \qquad (10.16)$$

where $m \neq 0$ is an integer, Z_i is supported in $W_i' \times W_i$ with $\dim W_i = i$ and $\dim W_i' = n - i$, and Z' is supported in $T \times X$, where $T \subset X$ is a closed algebraic subset of codimension $\geq k_0 + 1$.

Corollary 10.21 is in fact the case $k_0 = 0$ of this theorem.

Remark 10.30 *There is a version of this theorem concerning the correspondences* $\Gamma \subset Y \times X$ *of codimension* n*. Under the hypothesis that* $\Gamma_* : \mathrm{CH}_k(Y) \otimes \mathbb{Q} \to \mathrm{CH}_k(X) \otimes \mathbb{Q}$ *factors through* $\mathrm{cl} : \mathrm{CH}_k(Y) \otimes \mathbb{Q} \to H^{2\dim Y - 2k}(Y, \mathbb{Q})$ *for* $k \leq k_0$*, we then obtain a decomposition of* $\Gamma \in \mathrm{CH}^n(Y \times X)$ *similar to the one given in (10.16).*

Proof of theorem 10.29 We use induction on k_0. The case $k_0 = 0$ was considered above. We may thus assume that we have a decomposition

$$m\Delta_X = Z_0 + \cdots + Z_{k_0-1} + Z' \in \mathrm{CH}^n(X \times X), \tag{10.17}$$

where $m \neq 0$ is an integer, Z_i is supported in $W_i' \times W_i$ with $\dim W_i = i$ and $\dim W_i' = n - i$, and Z' is supported in $T \times X$, where $T \subset X$ is a closed algebraic subset of codimension $\geq k_0$. We may assume that T is of codimension k_0. Let $\tau : \widetilde{T} \to X$ be a desingularisation. Let $\widetilde{Z}' \subset \widetilde{T} \times X$ be a cycle such that $(\tau, \mathrm{Id})_* \widetilde{Z}' = Z'$. The cycle $\widetilde{Z}' \subset \widetilde{T} \times X$ is of codimension $n - k_0$ and induces a morphism $\widetilde{Z}'_* : \mathrm{CH}_0(\widetilde{T}) \to \mathrm{CH}_{k_0}(X)$. Now, we know that the kernel of $\mathrm{cl} : \mathrm{CH}_{k_0}(X) \to H^{2n-2k_0}(X, \mathbb{Z})$ is torsion. Thus, the map $\widetilde{Z}'_*$ maps $\mathrm{CH}_0(\widetilde{T})_{\mathrm{hom}}$ to the torsion of $\mathrm{CH}_{k_0}(X)$. By a countability argument, it follows that there exists an integer M such that $M\widetilde{Z}'_* = 0$ on $\mathrm{CH}_0(\widetilde{T})_{\mathrm{hom}}$. Indeed, the fact that $\widetilde{Z}'_*(\mathrm{CH}_0(\widetilde{T})_{\mathrm{hom}})$ is contained in the torsion of $\mathrm{CH}_{k_0}(X)$ means that each connected component $\widetilde{T}_i$ of $\widetilde{T}$ is a countable union

$$\widetilde{T}_i = \bigcup_{m>0} \widetilde{T}_{i,m},$$
$$\widetilde{T}_{i,m} = \{t \in \widetilde{T}_i \mid m\widetilde{Z}'_*(t - t_i) = 0 \quad \text{in} \quad \mathrm{CH}_{k_0}(X)\},$$

where t_i is a fixed point of $\widetilde{T}_i$. But we can show that each $\widetilde{T}_{i,m}$ is a countable union of closed algebraic subsets of $\widetilde{T}_i$. By Baire, we then conclude that one of these sets has non-empty interior, so it is equal to $\widetilde{T}_i$. Thus, we must have $\widetilde{T}_{i,m_i} = \widetilde{T}_i$ for some integer m_i. Doing this for each connected component of $\widetilde{T}$, and taking M to be the product of the integers m_i constructed above, we have then shown that $M\widetilde{Z}'_* = 0$ on $\mathrm{CH}_0(\widetilde{T})_{\mathrm{hom}}$.

For each component $\widetilde{T}_j$ of $\widetilde{T}$, choose a point $t_j \in \widetilde{T}_j$ as above, and let $W_j = \widetilde{Z}'_*(t_j)$. The cycle

$$Z'' = M\left(\widetilde{Z}' - \sum_j \widetilde{T}_j \times W_j\right) \subset \widetilde{T} \times X$$

then satisfies the property that its restriction to each $t \times X$ is trivial in $\mathrm{CH}_{k_0}(X)$. We then apply theorem 10.19 to conclude that there exists a cycle $\widetilde{Z}'' \subset \widetilde{T}' \times X$, where $\widetilde{T}' \subset \widetilde{T}$ is a closed algebraic subset with empty interior, and an integer M', such that

$$M'M\left(\widetilde{Z}' - \sum_i \widetilde{T}_i \times W_i\right) = \widetilde{Z}'' \text{ in } \mathrm{CH}_n(\widetilde{T} \times X).$$

Setting $T' = \tau(\widetilde{T}')$ and $Z'' = (\tau, \mathrm{Id})_* \widetilde{Z}''$, we obtain

$$M'M\left(Z' - \sum_i T_i \times W_i\right) = Z'' \text{ in } \mathrm{CH}_n(X \times X),$$

with Z'' supported on $T' \times X$ and $\operatorname{codim} T' \geq k_0 + 1$. Combining this equality with (10.17), we obtain the desired decomposition. □

10.3.2 An application

As a consequence of theorem 10.29, we obtain the following result, due to Schoen (1993), and Lewis (1995), which generalises corollary 10.18.

Theorem 10.31 *If a smooth projective complex variety X satisfies the condition that for every $k \leq k_0$, the map* $\mathrm{cl} : \mathrm{CH}_k(X) \otimes \mathbb{Q} \to H^{2n-2k}(X, \mathbb{Q})$ *is injective, then*

$$H^{p,q}(X) = 0, \quad \forall p \neq q, \quad q \leq k_0.$$

Proof Let us write the generalised decomposition of the diagonal given in theorem 10.29:

$$m\Delta_X = Z_0 + \cdots + Z_k + Z' \in \mathrm{CH}^n(X \times X)$$

By hypothesis, the Z_i are of the form

$$Z_i = \sum_j n_{i,j} W'_{i,j} \times W_{i,j}, \tag{10.18}$$

where the $W_{i,j}$ (resp. $W'_{i,j}$) are the irreducible components of W_i (resp. W'_i). Moreover, Z' is supported in $T \times X$ with $\operatorname{codim} T \geq k+1$. Let $\tau : \widetilde{T} \to X$ be a desingularisation of T, and let $\widetilde{Z}'$ in $\mathrm{CH}_n(\widetilde{T} \times X)$ be such that $(\tau, \mathrm{Id})_*(\widetilde{Z}') = Z'$. The above decomposition gives a decomposition of the corresponding cohomology classes

$$m\,\mathrm{cl}(\Delta_X) = \mathrm{cl}(Z_0) + \cdots + \mathrm{cl}(Z_k) + \mathrm{cl}(Z') \text{ in } \mathrm{CH}^n(X \times X),$$

and of the corresponding morphisms of Hodge structures for every l:

$$m[\Delta_X]^* = m\mathrm{Id} = [Z_0]^* + \cdots + [Z_k]^* + [Z']^* : H^l(\mathbb{Z}) \to H^l(X, \mathbb{Z}).$$

Now, by (10.18), we clearly have

$$[Z_i]^*(\alpha) = \sum_j n_{i,j} \langle \alpha, [W_{i,j}] \rangle [W'_{i,j}],$$

where $\langle\, , \rangle$ is the intersection form on $H^*(X)$. In particular, we have

$$Z_i^*(\alpha) = 0, \quad \forall \alpha \in H^{p,q}(X), \quad p \neq q.$$

For α satisfying this hypothesis, we thus have

$$m\alpha = [Z']^*(\alpha) = \tau_*([\widetilde{Z}']^*(\alpha)).$$

Thus, if $\alpha \in H^{p,q}(X)$ with $p \neq q$, we have

$$m\alpha \in \mathrm{Im}\, \tau_* \cap H^{p,q}(X).$$

Now, as $\mathrm{codim}\, T = k_0 + 1$, the Gysin morphism

$$\tau_* : H^{p+q-2k_0-2}(\widetilde{T}, \mathbb{Z}) \to H^{p+q}(X, \mathbb{Z})$$

is a morphism of Hodge structures of bidegree $(k_0 + 1, k_0 + 1)$, so that

$$\mathrm{Im}\, \tau_* \cap H^{p,q}(X) = 0, \quad q \leq k_0.$$

Hence, we have proved that $m\alpha = 0$ for $\alpha \in H^{p,q}(X), \quad p \neq q, \quad q \leq k_0.$ □

Exercises

1. *A lemma on the* 0*-cycles of surfaces.* Let S be a smooth connected projective surface, and let L be a very ample line bundle on S. For a smooth curve $C \in |L|$, write $j : C \hookrightarrow S$ for inclusion, and set

$$H := \mathrm{Ker}\,(j_* : H^1(C, \mathbb{Z}) \to H^3(S, \mathbb{Z})).$$

We also write $A \subset J(C)$ for the abelian subvariety corresponding to the Hodge substructure H.

(a) Show that for every smooth curve $C \in |L|$, the kernel

$$\mathrm{Ker}\,(j_* : \mathrm{CH}_0(C)_{\mathrm{hom}} = J(C) \to \mathrm{CH}_0(S))$$

is a countable union of translates of an abelian subvariety of A.

(b) Show that if this abelian subvariety were non-zero and different from A for general $C \in |L|$, this would contradict theorem 3.27.

(c) Deduce from the above that if $\mathrm{alb} : \mathrm{CH}_0(S)_{\mathrm{hom}} \to \mathrm{Alb}\, S$ is not an isomorphism, for C a general curve in $|L|$, then the kernel $\mathrm{Ker}\, j_*$ is countable. (Use Roitman's theorem 10.11 and the fact that since C moves in a linear system, for fixed $z \in \mathrm{Alb}\, S$, the cycles $j_*\tilde{z}$, where $\tilde{z}$ is a preimage of z under the isogeny $A^\perp \cong \mathrm{Alb}\, S$, are all rationally equivalent up to torsion in S to deal with the case $A = \mathrm{Ker}\, j_*$.)

2. *Standard conjectures.* Recall (see vI.6.2) that if (X, H) is a smooth polarised projective variety, i.e. if $H \in \mathrm{Pic}\, X$ is ample, we have Lefschetz operators

$$L = c_1(H)\cup : H^*(X, \mathbb{Q}) \to H^{*+2}(X, \mathbb{Q})$$

and

$$\Lambda : H^*(X, \mathbb{Q}) \to H^{*-2}(X, \mathbb{Q})$$

which are morphisms of Hodge structures of weight 2 and -2 respectively, satisfying the relations

$$[L, \Lambda] = (k - n)\mathrm{Id} \ \text{ on } \ H^k(X, \mathbb{Q}) \ \text{ for } \ n = \dim X.$$

We say that an operator $\gamma : H^*(X, \mathbb{Q}) \to H^{*+2r}(X, \mathbb{Q})$ of degree $2r$ is algebraic if there exists a cycle $\Gamma \in \mathrm{CH}^{n+r}(X \times X)_{\mathbb{Q}}$ such that

$$[\Gamma]^* = \gamma \quad \text{on} \quad H^*(X, \mathbb{Q}).$$

(a) Show that the operator L is algebraic. (Consider

$$j_*(\Delta_Y) \in \mathrm{CH}^{n+1}(X \times X),$$

where j is the inclusion of $Y \in |H|$ and Δ_Y is the diagonal of Y.)

(b) Show that the Künneth components $[\Delta]_i$ of the diagonal Δ of X can be expressed as combinations with rational coefficients of $[\Delta]$ and the $[L, \Lambda]^k$. Deduce that if the operator Λ is algebraic, then the $[\Delta]_i$ are algebraic.

(c) Show that if Λ is algebraic, then for every $i \leq n$, there exists a cycle $\Gamma_i \in \mathrm{CH}^{n-i}(X \times X)_{\mathbb{Q}}$ such that

$$[\Gamma_i]^* : H^{n+i}(X, \mathbb{Q}) \to H^{n-i}(X, \mathbb{Q})$$

is the inverse of the Lefschetz isomorphism

$$L^i : H^{n+i}(X, \mathbb{Q}) \to H^{n-i}(X, \mathbb{Q}).$$

NB. That the operator Λ is algebraic is the subject of the '$B(X, H)$ conjecture' (see Kleiman 1968). It is, of course, simply a special case of the Hodge conjecture.

11

The Bloch Conjecture and its Generalisations

In this final chapter, we assemble the evidence in favour of the conjectural converse to Mumford's theorem and its generalisations. We first consider Bloch's conjecture for surfaces satisfying the condition $p_g := \dim H^{2,0}(S) = 0$. This conjecture, which is the converse to theorem 10.15, is given in the following statement.

Conjecture 11.1 *Let S be a smooth complex projective surface such that $H^{2,0}(S) = 0$. Then the Albanese map*

$$\mathrm{alb}_S : \mathrm{CH}_0(S)_{\mathrm{hom}} \to \mathrm{Alb}\, S$$

is an isomorphism.

This conjecture was proved by Bloch, Kas & Lieberman in the case of surfaces which are not of general type, and we sketch their proof, which uses the classification of these surfaces. We also consider the example of certain types of Godeaux surfaces, which are surfaces of general type with $p_g = 0$, for which the conjecture is also proved.

In the second section, we turn to the the problem of formulating a general conjecture which includes the converse to theorem 10.31 and its functorial versions, where a variety and its Chow groups are replaced by a correspondence between varieties and the induced morphisms between their Chow groups. The conjecture which emerges is then a statement on the existence of a decreasing filtration $F^i\mathrm{CH}^k(X)_{\mathbb{Q}}$ on the Chow groups with rational coefficients of a variety X, which satisfies the following properties:

(i) The filtration is stable under correspondences.
(ii) $F^{k+1}\mathrm{CH}^k(X)_{\mathbb{Q}} = 0$.

(iii) If $\Gamma \subset X \times Y$, the map

$$\Gamma_* : \mathrm{Gr}^i_F \mathrm{CH}^k(X)_{\mathbb{Q}} \to \mathrm{Gr}^i_F \mathrm{CH}^l(Y)_{\mathbb{Q}}, \quad l = k + \operatorname{codim} \Gamma - \dim X$$

induced by Γ_* is zero if the morphism of Hodge structures

$$[\Gamma]_* : H^{2k-i}(X, \mathbb{Q}) \to H^{2l-i}(Y, \mathbb{Q})$$

is zero.

Shuji Saito proposed a construction of such a filtration. His filtration satisfies the first and the last properties, but the condition $F^{k+1}\mathrm{CH}^k(X)_{\mathbb{Q}} = 0$ remains a fundamental open problem. Murre also proposed a construction, based on the conjectural existence of a decomposition of the diagonal as a sum of idempotents acting on the Chow groups. This last point is particularly conjectural, in that even its cohomological version (the Hodge conjecture for the Künneth components of the diagonal) is not known.

The chapter ends with the example of abelian varieties, for which the existence of the Pontryagin product enabled Bloch to construct a filtration on the group CH_0 satisfying properties (i) and (ii) above. Finally, Beauville proved a theorem of decomposition of the Chow groups as direct sums of characteristic spaces for the action of homotheties, which is almost exactly what is predicted by the conjecture. For this, he used the 'Fourier transform' considered by Mukai in a different context, which passes from the Chow groups of an abelian variety A to the Chow groups of the dual abelian variety $\widehat{A}$.

11.1 Surfaces with $p_g = 0$

11.1.1 Statement of the conjecture

Let X be a smooth complex projective surface. Mumford's theorem 10.15 states that if $H^{2,0}(X) \neq 0$, then the group $\mathrm{CH}_0(X)$ is not representable, and by Roitman's theorem 10.11, this is equivalent to the fact that the Albanese map

$$\mathrm{alb}_X : \mathrm{CH}_0(X)_{\mathrm{hom}} \to \mathrm{Alb}\, X$$

is not injective.

Bloch (1980) conjectures the converse to Mumford's theorem.

Conjecture 11.2 *Let X be a smooth projective surface such that $H^{2,0}(X) = 0$. Then the map*

$$\mathrm{alb}_X : \mathrm{CH}_0(X)_{\mathrm{hom}} \to \mathrm{Alb}\, X$$

is an isomorphism.

In particular, if the surface X is also regular, i.e. satisfies the condition that the irregularity

$$q(X) := \dim H^{1,0}(X)$$

is zero, we should have $\mathrm{CH}_0(X)_{\mathrm{hom}} = 0$, since then $\mathrm{Alb}\, X = 0$.

The simplest examples of surfaces X satisfying the condition that the geometric genus

$$p_g(X) := \dim H^{2,0}(X)$$

is zero are ruled surfaces (or equivalently, by theorem 11.3 below, uniruled surfaces, i.e. those which are covered by rational curves (see Beauville 1978)). Such surfaces necessarily satisfy $H^{2,0}(X) = 0$, and more generally, $H^0(X, K_X^{\otimes n}) = 0$ for all $n > 0$. Indeed, for a non-constant morphism $\phi : \mathbb{P}^1 \cong C \to X$ from a rational curve to X, we have the adjunction formula

$$\phi^* K_X \otimes \det N_\phi = K_C = \mathcal{O}_{\mathbb{P}^1}(-2), \tag{11.1}$$

where $N_\phi = \phi^* T_X / \phi_* T_C$ is the normal sheaf. If moreover the deformations of $\phi(C)$ cover X, then $\deg \det N_\phi \geq 0$, since the normal bundle admits non-zero sections corresponding to the first-order deformations of ϕ, so formula (11.1) shows that $\deg \phi^* K_X < 0$. It follows that

$$H^0\left(\mathbb{P}^1, \phi^* K_X^{\otimes n}\right) = 0 \quad \text{for all} \quad n > 0,$$

which of course implies that $H^0(X, K_X^{\otimes n}) = 0$, since X is covered by the deformations of $\phi(C)$.

These surfaces satisfy conjecture 11.2. Indeed, since they are covered by rational curves, every point $x \in X$ of such a surface belongs to a rational curve which intersects a fixed ample curve $D \subset X$ in at least one point y. Then x is rationally equivalent to y, which shows that the inclusion $j : D \to X$ satisfies the property that the map

$$j_* : \mathrm{CH}_0(D) \to \mathrm{CH}_0(X)$$

is surjective, which implies the result by Roitman's theorem 10.11.

However, there exist surfaces which satisfy the condition $p_g = 0$ and which are not covered by rational curves, so that that the representability of the group CH_0 is not immediately visible geometrically.

Note, finally, that conjecture 11.2 is compatible with the fact that the group $\mathrm{CH}_0(X)$ and the space $H^{2,0}(X)$ are both birational invariants of a surface. For the Chow groups, this follows from the fact that if $\phi : X \dashrightarrow Y$ is a birational transformation then, by Hironaka, we can use successive blowups to construct a variety $\widetilde{X}$ which is birationally equivalent to X, and a diagram of morphisms

$$\begin{array}{ccc} \widetilde{X} & \xrightarrow{\tilde{\phi}} & Y \\ \downarrow{\scriptstyle \tau} & & \\ X & & \end{array}$$

such that $\phi = \tilde{\phi} \circ \tau^{-1}$ on a Zariski open set of X. Then, as ϕ is birational, $\tilde{\phi}$ is of degree 1, and the map

$$\tilde{\phi}^* : \mathrm{CH}_0(Y) \to \mathrm{CH}_0(\widetilde{X})$$

admits $\tilde{\phi}_*$ as a left inverse. Now, by theorem 9.27, the map

$$\tau_* : \mathrm{CH}_0(\widetilde{X}) \to \mathrm{CH}_0(X)$$

is an isomorphism. Thus, we have an injective map

$$\tau_* \circ \tilde{\phi}^* : \mathrm{CH}_0(Y) \to \mathrm{CH}_0(X).$$

Using the fact that $\tau_* \circ \tau^* = \mathrm{Id}$, we also see that

$$\tilde{\phi}_* \circ \tau^* : \mathrm{CH}_0(X) \to \mathrm{CH}_0(Y)$$

is surjective. Exchanging the roles of X and Y, and noting that the morphisms above do not depend on the choice of desingularisation of ϕ, we conclude that $\tau_* \circ \tilde{\phi}^*$ is an isomorphism.

As for the spaces $H^0(X, K_X)$, and more generally $H^0(X, \Omega_X^q)$, their birational invariance follows from Hartogs' theorem (see vI.1.2.2), which says that one can extend the sections of a vector bundle (here Ω_X^q) defined on the complement of a Zariski closed set of codimension at least 2, and from the fact that a rational map with values in a projective variety is defined on the complement of a Zariski closed subset of codimension at least 2.

11.1.2 Classification

Let X be a projective surface. The Kodaira dimension $\kappa(X)$ of X is defined in the following way. For each integer $n \geq 0$, if the linear system $|nK_X|$ is non-zero, it gives a rational map

$$\Phi_{nK_X} : X \dashrightarrow \mathbb{P}^r, \quad r = h^0(X, nK_X) - 1.$$

We then set

$$\kappa(X) = -\infty \quad \text{if} \quad h^0(X, nK_X) = 0 \quad \text{for all} \quad n > 0,$$
$$\kappa(X) = \overline{\lim_{n\to\infty}} \ \dim \operatorname{Im} \Phi_{nK_X} \ \text{otherwise.}$$

The classification of surfaces up to birational equivalence gives quite a good description of surfaces with Kodaira dimension ≤ 1. (The other surfaces are said to be 'of general type'.)

Theorem 11.3 (see Beauville 1978) *A surface whose Kodaira dimension is equal to $-\infty$ is birationally ruled, i.e. birationally equivalent to a surface X equipped with a morphism*

$$X \to C$$

all of whose fibres are isomorphic to $\mathbb{P}^1$.

When the irregularity $q = \dim \operatorname{Alb} X$ is non-zero, we obtain such a morphism by considering the Albanese map of X. When $q = 0$, the curve C above should be of genus 0, i.e. isomorphic to $\mathbb{P}^1$. Such surfaces are clearly rational, i.e. birationally equivalent to $\mathbb{P}^2$. These surfaces are characterised by the following theorem, due to Castelnuovo (see Beauville 1978).

Theorem 11.4 *A smooth projective surface is rational if and only if it satisfies the equalities $q = p_g = p_2 = 0$, where $p_2 := \dim H^0(X, K_X^{\otimes 2})$.*

When a surface X has Kodaira dimension ≥ 0, we can show that it has a unique minimal model, i.e. a smooth projective surface Y such that there exists a birational morphism $X \to Y$, but which has no smooth rational curve of self-intersection -1. (By a result due to Castelnuovo, such a curve can be contracted to a point, which gives a smooth projective surface.) From now on, we will only consider minimal surfaces, i.e. surfaces which have no rational curve of self-intersection -1.

Theorem 11.5 *Let X be a minimal surface with $\kappa(X) = 0$. Then we have $q(X) \leq 2$, and also the following results:*

(i) If $q(X) = 2$, then X is an abelian surface, i.e. a projective complex torus of dimension 2. In particular, $H^0(X, K_X) \neq 0$.
(ii) If $q(X) = 1$, then X is a quotient of the product of two elliptic curves E and F (i.e. of 1-dimensional complex tori) by a finite group G acting with no fixed points, via an action which is compatible with the product

structure (i.e. $G \subset \operatorname{Aut} F \times \operatorname{Aut} E$). If $p_g(X) = 0$, then G acts on E via translations, and the action on F is such that the quotient F/G is a rational curve.

(iii) *If $q(X) = 0$, then X is a $K3$ surface (see vI.7.2.3; Barth et al. 1984), or an Enriques surface, i.e. the quotient of a $K3$ surface Y by an involution without fixed point acting by -1 on $H^0(Y, K_Y)$. In the second case only, we have $p_g(X) = 0$.*

Remark 11.6 *The inequality $q \leq 2$ follows easily from the fact that $K_X^2 := c_1(K_X)^2 = 0$ for a minimal surface of Kodaira dimension 0. To show this, one uses the symmetry of the Hodge numbers and Woether's Noether is equality*

$$\chi(X, \mathcal{O}_X) := \sum_i (-1)^i h^i(X, \mathcal{O}_X) = \frac{c_1(K_X)^2 + c_2(X)}{12}$$

with $c_2(X) := c_2(T_X) = \chi_{\text{top}}(X) = \sum_i (-1)^i b_i(X)$ (Hopf's theorem).

The same argument also shows that if $\kappa(X) = 0$ and $H^0(X, K_X) = 0$, then $q \leq 1$.

Now consider minimal surfaces with Kodaira dimension equal to 1.

Theorem 11.7 *If $\kappa(X) = 1$ and $q > 0$, then X admits a morphism $\phi : X \to C$ whose fibre is an elliptic curve.*

If $\kappa(X) = 1$ and $q(X) = 0$, then X admits a pencil of elliptic curves $(E_t)_{t \in \mathbb{P}^1}$.

Remark 11.8 *Conversely, a surface which is fibred in elliptic curves always satisfies the condition that $\kappa(X) \leq 1$. Indeed, if E is the generic fibre of the fibration $\phi : X \to C$, then the normal bundle of E in X is trivial, as is always the case for the fibre of a fibration, and the adjunction formula then shows that $K_{X|E} = K_E$, where K_E is the trivial bundle if E is an elliptic curve. But then it is clear that the pluricanonical maps Φ_{nK_X} factor through ϕ.*

Finally, let us consider the surfaces X satisfying $\kappa(X) = 1$ and $p_g(X) = 0$.

Theorem 11.9 *If X is a surface satisfying $\kappa(X) = 1$ and $p_g(X) = 0$, then $q(X) \leq 1$. If $q(X) = 1$, then the Albanese map*

$$\text{alb} : X \to \text{Alb}\, X = E$$

is an isotrivial fibration if the fibres are of genus >1, or an elliptic fibration with smooth isomorphic fibres if the fibres are of genus 1. In the latter case, *replacing $X \to E$ by the associated Jacobian fibration $J \to E$ (see section 7.1), one can show that $\kappa(J) = 0$, $q(J) = 1$, so that J is of the type described in theorem 11.5(ii).*

The inequality $q(X) \leq 1$ again follows from Noether's formula and from the equality $K_X^2 = 0$. The isotriviality of the fibration can be deduced from the positivity of the bundles $R^0\mathrm{alb}_{X*}K_{X/E}$.

11.1.3 Bloch's conjecture for surfaces which are not of general type

For surfaces with Kodaira dimension equal to $-\infty$, Bloch's conjecture is satisfied; this follows from theorem 11.3 together with the observations at the end of subsection 11.1.1.

Let us now sketch the arguments of Bloch *et al.* (1976) concerning conjecture 11.2 for surfaces satisfying $\kappa(X) \leq 1$.

Theorem 11.10 *Conjecture 11.2 holds for surfaces which are not of general type.*

Proof Consider the surfaces X whose Kodaira dimension is 0 or 1 and which satisfy the condition $p_g(X) = 0$. By theorems 11.5 and 11.7, they can be of two possible types, according to their irregularity.

(i) $q(X) = 1$. Then the Albanese map of X is an isotrivial fibration whose fibres are curves of genus ≥ 1, and if $g \geq 2$, then by theorem 11.9, X is the quotient of a product $C \times \widetilde{E}$ by a finite group G acting with no fixed points, via an inclusion $G \subset \mathrm{Aut}\, C \times \mathrm{Aut}\, \widetilde{E}$. Here $\widetilde{E}$ is an étale cover of $E = \mathrm{Alb}\, X$ giving a trivialisation of the fibration $\mathrm{alb}_X : X \to E$.

In particular the group G acts by translations on $\widetilde{E}$, which is also an elliptic curve.

When the fibres of alb_X are of genus 1, the map alb_X can have multiple fibres, but these disappear when the elliptic fibration $X \to E$ is replaced by its associated Jacobian fibration $J \to E$. Now, it is easy to see that X and J have the same Chow groups CH_0 with rational coefficients. Indeed, if the fibration $X \to E$ has a section σ, then clearly X and J are birationally equivalent via the map

$$x \in X_b \mapsto \mathrm{alb}_{X_b}(x - \sigma(b)),$$

where X_b is the elliptic curve $\mathrm{alb}_X^{-1}(b) \subset X$. In general, let $C \subset X$ be a curve which is not contained in the fibres of alb_X and is not ramified over E over the critical points of alb_X. Then the fibred product $X_C := X \times_E C$ is a an elliptic curve fibration alb_X which has a section, so it is birationally equivalent to the associated Jacobian fibration $J_C = J \times_E C$. If $\Gamma \subset X_C \times J_C$ is the graph of this isomorphism, we easily check that the image Γ' of Γ in $X \times J$ induces an isomorphism

$$\Gamma'_* : \mathrm{CH}_0(X)_{\mathrm{hom}} \otimes \mathbb{Q} \to \mathrm{CH}_0(J)_{\mathrm{hom}} \otimes \mathbb{Q},$$

using the fact that it does on fibres. Replacing X by J, theorem 11.9 then allows us to reduce to the preceding situation, where we are dealing with the quotient of a product of two curves C and $\widetilde{E}$ by a finite group.

Note that as G acts by translations on $\widetilde{E}$, G acts trivially on $H^0(\widetilde{E}, \Omega_{\widetilde{E}})$, and the condition

$$H^0(X, K_X) = 0 = H^0(C \times \widetilde{E}, K_{C \times \widetilde{E}})^G$$

then implies that $H^0(C, \Omega_C)^G = 0$, where the upper index G denotes the G-invariant part. Thus, the action of G on C satisfies $C/G \cong \mathbb{P}^1$, and

$$\mathrm{CH}_0(C)^G_{\mathrm{hom}} = 0.$$

Furthermore, up to torsion, the group G acts trivially on $\mathrm{CH}_0(\widetilde{E})$. Indeed, as it acts trivially on $H^0(\widetilde{E}, \Omega_{\widetilde{E}})$, it acts trivially on $\mathrm{Alb}\,\widetilde{E}$, and thus it also acts trivially on $\mathrm{CH}_0(\widetilde{E})_{\mathrm{hom}}$. Moreover, it obviously acts trivially on the quotient

$$\mathrm{CH}_0(\widetilde{E})/\mathrm{CH}_0(\widetilde{E})_{\mathrm{hom}} \cong H^2(\widetilde{E}, \mathbb{Z}) = \mathbb{Z},$$

so the action on $\mathrm{CH}_0(\widetilde{E})$ is of the form

$$g^*z = z + (\deg z)z_g$$

with $z_g \in \mathrm{CH}_0(\widetilde{E})_{\mathrm{hom}}$. As G is torsion, iterating this formula shows that z_g is also torsion.

Let now $(c, e) \in C \times \widetilde{E}$; then for $g \in G$, writing $g = (g_1, g_2)$ in $\mathrm{Aut}\,C \times \mathrm{Aut}\,\widetilde{E}$, we have

$$g(c, e) = (g_1c, g_2e) \equiv_{\mathbb{Q}} (g_1c, e),$$

where the equivalence relation is that of rational equivalence modulo torsion in $C \times \widetilde{E}$. We thus obtain

$$\sum_{g \in G} g(c, e) \equiv_{\mathbb{Q}} \sum_{g \in G} (g_1c, e) \quad \text{in} \quad \mathrm{CH}_0(C \times \widetilde{E}) \otimes \mathbb{Q}.$$

Now, we know that $\sum_{g\in G} g_1 c \in \mathrm{CH}_0(C)$ is a constant independent of c. To conclude, in the group

$$\mathrm{CH}_0(X) \otimes \mathbb{Q} = \mathrm{CH}_0(C \times \widetilde{E})^G \otimes \mathbb{Q},$$

we have the equality

$$\sum_{g\in G} g(c, e) = \sum_{g\in G} (g_1 c_0, e),$$

where c_0 is a fixed point of C. As $\pi^*\mathrm{CH}_0(X)$ is generated by cycles of the form $\sum_{g\in G} g(c, e)$, where $\pi : C \times \widetilde{E} \to X$ is the quotient map, and furthermore $\pi_* \circ \pi^* = \deg \pi \, \mathrm{Id}_{\mathrm{CH}_0(X)}$, we conclude that $\mathrm{CH}_0(X)_{\mathbb{Q}}$ is supported on the image of the curve $c_0 \times \widetilde{E}$ in X. This shows that $\mathrm{CH}_0(X)$ is finite-dimensional, so alb_X is an isomorphism by Roitman's theorem 10.11.

(ii) $q(X) = 0$. By theorems 11.5 and 11.7, we see that X has a pencil of elliptic curves. (In the case of Enriques surfaces, this follows from a study of the Picard group.) After blowing up the base locus, such a pencil gives a fibration

$$X \to \mathbb{P}^1$$

into elliptic curves. Let J be the associated Jacobian fibration. We have already seen that $\mathrm{CH}_0(X)_{\mathbb{Q}} = \mathrm{CH}_0(J)_{\mathbb{Q}}$, and this even implies that $\mathrm{CH}_0(X) = \mathrm{CH}_0(J)$ by Roitman's theorem 10.14 and the fact that the considered surfaces have trivial Albanese varieties. We have now the following.

Proposition 11.11 *If $p_g(X) = q(X) = 0$, then $p_2(J) = q(J) = 0$.*

Proof This proposition is proved using results on the canonical bundle of an elliptic fibration due to Kodaira (1963a). The disappearance of the multiple fibres in the Jacobian fibration then shows that $p_g(J) = 0 \Rightarrow p_2(J) = 0$. □

Castelnuovo's theorem 11.4 then applies to J and shows that J is rational. Thus, we have $\mathrm{CH}_0(J) = \mathbb{Z} = \mathrm{CH}_0(X)$, and X satisfies Bloch's conjecture. □

11.1.4 Godeaux surfaces

Let us now consider the two following families of surfaces of general type satisfying the condition $p_g = 0$. (Other constructions can be found, for example in Peters (1977).)

(i) Consider the projective space $\mathbb{P}^3$ equipped with the following action of the group $G = \mathbb{Z}/5\mathbb{Z}$. Let g be a generator of G, and choosing an inclusion $G \subset \mathbb{C}^*$, let $\eta \in \mathbb{C}^*$ be the 5th root of unity corresponding to g. The action of $g \in G$ on the homogeneous coordinates $X_0, \ldots, X_3$ is then given by

$$g^* X_i = \eta^i X_i, \quad i = 0, \ldots, 3. \tag{11.2}$$

Lemma 11.12 *If $F \in H^0(\mathbb{P}^3, \mathcal{O}_{\mathbb{P}^3}(5))$ is a generic polynomial invariant under G, then the surface $S = V(F)$ is smooth, the action of G on S has no fixed points, and the surface $\Sigma = S/G$ is smooth of general type and satisfies $H^0(\Sigma, K_\Sigma) = 0$.*

Proof As the space $H := H^0(\mathbb{P}^3, \mathcal{O}_{\mathbb{P}^3}(5))^G$ of polynomials invariant under G contains the X_i^5, it has no base points, so by Bertini's lemma, the surface S is smooth for general F. A local computation then shows that if S is smooth, F cannot vanish at the fixed points $(0, \ldots, 1, \ldots, 0)$ of the action of G on $\mathbb{P}^3$, so the action of G on S must be free. Like S, the surface Σ is thus smooth of general type, and satisfies

$$H^0(\Sigma, K_\Sigma) = H^0(S, K_S)^G.$$

Finally, the action of $g \in G$ on the generator

$$\Omega = \sum_i (-1)^i X_i dX_0 \wedge \cdots \wedge d\widehat{X}_i \wedge \cdots \wedge dX_3$$

of $H^0(\mathbb{P}^3, K_{\mathbb{P}^3}(4))$ is given by $g^*\Omega = \eta^6\Omega = \eta\Omega$. Thus, if

$$\omega = \mathrm{Res}_S\, P\Omega/F \in H^0(S, K_S),$$

where P is a homogeneous polynomial of degree 1, then

$$g^*\omega = \eta\, \mathrm{Res}_S g^* P \cdot \Omega/F.$$

But by (11.2), g^* does not admit the eigenvalue $\eta^{-1} = \eta^4$ on $H^0(\mathbb{P}^3, \mathcal{O}_{\mathbb{P}^3}(1))$, so we must have $H^0(S, K_S)^G = 0$. □

(ii) A similar example can be obtained by considering the action of $G = \mathbb{Z}/8\mathbb{Z}$ on $\mathbb{P}^6$ given by

$$g^* X_i = \eta^i X_i,\ 0 \le i \le 6,$$

where η is a primitive 8th root of unity. We then consider surfaces S which are complete intersections of four quadrics Q_i, $i = 1, \ldots, 4$, satisfying

$$g^*Q_1 = Q_1, \quad g^*Q_2 = \eta^2 Q_2, \quad g^*Q_3 = \eta^4 Q_3, \quad g^*Q_4 = \eta^6 Q_4.$$

As before, we have the following lemma.

Lemma 11.13 *If the Q_i are generic, then the surface $S = \bigcap_i V(Q_i)$ is smooth, the action of G on S has no fixed points, and the surface $\Sigma = S/G$ is smooth of general type and satisfies $H^0(\Sigma, K_\Sigma) = 0$.*

By Lefschetz' theorem on hyperplane sections, complete intersection surfaces in projective space satisfy the condition $H^1(S, \mathbb{Z}) = 0$, so the surfaces Σ constructed above satisfy the condition $\mathrm{Alb}\,\Sigma = 0$. Bloch's conjecture thus predicts that these surfaces satisfy $\mathrm{CH}_0(\Sigma) = \mathbb{Z}$. This was proved by Inoze & Mizukami (1979) for certain special cases of these surfaces, and by Voisin (1992) in general.

Theorem 11.14 *The Godeaux surfaces constructed in examples (i) and (ii) above satisfy* $\mathrm{CH}_0(\Sigma) = \mathbb{Z}$.

Proof We will give the proof in case (i), the other case being similar. Note first that we have a natural isomorphism

$$\pi^* : \mathrm{CH}_0(\Sigma) \otimes \mathbb{Q} \rightarrow \mathrm{CH}_0(S)^G_{\mathbb{Q}},$$

where $\pi : S \to \Sigma$ is the quotient map. Indeed, π^* is injective on the Chow groups with rational coefficients, since $\pi_* \circ \pi^*$ is equal to multiplication by $\deg \pi = 5$. Furthermore, if $z = \sum_i n_i x_i \in \mathrm{CH}_0(S)^G$, then

$$5z = \sum_{g \in G} g_* z = \sum_i n_i G x_i \in \pi^* \mathrm{CH}_0(\Sigma),$$

where Gx_i is the cycle of S given by the orbit of x_i. Thus, π^* is also surjective on the rational Chow groups.

Moreover, if

$$\mathrm{cl} : \mathrm{CH}_0(\Sigma)_{\mathbb{Q}} \to \mathbb{Q} = H^4(\Sigma, \mathbb{Q})$$

is an isomorphism, then by Roitman's theorem 10.14, so is the map

$$\mathrm{cl} : \mathrm{CH}_0(\Sigma) \to \mathbb{Z} = H^4(\Sigma, \mathbb{Z}).$$

To conclude, it thus suffices to show that $\mathrm{CH}_0(S)^G_{\mathrm{hom}} \otimes \mathbb{Q} = 0$.

Note that if x, y are two generic points of S, there exists a G-invariant element $F' \in H$ defining a smooth, G-invariant surface S' such that the curve $C := S \cap S'$ is smooth (and G-invariant) and contains x and y. This follows immediately from Bertini's lemma, together with the fact that the map from $\mathbb{P}^3$ to $\mathbb{P}^M$, where $M+1 = \mathrm{rank}\, H$, given by the linear system $H \subset H^0(\mathbb{P}^3, \mathcal{O}_{\mathbb{P}^3}(5))$ is finite on its image.

Let $j : C \to S$ be inclusion. Clearly, the cycle $Gx - Gy$ belongs to

$$\mathrm{Im}\left(j_* : \mathrm{CH}_0(C)^G_{\mathrm{hom}} \to \mathrm{CH}_0(S)^G_{\mathrm{hom}}\right).$$

As the cycles of the form $Gx - Gy$, with $x,\ y$ generic in S, generate

$$\mathrm{CH}_0(S)^G_{\mathrm{hom}} \otimes \mathbb{Q} = \pi^*\mathrm{CH}_0(\Sigma)_{\mathbb{Q}},$$

it suffices to show that for C as above, the map

$$j_* : \mathrm{CH}_0(C)^G_{\mathrm{hom}} \to \mathrm{CH}_0(S)^G_{\mathrm{hom}}$$

is zero.

Proposition 11.15 *Let $(S_t)_{t\in\mathbb{P}^1}$ denote the pencil of surfaces generated by S and S'. For $t \in \mathbb{P}^1$, let $j_t : C \hookrightarrow S_t$ denote inclusion. Then $\mathrm{CH}_0(C)^G_{\mathrm{hom}}$ is generated by the cycles of the form $j_t^* D_t$, where D_t is a G-invariant divisor of S_t which is of degree 0, i.e. such that $\deg D_t|_C = 0$.*

Remark 11.16 *The surface S_t is not necessarily smooth, but the map j_t^* is well-defined by the fact that $C \subset S_t$ is a Cartier divisor, or more simply, by the fact that S_t is smooth along C, so that we can replace $\mathrm{Im}\, j_t^*$ by $\mathrm{Im}\, \tilde{j}_t^*$, where $\tilde{j}_t : C \subset \widetilde{S}_t$ is the inclusion of C into a desingularisation of S_t.*

Admitting proposition 11.15, we conclude by observing that if

$$k : S \hookrightarrow \mathbb{P}^3, \quad k_t : S_t \hookrightarrow \mathbb{P}^3$$

are the natural inclusions, we have a Cartesian diagram

$$\begin{array}{ccc} C & \xrightarrow{j} & S \\ \downarrow{\scriptstyle j_t} & & \downarrow{\scriptstyle k} \\ S_t & \xrightarrow{k_t} & \mathbb{P}^3 \end{array}$$

which identifies the normal bundles of C in S and S_t in $\mathbb{P}^3$, so that we have the equality (see Fulton 1984)

$$j_* \circ j_t^* = k^* \circ k_{t*} : \mathrm{CH}_1(S_t) \to \mathrm{CH}_0(S).$$

But by theorem 9.25, we have $\mathrm{CH}_1(\mathbb{P}^3)_{\mathrm{hom}} = 0$, so $k_{t*}(D_t) = 0$ in $\mathrm{CH}_1(\mathbb{P}^3)$ for every divisor D_t of degree 0 on S_t. Thus, $j_* \circ j_t^* D_t = 0$ for such a divisor, and by proposition 11.15, it follows that

$$j_* : \mathrm{CH}_0(C)^G_{\mathrm{hom}} \to \mathrm{CH}_0(S)^G_{\mathrm{hom}}$$

is zero. □

Proof of proposition 11.15 Let $\widetilde{\mathbb{P}^3}$ be the blowup of $\mathbb{P}^3$ along C, and

$$f : \widetilde{\mathbb{P}^3} \to \mathbb{P}^1$$

the map given by the pencil of surfaces $(S_t)_{t\in\mathbb{P}^1}$. By theorem 9.27, we know that we have an isomorphism

$$\mathrm{CH}_0(C)^G_{\mathrm{hom}} \cong \mathrm{CH}_1(\widetilde{\mathbb{P}^3})^G_{\mathrm{hom}}, \tag{11.3}$$

where by the functoriality of blowups, G acts naturally on $\widetilde{\mathbb{P}^3}$. It is not hard to see that if D_t is a G-invariant divisor of degree 0 on S_t, this isomorphism maps $j_t^* D_t$ to the class of $\tilde{k}_{t\,*}D_t$, where $\tilde{k}_t$ is the inclusion of S_t into $\widetilde{\mathbb{P}^3}$. Furthermore, via the Abel–Jacobi map, the isomorphism (11.3) can be identified with the corresponding isomorphism

$$J(C)^G \cong J^3(\widetilde{\mathbb{P}^3})^G$$

between the intermediate Jacobians. It thus suffices to prove that the Abel–Jacobi invariants

$$\Phi^2_{\widetilde{\mathbb{P}^3}}(\tilde{k}_{t\,*}D_t) \in J^3(\widetilde{\mathbb{P}^3})_G,$$

where the D_t are the G-invariant divisors of degree 0 on S_t, $t \in \mathbb{P}^1$, generate $J^3(\widetilde{\mathbb{P}^3})^G$.

The smooth surfaces S_t of the pencil satisfy the property that $H^{2,0}(S_t)^G = 0$. By Lefschetz theorem on (1, 1)-classes, the cohomology $H^2(S_t, \mathbb{Z})^G$ is thus generated by the classes of G-invariant divisors on S_t. As this group is of finite type, it is generated by a finite number of classes of such divisors. Now, let $U \subset \mathbb{P}^1$ be the open set parametrising the smooth surfaces S_t. Via an argument using relative Hilbert schemes, one can show that there exists a smooth curve

$$D \xrightarrow{r} \mathbb{P}^1$$

such that the fibre $r^{-1}(t) = \{d_{1,t}, \ldots, d_{N,t}\}$ parametrises G-invariant divisors $D_{1,t}, \ldots, D_{N,t}$ on the fibre S_t, whose cohomology classes generate the invariant cohomology $H^2(S_t, \mathbb{Z})^G$ for every $t \in U$.

Let

$$\alpha : R^0 r_* \mathbb{Z} \to R^2 f_* \mathbb{Z}^G$$

be the morphism of local systems on U which maps the class $[d_{i,t}] \in H^0(D_t, \mathbb{Z})$ of the point $d_{i,t} \in D_t = r^{-1}(t)$ to the class $[D_{i,t}] \in H^2(S_t, \mathbb{Z})^G$ of the cycle $D_{i,t} \subset S_t = f^{-1}(t)$. By definition, α is surjective on U, so it induces a surjective map

$$H^1(\alpha) : H^1(U, R^0 r_*\mathbb{Z}) \to H^1(U, R^2 f_*\mathbb{Z}^G),$$

by the fact that U has the homotopy type of a 1-dimensional CW-complex (see theorem 1.22).

Note that we have a codimension 1 cycle

$$\mathcal{D} \subset D \times_{\mathbb{P}^1} \widetilde{\mathbb{P}^3},$$

with fibre $D_{i,t} \subset S_t$ over the point $d_{i,t} \in D$, and that α can be identified with the image of the cohomology class

$$[\mathcal{D}] \in H^2(D \times_{\mathbb{P}^1} \widetilde{\mathbb{P}^3}, \mathbb{Z})$$

in $H^0(U, R^2 f'_*\mathbb{Z})$, where $f' : D \times_{\mathbb{P}^1} \widetilde{\mathbb{P}^3} \to \mathbb{P}^1$ is the natural map with fibre $r^{-1}(t) \times f^{-1}(t)$. By the inclusion

$$D \times_{\mathbb{P}^1} \widetilde{\mathbb{P}^3} \subset D \times \widetilde{\mathbb{P}^3},$$

$\mathcal{D}$ also gives a cycle $\mathcal{D} \subset D \times \widetilde{\mathbb{P}^3}$ of codimension 2. The following result can easily be deduced from this.

Lemma 11.17 *Let $0 \in D$ be a reference point, $0' = r(0) \in \mathbb{P}^1$, and let*

$$\beta : JD \to J^3(\widetilde{\mathbb{P}^3})^G$$

be the morphism

$$\beta(\mathrm{alb}\,(d - 0)) = \Phi^2_{\widetilde{\mathbb{P}^3}}(k_{t*}D_d - k_{0'*}D_0), \quad d \in D, \quad t = r(d)$$

induced by the Abel–Jacobi map of $\widetilde{\mathbb{P}^3}$ and the cycle $\mathcal{D} \subset D \times \widetilde{\mathbb{P}^3}$. Then the corresponding morphism of Hodge structures

$$\beta_{\mathbb{Z}} : H^1(D, \mathbb{Z}) \to H^3(\widetilde{\mathbb{P}^3}, \mathbb{Z})^G$$

makes the following diagram commute:

$$
\begin{array}{llcl}
\beta_{\mathbb{Z}} : & H^1(D,\mathbb{Z}) & \longrightarrow & H^3(\widetilde{\mathbb{P}}^3,\mathbb{Z})^G \\
 & \downarrow & & \downarrow \\
 & H^1(r^{-1}(U),\mathbb{Z}) & \longrightarrow & H^3(f^{-1}(U),\mathbb{Z})^G \\
 & \| & & \| \\
H^1(\alpha) : & H^1(U, R^0 r_*\mathbb{Z}) & \longrightarrow & H^1(U, R^2 f_*\mathbb{Z})^G .
\end{array}
\tag{11.4}
$$

Here, the upper vertical arrows are the restrictions, and the vertical equalities are given by the Leray spectral sequences for r and f and the fact that on U, we have

$$R^1 r_*\mathbb{Z} = 0, \quad R^3 f_*\mathbb{Z} = 0.$$

Proof of lemma 11.17 We know that $\beta_{\mathbb{Z}}$ is the morphism of Hodge structures induced by the Künneth component of type $(1, 3)$ of the Hodge class

$$[\mathcal{D}] \in H^4(D \times \widetilde{\mathbb{P}}^3, \mathbb{Z})$$

(see vI.12.2.1). The compatibility of the diagram 11.4 then follows from the fact that the morphism α is induced by the image of

$$[\mathcal{D}] \in H^2(D \times_{\mathbb{P}^1} \widetilde{\mathbb{P}}^3, \mathbb{Z})$$

in

$$H^0(R^2 f'_*\mathbb{Z}) = \mathrm{Hom}\,(R^0 r_*\mathbb{Z}, R^2 f_*\mathbb{Z}).$$

□

It is easy to see that $H^1(\alpha)$ is a morphism of mixed Hodge structures, and as $H^1(\alpha)$ is surjective, we conclude that the composed map

$$H^1(D,\mathbb{Q}) \stackrel{\beta_{\mathbb{Q}}}{\to} H^3(\widetilde{\mathbb{P}}^3,\mathbb{Q})^G \stackrel{j^*}{\to} H^3(f^{-1}(U),\mathbb{Q})^G$$

has the same image as the restriction map j^*, where j is the inclusion of $f^{-1}(U)$ in $\widetilde{\mathbb{P}}^3$. Indeed, this follows from the strictness of the morphisms of mixed Hodge structures (theorem 4.20) and from the fact that

$$\mathrm{Im}\,(H^1(D,\mathbb{Q}) \to H^1(r^{-1}(U),\mathbb{Q})) = W_0 H^1(r^{-1}(U),\mathbb{Q}),$$
$$\mathrm{Im}\,(j^* : H^3(\widetilde{\mathbb{P}}^3,\mathbb{Q})^G \to H^3(f^{-1}(U),\mathbb{Q})^G) = W_0 H^3(f^{-1}(U),\mathbb{Q})^G.$$

But the kernel of the restriction map

$$H^3(\widetilde{\mathbb{P}}^3, \mathbb{Q})^G \to H^3(f^{-1}(U), \mathbb{Q})^G$$

is generated by the images of the $\tilde{k}_{t_i *} : H^1(\widetilde{S}_{t_i}, \mathbb{Q})^G \to H^3(\widetilde{\mathbb{P}}^3, \mathbb{Q})^G$, where t_i runs through the set of critical values of the pencil. It follows that the map

$$\beta_{\mathbb{Q}} \oplus_i \tilde{k}_{t_i *} : H^1(D, \mathbb{Q}) \oplus_i H^1(\widetilde{S}_{t_i}, \mathbb{Q})^G \to H^3(\widetilde{\mathbb{P}}^3, \mathbb{Q})^G$$

is surjective. Now, $\beta_{\mathbb{Q}} \oplus_i \tilde{k}_{t_i *}$ is the morphism of Hodge structures corresponding to the morphism of abelian varieties

$$\beta \oplus_i \Phi^2_{\widetilde{\mathbb{P}}^3} \circ \tilde{k}_{t_i *} : J(D) \oplus_i \mathrm{Pic}^0(\widetilde{S}_{t_i}) \to J^3(\widetilde{\mathbb{P}}^3)^G.$$

Thus, $\beta \oplus_i \Phi^2_{\widetilde{\mathbb{P}}^3} \circ \tilde{k}_{t_i *}$ is surjective, which means that the Abel–Jacobi invariants of the combinations homologous to 0 of 1-cycles of the form $k_{t*}D_t$ or $\tilde{k}_{t_i *}D_{t_i}$, where D_t and D_{t_i} are G-invariant divisors, generate $J^3(\widetilde{\mathbb{P}}^3)^G$. This proves proposition 11.15. □

11.2 Filtrations on Chow groups

11.2.1 The generalised Bloch conjecture

If S, T are smooth projective surfaces and Γ is a cycle of codimension 2 in $S \times T$, we have the morphism

$$\Gamma_* : \mathrm{CH}_0(S) \to \mathrm{CH}_0(T),$$

which is compatible with the map cl, and more generally cl_D (see subsection 9.2.4), so that in fact we have commutative diagrams

$$\begin{array}{lccc} \Gamma_* : & \mathrm{CH}_0(S) & \longrightarrow & \mathrm{CH}_0(T) \\ & \downarrow{\scriptstyle \mathrm{cl}} & & \downarrow{\scriptstyle \mathrm{cl}} \\ [\Gamma]_* : & H^4(S, \mathbb{Z}) & \longrightarrow & H^4(T, \mathbb{Z}) \end{array}$$

and

$$\begin{array}{lccc} \Gamma_* : & \mathrm{CH}_0(S)_{\mathrm{hom}} & \longrightarrow & \mathrm{CH}_0(T)_{\mathrm{hom}} \\ & \downarrow{\scriptstyle \mathrm{alb}_S} & & \downarrow{\scriptstyle \mathrm{alb}_T} \\ [\Gamma]_* : & \mathrm{Alb}\, S = J^3(S) & \longrightarrow & \mathrm{Alb}\, T = J^3(T). \end{array}$$

Here, the maps $[\Gamma]_*$ are induced by the Hodge class of Γ. Another way to formulate this is by introducing the filtration

$$F^0\mathrm{CH}_0(W) = \mathrm{CH}_0(W), \quad F^1\mathrm{CH}_0(W) = \mathrm{CH}_0(W)_{\mathrm{hom}},$$
$$F^2\mathrm{CH}_0(W) = \mathrm{Ker}\,\mathrm{alb}_W \subset F^1\mathrm{CH}_0(W),$$

where W is a smooth projective complex variety. Then the morphism Γ_* is compatible with the filtrations F on $\mathrm{CH}_0(S)$ and $\mathrm{CH}_0(T)$. Proposition 10.24 then implies the following result.

Proposition 11.18 *If Γ_* vanishes on $F^2\mathrm{CH}_0(S)$, then the map $[\Gamma]^* : H^2(T, \mathbb{C}) \to H^2(S, \mathbb{C})$ vanishes on $H^{2,0}(T)$.*

Proof Let $j : C \hookrightarrow S$ be an ample smooth curve. By the Lefschetz theorem on hyperplane sections, we know that the map $j_* : \mathrm{Alb}\,C \to \mathrm{Alb}\,S$ is surjective. If the map Γ_* vanishes on $F^2\mathrm{CH}_0(S)$, we thus have $\mathrm{Ker}\,\Gamma_* + \mathrm{Im}\,j_* = \mathrm{CH}_0(S)$ and thus

$$\mathrm{Im}\,\Gamma_* = \Gamma_*(j_*\mathrm{CH}_0(C)) = \mathrm{Im}\,\Gamma_{C*},$$

where $\Gamma_C = \Gamma_{|C\times T}$ is the restriction of Γ to C. But Γ_C is a 1-cycle in $C \times T$, so its support is a curve which projects to a curve C' of T. Clearly, $\mathrm{Im}\,\Gamma_{C*}$ consists of cycles supported on C'. Thus, $\mathrm{Im}\,\Gamma_*$ consists of cycles supported on C', and we can apply proposition 10.24, which shows that

$$[\Gamma]^* : H^{2,0}(T) \to H^{2,0}(S)$$

is the zero map. □

Conjecture 11.2 is a special case of the following conjecture, which is the converse to proposition 11.18.

Conjecture 11.19 *If the codimension 2 cycle $\Gamma \subset S \times T$ satisfies the condition that $[\Gamma]^* = 0$ on $H^0(T, \Omega^2_T)$, then the morphism*

$$\Gamma_* : \mathrm{CH}_0(S) \to \mathrm{CH}_0(T)$$

vanishes on the subgroup $F^2\mathrm{CH}_0(S)$.

Remark 11.20 *Define $H^2(T, \mathbb{Q})_{\mathrm{tr}}$ to be the smallest Hodge substructure of $H^2(T, \mathbb{Q})$ containing $H^{2,0}(T)$. (By the fact that the intersection form on $H^2(T)$ is non-degenerate on $H^{2,0}(T) \oplus H^{0,2}(T)$ (see vI.6.3.2), $H^2(T, \mathbb{Q})_{\mathrm{tr}}$ can also be identified with the orthogonal complement of the subgroup $\mathrm{NS}(T) = \mathrm{Hdg}^2(T)$ in $H^2(T, \mathbb{Q})$.)*

Then the condition $[\Gamma]^* = 0$ on $H^0(T, \Omega_T^2)$ is equivalent to $\Gamma^* = 0$ on $H^2(T, \mathbb{Q})_{\mathrm{tr}}$, by the definition of $H^2(T, \mathbb{Q})_{\mathrm{tr}}$ together with the fact that Γ^* is a morphism of Hodge structures, so that its kernel is a Hodge substructure of $H^2(T, \mathbb{Q})$.

Using the Künneth decomposition of $[\Gamma]$ and the Lefschetz theorem on $(1,1)$-classes, one can prove that this condition is also equivalent to the fact that there exist cycles Z_1, Z_2, supported in $C \times T$ and $S \times D$ respectively, where C is a curve contained in S and D is a curve contained in T, such that a multiple of Γ is cohomologous to $Z_1 + Z_2$. For this, one must use the fact that the Künneth components of type $(1, 3)$ and $(3, 1)$ of $[\Gamma]$ are classes of algebraic cycles supported on subvarieties of the form $S \times D$ and $C \times T$ respectively (see Murre 1990; and chapter 9, exercise 2).

The Bloch conjecture for surfaces S such that $p_g(S) = 0$ can then be obtained by taking

$$T = S, \quad \Gamma = \operatorname{diag} S \cong S \subset S \times S$$

in conjecture 11.19. Then $\Gamma_* = \mathrm{Id}$ and $\Gamma_* = 0$ on $F^2\mathrm{CH}_0(S)$ is equivalent to $F^2\mathrm{CH}_0(S) = 0$.

11.2.2 Conjectural filtration on the Chow groups

More generally, Bloch and Beilinson formulated the following conjecture (see Janssen 1994).

Conjecture 11.21 *If X is a smooth projective variety, then for every integer k, there exists a decreasing filtration $F^i\mathrm{CH}^k(X)_{\mathbb{Q}}$ on the Chow groups with rational coefficients $\mathrm{CH}^k(X)_{\mathbb{Q}} := \mathrm{CH}^k(X) \otimes \mathbb{Q}$, satisfying the following properties:*

(i) $F^{k+1}\mathrm{CH}^k(X)_{\mathbb{Q}} = 0$.

(ii) The filtration F^i is stable under correspondences: if $\Gamma \in \mathrm{CH}^l(X \times Y)$, then the maps

$$\Gamma_* : \mathrm{CH}^k(X)_{\mathbb{Q}} \to \mathrm{CH}^{k+l-\dim X}(Y)_{\mathbb{Q}}$$

satisfy

$$\Gamma_*(F^i\mathrm{CH}^k(X)_{\mathbb{Q}}) \subset F^i\mathrm{CH}^{k+l-\dim X}(Y)_{\mathbb{Q}}.$$

(iii) The induced map

$$\mathrm{Gr}_F^i\Gamma_* : \mathrm{Gr}_F^i\mathrm{CH}^k(X)_{\mathbb{Q}} \to \mathrm{Gr}_F^i\mathrm{CH}^{k+l-n}(Y)_{\mathbb{Q}}, \quad n = \dim X,$$

vanishes if the Künneth component of bidegree $(2n-2k+i, 2l-2n+2k-i)$ *of the class of* Γ *is zero in* $H^{2l}(X\times Y,\mathbb{Q})$, *i.e.*

$$0=[\Gamma]_*: H^{2k-i}(X,\mathbb{Q})\to H^{2k+2l-2n-i}(Y,Q).$$

This conjecture can be further refined by replacing condition (iii) by

(iii)′ The induced map

$$\mathrm{Gr}_F^i\Gamma_*:\mathrm{Gr}_F^i\mathrm{CH}^k(X)_{\mathbb{Q}}\to \mathrm{Gr}_F^i\mathrm{CH}^{k+l-n}(Y)_{\mathbb{Q}},\quad n=\dim X$$

vanishes if the map

$$[\Gamma]_*: H^{2k-i}(X,\mathbb{Q})\to H^{2k+2l-2n-i}(Y,Q)$$

vanishes on $H^{r,s}(X)$ *for* $s\le k-i$.

For example, in the case of the group CH_0, conjecture 11.21 predicts that the graded object associated to the filtered group $\mathrm{CH}_0(X)$ is controlled by the holomorphic forms on X in the following way.

Conjecture 11.22 *There exists a filtration* $F^i\mathrm{CH}_0(X)_{\mathbb{Q}}$ *such that*

$$F^{n+1}\mathrm{CH}_0(X)_{\mathbb{Q}}=0,\ n=\dim X,$$

for every smooth projective variety X stable under correspondences and satisfying the following property:

(*) *For a correspondence* $\Gamma\subset X\times Y$ *of codimension* $m=\dim Y$, *the map*

$$Gr_F^i\Gamma_*:\mathrm{Gr}_F^i\mathrm{CH}_0(X)_{\mathbb{Q}}\to \mathrm{Gr}_F^i\mathrm{CH}_0(Y)_{\mathbb{Q}}$$

is zero if and only if

$$[\Gamma]^*: H^0(Y,\Omega_Y^i)\to H^0(X,\Omega_X^i)$$

is zero.

(To see how condition (iii)′ is related to (*), note that $[\Gamma]^*$ is the dual of

$$[\Gamma]_*: H^n(X,\Omega_X^{n-i})\to H^m(Y,\Omega_Y^{m-i})\ \text{ for } m=\dim Y,$$

which is also the complex conjugate of

$$[\Gamma]_*: H^{n,n-i}(X)\to H^{m,m-i}(Y).)$$

Conjecture 11.22 would imply the following result, which is the converse to theorem 10.31.

Conjecture 11.23 *If the smooth projective variety X satisfies $H^{p,q}(X)=0$ for $p\neq q,\ q\leq k$, then the map*

$$\mathrm{cl}:\mathrm{CH}_l(X)_{\mathbb{Q}}\to H^{2n-2l}(X,\mathbb{Q})$$

is injective for $l\leq k$.

Proof of the implication 11.22 ⇒ 11.23 The proof is by induction on k. For $k=0$, conjecture 11.22 applied to the diagonal diag $X\subset X\times X$ says that if $H^0(X,\Omega_X^i)=0$ for all $i\neq 0$, then the groups $\mathrm{Gr}_F^i\mathrm{CH}_0(X)_{\mathbb{Q}}$ are zero for $i\neq 0$, and as the filtration is finite, this implies that $\mathrm{cl}:\mathrm{CH}_0(X)_{\mathbb{Q}}\to H^{2n}(X,\mathbb{Q})$ is an isomorphism, since clearly by condition (*) on the filtration F, we have $\mathrm{CH}_0(X)_{\mathrm{hom}}\subset F^1\mathrm{CH}_0(X)$.

So assume that conjecture 11.23 holds for $k-1$, and let X satisfy the hypothesis $H^{p,q}(X)=0$ for $p\neq q,\ q\leq k$. By the induction hypothesis,

$$\mathrm{cl}:\mathrm{CH}_l(X)_{\mathbb{Q}}\to H^{2n-2l}(X,\mathbb{Q})$$

is injective for $l\leq k-1$. Thus, we can apply theorem 10.29, which gives a decomposition

$$m\Delta=\sum_{i\leq k-1}Z_i\oplus Z\in\mathrm{CH}^n(X\times X),\qquad(11.5)$$

where the Z_i are of the form $\sum_j n_jW_{ij}\times W'_{ij}$ with $\dim W_{ij}=i$ and $\dim W'_{ij}=n-i$, and Z is supported on $T\times X$ with $\operatorname{codim}T=k$.

Let $\tau:\widetilde{T}\to T\to X$ be a desingularisation of T, and let $\widetilde{Z}$ be an n-dimensional cycle such that $(\tau,\mathrm{Id})_*\widetilde{Z}=Z$. Formula (11.5) shows that for $W\in\mathrm{CH}_k(X)$, we have

$$mW=m\Delta_*W=(\widetilde{Z}_*\circ\tau^*)W.\qquad(11.6)$$

Consider the correspondence of dimension $n-k=\dim\widetilde{T}$ given by

$$Z'=(\mathrm{Id},\tau^*)\widetilde{Z}\subset\widetilde{T}\times\widetilde{T}.$$

Its action on 0-cycles of $\widetilde{T}$ is given by

$$Z'_*\alpha=\tau^*\widetilde{Z}_*\alpha.$$

Thus, by (11.6), we have

$$Z'_*\tau^*W=m\tau^*W,\quad W\in\mathrm{CH}_k(X),$$

so

$$\widetilde{Z}_*(Z'_*\tau^*W)=m(\widetilde{Z}\circ\tau^*)W=m^2W,\quad W\in\mathrm{CH}_k(X).$$

Similarly, one can show that if $Z'^{(l)} := Z'^{\circ l} \in \mathrm{CH}_{n-k}(\widetilde{T} \times \widetilde{T})$, then

$$\widetilde{Z}_*(Z'^{(l)}_* \tau^* W) = m^{l+1} W, \quad W \in \mathrm{CH}_k(X).$$

To conclude that

$$\mathrm{cl} : \mathrm{CH}_k(X)_{\mathbb{Q}} \to H^{2n-2k}(X, \mathbb{Q})$$

is injective, it thus suffices to prove that for an integer l, $Z'^{(l)}_*$ annihilates $\mathrm{CH}_0(\widetilde{T})_{\mathrm{hom}}$. But the hypothesis $H^{p,q}(X) = 0$ for $p > q$, $q \leq k$ shows immediately that the correspondence Z' acts trivially on $H^{l,0}(\widetilde{T})$ for every $l \neq 0$, so conjecture 11.22 implies that Z'_* acts trivially on the associated graded group $\mathrm{Gr}^i_F \mathrm{CH}_0(T)$, $i > 0$. But then Z'_* is nilpotent on $\mathrm{CH}_0(T)_{\mathrm{hom}}$, since the filtration is finite. Thus, there exists an l such that $Z'^{(l)}_* = (Z'_*)^{\circ l}$ annihilates $\mathrm{CH}_0(\widetilde{T})_{\mathrm{hom}}$.

□

11.2.3 The Saito filtration

Shuji Saito (1996) constructed a filtration $F^i\mathrm{CH}^k(X)$ on the Chow groups of a projective variety X, which satisfies properties (ii) and (iii) of conjecture 11.21. Unfortunately, the finiteness property $F^{k+1}\mathrm{CH}^k(X)_{\mathbb{Q}} = 0$ remains completely mysterious. We can only say that the subgroup $F^{k+1}\mathrm{CH}^k(X)_{\mathbb{Q}}$ is the part of the Chow group $\mathrm{CH}^k(X)_{\mathbb{Q}}$ which in no way governs the Hodge theory of X. Its construction is roughly as follows. Assume that we have constructed the groups $F^{i-1}\mathrm{CH}^p(Y)_{\mathbb{Q}}$ for every p and every smooth projective variety Y. Define $F^i\mathrm{CH}^k(X)$ to be the subgroup of $F^{i-1}\mathrm{CH}^k(X)$ generated by the images $\Gamma_*(F^{i-1}\mathrm{CH}^{\dim Y - r + k}(Y)_{\mathbb{Q}})$ for the correspondences $\Gamma \subset \mathrm{CH}^r(Y \times X)$ satisfying the property:

The Künneth component of bidegree $(2r - 2k + i - 1, 2k - i + 1)$ of the class of Γ is zero in $H^{2r}(Y \times X, \mathbb{Q})$, i.e.

$$\Gamma_* : H^{2\dim Y - 2r + 2k - i + 1}(Y, \mathbb{Q}) \to H^{2k-i+1}(X, Q)$$

is zero.

This definition could be refined by requiring only that Γ_* vanish on the $H^{p,q}$ with $q \leq \dim Y - r + k - i + 1$, but conjecturally, the resulting filtration should be the same.

Saito shows that his filtration is the minimal filtration satisfying properties (ii) and (iii) of conjecture 11.21.

Murre (1993) also proposes a definition of the filtration F on the Chow groups $\mathrm{CH}(X)$, using a (conjectural) decomposition of the diagonal

$$\mathrm{diag}\, X \in \mathrm{CH}(X \times X)_{\mathbb{Q}}$$

as a sum of projectors. Such a decomposition is known to exist for curves, surfaces and certain 3-dimensional varieties (see chapter 9, exercise 2).

11.3 The case of abelian varieties

11.3.1 The Pontryagin product

If A is a g-dimensional abelian variety, i.e. a complex torus which is a projective variety (see vI.7.2.2), the group structure on A induces additional structure on the Chow groups of A, which in turn helps to obtain results in the direction of conjecture 11.21.

Let $\mu : A \times A \to A$ denote the sum map

$$\mu(z, z') = z + z'.$$

If $Z = \sum_i n_i Z_i \in \mathcal{Z}_k(A)$ and $Z' = \sum_j m_j Z'_j \in \mathcal{Z}_l(A)$, we set

$$Z \times Z' := \sum_{i,j} n_i m_j Z_i \times Z_j \in \mathcal{Z}_{k+l}(A \times A).$$

It is easy to see that the bilinear map

$$\mathcal{Z}_k(A) \times \mathcal{Z}_l(A) \to \mathcal{Z}_{k+l}(A \times A), \quad (Z, Z') \to Z \times Z'$$

factors through rational equivalence (indeed this is the same thing as $\mathrm{pr}_1^* Z \cdot \mathrm{pr}_2^* Z'$), and thus gives a bilinear map

$$\mathrm{CH}_k(A) \times \mathrm{CH}_l(A) \to \mathrm{CH}_{k+l}(A \times A).$$

Definition 11.24 *The Pontryagin product* $(Z, Z') \mapsto Z * Z'$ *is defined by*

$$Z * Z' = \mu_*(Z \times Z') \in \mathrm{CH}_{k+r}(A) \text{ for } Z \in \mathrm{CH}_k(A), \quad Z' \in \mathrm{CH}_l(A).$$

Lemma 11.25 *The Pontryagin product is commutative and associative on* $\mathrm{CH}(A) := \bigoplus_k \mathrm{CH}_k(A)$.

Proof The commutativity follows from the commutativity of μ, and the associativity follows from the equality

$$Z * (Z' * Z'') = \mu'_*(Z \times Z' \times Z'') \in \mathrm{CH}_{k+l+r}(A),$$

with $Z \in \mathrm{CH}_k(A)$, $Z' \in \mathrm{CH}_l(A)$, and $Z'' \in \mathrm{CH}_r(A)$, where $\mu' : A \times A \times A \to A$ is the sum map $\mu'(z, z', z'') = z + z' + z''$. Here, as above, the product $Z \times Z' \times Z''$

is defined by trilinearity starting from the products $Z \times Z' \times Z''$ for three subvarieties of A. □

11.3.2 Results of Bloch

In the case where A is an abelian variety, Bloch (1976) used the Pontryagin product to put a filtration $F^i\mathrm{CH}_0$ on the Chow groups $\mathrm{CH}_0(A)$, which satisfies some of the properties of conjecture 11.22. To avoid confusion between addition in A and addition in $\mathrm{CH}_0(A)$, we will write $z \in A$ for the points of A and $\{z\} \in \mathrm{CH}_0(A)$ for the corresponding cycles.

Let $I = \mathrm{CH}_0(A)_{\mathrm{hom}} \subset \mathrm{CH}_0(A)$. Bloch sets

$$F^i\mathrm{CH}_0(A) = I^{*i} \subset \mathrm{CH}_0(A).$$

The following lemma relates this filtration with Hodge theory.

Lemma 11.26 *Let $\Gamma \subset W \times A$ be a correspondence of codimension g. Then if*

$$\mathrm{Im}\,(\Gamma_* : \mathrm{CH}_0(W) \to \mathrm{CH}_0(A)) \subset F^i\mathrm{CH}_0(A), \tag{11.7}$$

we have

$$[\Gamma]^* = 0 : H^0\big(A, \Omega_A^l\big) \to H^0\big(W, \Omega_W^l\big) \text{ for } l < i.$$

Proof Let $\Gamma'' = \Gamma_{\mu_i} \circ \Gamma_1^i \subset A^i \times A$ be the correspondence consisting of the graph $\Gamma_{\mu_i} \subset A^i \times A$ of the sum map $\mu_i : A^i \to A$, composed with the correspondence $\Gamma_1^i \subset (A \times A)^i \cong A^i \times A^i$, where $\Gamma_1 := \Delta_A - A \times 0$. In other words, we have

$$\Gamma_{1*}(\{z\}) = \{z\} \quad \{0\} \in \mathrm{CH}_0(A)_{\mathrm{hom}}, \quad z \in A,$$
$$\Gamma_{1*}^i(\{(z_1, \ldots, z_i)\}) = (\{z\}_1 - \{0\}) \times \cdots \times (\{z\}_i - \{0\}) \in \mathrm{CH}_0(A^i),$$
$$\Gamma_{\mu_i *}(\{(z_1, \ldots, z_i)\}) = \{z_1 + \cdots + z_i\},$$

so

$$\Gamma''_*(\{(z_1, \ldots, z_i)\}) = (\{z_1\} - \{0\}) * \cdots * (\{z_i\} - \{0\}).$$

We will prove the following lemma below.

Lemma 11.27 *Under the condition (11.7), there exists a correspondence $\Gamma' \subset W \times A^i$, an integer $N \neq 0$, and a Zariski open dense set $W_0 \subset W$, such that*

$$N\Gamma = \Gamma'' \circ \Gamma' \text{ in } \mathrm{CH}^g(W_0 \times A). \tag{11.8}$$

Admitting this lemma, we then have

$$N[\Gamma]^* = [\Gamma']^* \circ [\Gamma'']^* : H^0(A, \Omega_A^l) \to H^0(W_0, \Omega_{W_0}^l).$$

It thus suffices to prove the result for $\Gamma = \Gamma''$.

Now, for $\omega = \alpha_1 \wedge \cdots \wedge \alpha_l \in H^0(A, \Omega_A^l)$, we have

$$\mu_i^* \omega = \sum_K p_{k_1}^* \alpha_1 \wedge \cdots \wedge p_{k_l}^* \alpha_l,$$

where K runs through the set of l-tuples $(k_1, \ldots, k_l)$ of elements of $\{1, \ldots, i\}$. If $l < i$, then each form $\alpha_K = p_{k_1}^* \alpha_1 \wedge \cdots \wedge p_{k_l}^* \alpha_l$ is a pullback $q_s^* \eta$, for at least one s, where $q_s : A^i \to A^{i-1}$ is the projection forgetting the sth factor, so we have

$$\alpha_K = q_s^*(\alpha_{K \,|\, A \times \cdots 0 \cdots \times A}),$$

where the 0 is on the sth factor. This immediately implies that α_K is annihilated by $[\Gamma_1^i]^*$, by a direct computation, or using the functorial version of Mumford's theorem and the fact that $q_{s*} \circ \Gamma_{1*}^i = 0$ on $\mathrm{CH}_0(A^i)$. Thus, $\mu_i^* \omega$ is annihilated by $[\Gamma_1^i]^*$ and ω is annihilated by $[\Gamma'']^*$. □

Proof of lemma 11.27 Hypothesis (11.7) says exactly that for every $w \in W$, there exists $z \in \mathrm{CH}_0(A^i)$ such that

$$\Gamma_*(w) = \Gamma'_*(z) \text{ in } \mathrm{CH}_0(A).$$

For each pair of integers (j, j'), let us introduce the subset $R_{j,j'} \subset W \times S^j A^i \times S^{j'} A^i$ defined by

$$R_{j,j'} = \{(w, z, z') \mid \Gamma_*(w) = \Gamma''_*(z - z') \text{ in } \mathrm{CH}_0(A)\}. \tag{11.9}$$

(Here, in order to avoid heavy notation, we write $S^j X$ for the jth symmetric product of X which we previously wrote as $X^{(j)}$.) We know that the $R_{j,j'}$ are countable unions of algebraic closed subsets of $W \times S^j A^i \times S^{j'} A^i$. Our hypothesis is that the union over j, j' of the projections onto W of the $R_{j,j'}$ is equal to W. Thus, there exists an algebraic component Z of a set $R_{j,j'}$ which projects surjectively onto W. Up to replacing Z by an algebraic closed subset, we may assume that the projection

$$p : Z \to W$$

is generically finite of degree N. Then $Z \subset W \times S^j A^i \times S^{j'} A^i$ gives two correspondences $Z_j \subset W \times S^j A^i$, resp. $Z_{j'} \subset W \times S^{j'} A^i$, of dimension equal to $\dim W$. Composing with the natural correspondences between $S^j A^i$, resp.

$S^{j'}A^i$, and A^i, we thus obtain

$$\Gamma'_j,\ \Gamma'_{j'} \in \mathrm{CH}_{\dim W}(W \times A^i).$$

Now set

$$\Gamma' = \Gamma'_j - \Gamma'_{j'}.$$

The fact that $Z \subset R_{j,j'}$, together with definition (11.9), then shows that for $w \in W$, we have

$$\Gamma''_*(\Gamma'_*(w)) = N\Gamma_*(w).$$

The fact that this implies equality (11.8) on an open subset of W is then a consequence of corollary 10.20. □

Lemma 11.26 shows in particular that $F^2\mathrm{CH}_0(A) = I^{*2}$ is contained in

$$\mathrm{Ker}\,(\mathrm{alb}_A : \mathrm{CH}_0(A)_{\mathrm{hom}} = F^1\mathrm{CH}_0(A) \to \mathrm{Alb}\,A = A).$$

Indeed, this is equivalent to the statement that for every correspondence

$$\Gamma \subset B \times A \ \text{ with } \ \dim \Gamma = \dim B$$

such that $\mathrm{Im}\,\Gamma_* \subset F^2\mathrm{CH}_0(A)$, we have $\mathrm{alb}_A \circ \Gamma_* = 0$. Clearly this is equivalent to the statement that the holomorphic map

$$\phi = \mathrm{alb}_A \circ \Gamma_* : B \to \mathrm{Alb}\,A = A$$

is constant. But ϕ admits for differential the map

$$\phi_* : T_B \to T_{A,0} \otimes \mathcal{O}_B,$$

which is the dual of the composed map

$$\Gamma^* : H^0(A, \Omega_A) \to H^0(B, \Omega_B)\Omega_B.$$

Thus, lemma 11.26 shows that ϕ is constant, which proves the result.

Lemma 11.28 *The Bloch filtration satisfies* $F^2\mathrm{CH}_0(A) = \mathrm{Ker}\,\mathrm{alb}_A$.

Proof The map alb_A sends $\{z\} - \{0\} \in \mathrm{CH}_0(A)_{\mathrm{hom}}$ to

$$z \in A = \mathrm{Alb}\,A,$$

and more generally, $\sum_i \{z_i\} - \{0\}$ to $\sum_i z_i \in A$. We want to show that $\mathrm{Ker}\,\mathrm{alb}_A$ is generated by the cycles of the form

$$\{z_1 + z_2\} - \{z_1\} - \{z_2\} + \{0\} = (\{z_1\} - \{0\}) * (\{z_2\} - \{0\}).$$

Now, if $z = \sum_{1\leq i\leq n}\{z_i\} - \sum_{1\leq i\leq n}\{z'_i\} \in \operatorname{Ker} \mathrm{alb}_A$, let

$$z' = \sum_{3\leq i\leq n}\{z_i\} - \sum_{3\leq i\leq n}\{z'_i\} + \{z_1+z_2\} - \{z'_1+z'_2\}.$$

Then $z = z' - (\{z_1+z_2\} - \{z_1\} - \{z_2\} + \{0\}) + (\{z'_1+z'_2\} - \{z'_1\} - \{z'_2\} + \{0\})$, and z' is of the form

$$\sum_{1\leq i\leq n-1}\{z''_i\} - \sum_{1\leq i\leq n-1}\{z'''_i\}.$$

As $z \in I^{*2}$ if and only if $z' \in I^{*2}$, the arguments above show that we can reduce to the case $n = 2$. But then

$$z = \{z_1\} + \{z_2\} - \{z'_1\} - \{z'_2\}$$

with $z_1 + z_2 = z'_1 + z'_2$. Therefore,

$$\begin{aligned} z &= \{z_1\} + \{z_2\} - \{z_1+z_2\} + \{z'_1+z'_2\} - \{z'_1\} - \{z'_2\} \\ &= (\{z_1\} - \{0\}) * (\{z_2\} - \{0\}) - (\{z'_1\} - \{0\}) * (\{z'_2\} - \{0\}). \end{aligned}$$

□

The second property satisfied by the Bloch filtration is the following.

Theorem 11.29 (Bloch 1976) *The Bloch filtration $F^i\mathrm{CH}_0(A)$ satisfies*

$$F^{g+1}\mathrm{CH}_0(A) = 0, \quad g = \dim A.$$

The proof of this theorem requires two lemmas.

Lemma 11.30 *For $l > g$, we have*

$$\mathrm{Gr}^l_F\mathrm{CH}_0(A) = 0.$$

Proof Let $j : C \hookrightarrow A$ be a smooth curve contained in A, obtained as the complete intersection of ample hypersurfaces $H_i \in |L|$. Then the map

$$\mathrm{alb}_A \circ j_* \circ j^* : \mathrm{Pic}^0(A) \to \mathrm{Alb}\, A$$

is surjective. Indeed, it suffices to check that the corresponding morphism of rational Hodge structures

$$j_* \circ j^* : H^1(A, \mathbb{Q}) \to H^{2g-1}(A, \mathbb{Q})$$

is surjective. But it is actually an isomorphism, by the hard Lefschetz theorem (see vI.6.2.3).

It follows from lemma 11.28 that the map

$$\mathrm{CH}_0(A)_{\mathrm{hom}}^{\otimes i} \to F^i\mathrm{CH}_0(A), \quad z_1 \otimes \cdots \otimes z_i \mapsto z_1 * \cdots * z_i$$

induces a surjective morphism

$$A^{\otimes i} \to \mathrm{Gr}_F^i \mathrm{CH}_0(A),$$

since this lemma identifies A with $F^1\mathrm{CH}_0(A)/F^2\mathrm{CH}_0(A)$. Thus, we conclude that $\mathrm{Gr}_F^i\mathrm{CH}_0(A)$ is generated by the cycles of the form

$$z = (j_* \circ j^* D_1) * \cdots * (j_* \circ j^* D_i), \quad D_l \in \mathrm{Pic}^0(A).$$

Note that furthermore, by the commutativity of the Pontryagin product, we have

$$z = \left(j_* \circ j^* D_{\sigma(1)}\right) * \cdots * \left(j_* \circ j^* D_{\sigma(i)}\right)$$

for every permutation σ of $\{1, \ldots, i\}$. Let $J : C^i \to A$ be the composition of $j^i : C^i \to A^i$ with $\mu : A^i \to A$.

The above arguments together with the definition of the Pontryagin product show that $\mathrm{Gr}_F^i\mathrm{CH}_0(A)$ is generated by cycles of the form

$$\begin{aligned} z &= J_*(\mathrm{pr}_1^* j^* D_1 \cdot \cdots \cdot \mathrm{pr}_i^* \circ j^* D_i) \\ &= J_*\left(\mathrm{pr}_{\sigma(1)}^* j^* D_{\sigma(1)} \cdot \cdots \cdot \mathrm{pr}_{\sigma(i)}^* j^* D_{\sigma(i)}\right), \quad D_l \in \mathrm{Pic}^0 A, \end{aligned} \tag{11.10}$$

where the pr_k denote the projections of C^i onto its factors. But the invariance of $\mathrm{Pic}^0 A$ under translations of A shows that

$$J^* D_l = \sum_k (\mathrm{pr}_k^* \circ j^*) D_l, \quad \forall D_l \in \mathrm{Pic}^0 A.$$

Thus, we have

$$\begin{aligned} J^*(D_1 \cdot \cdots \cdot D_i) &= J^* D_1 \cdot \cdots \cdot J^* D_i \\ &= \sum_{\sigma \in S_i} \left(\mathrm{pr}_{\sigma(1)}^* j^* D_1\right) \cdot \cdots \cdot \left(\mathrm{pr}_{\sigma(i)}^* \circ j^* D_i\right), \end{aligned}$$

where S_i denotes the symmetric group of order i, so by (11.10), we have

$$i!\, z = J_* J^*(D_1 \cdot \cdots \cdot D_i).$$

Therefore, for $i > g$, we have $i!\, z = 0$, i.e. the group $\mathrm{Gr}_F^i\mathrm{CH}_0(A)$ is annihilated by $i!$. As it is clearly a divisible group, it must be trivial. □

To conclude the proof of the theorem, we show the following result.

Lemma 11.31 *For sufficiently large N, we have $F^N\mathrm{CH}_0(A) = 0$.*

Proof If $\phi : B \to A$ is a morphism of abelian varieties, then clearly $\phi_* : \mathrm{CH}_0(B) \to \mathrm{CH}_0(A)$ satisfies $\phi_* F^i\mathrm{CH}_0(B) \subset F^i\mathrm{CH}_0(A)$, since ϕ_* is a morphism of rings for the Pontryagin product. Moreover, if ϕ is surjective, we actually have $\phi_* F^i\mathrm{CH}_0(B) = F^i\mathrm{CH}_0(A)$, since $\phi_* : \mathrm{CH}_0(B)_{\mathrm{hom}} \to \mathrm{CH}_0(A)_{\mathrm{hom}}$ is surjective. Thus, if the result holds for B, it also holds for A.

Now, if $C \subset A$ is a curve as above, we know that $\mathrm{alb}_A \circ j_* : J(C) = \mathrm{CH}_0(C)_{\mathrm{hom}} \to A$ is surjective. We can thus restrict ourselves to proving the lemma for $A = J(C)$, the embedding $j : C \hookrightarrow A$ being given by the Albanese map of C relative to the choice of a reference point $c_0 \in C$. For $g = g(C) = \dim A$, consider the map

$$\Phi : C^{(g)} \to A, \quad z \mapsto \mathrm{alb}_A(j_* z - g j(c_0) c_o).$$

We know that this map is surjective (see vI.12.1.3). Let

$$\Phi' : C^{(g+1)} \to A, \quad z \mapsto \mathrm{alb}_A(j_* z - (g+1) j(c_0) c_0),$$

and let us introduce the following notation: for $L_i \in \mathrm{CH}_0(C) = \mathrm{Pic}\, C$, $i = 1, \ldots, k$, define $L_1 \cdot \cdots \cdot L_k \in \mathrm{CH}_0(C^{(k)})$ by

$$L_1 \cdot \cdots \cdot L_k = p_*(\mathrm{pr}_1^* L_1 \cdot \cdots \cdot \mathrm{pr}_k^* L_k),$$

where $p : C^k \to C^{(k)}$ is the quotient map, and the $\mathrm{pr}_i : C^k \to C$ are the projections onto the different factors.

Lemma 11.32 *The group $F^{g+1}\mathrm{CH}_0(A)$ is generated by cycles of the form*

$$\{u\} * \Phi'_*(D_1 \cdot \cdots \cdot D_{g+1}), \quad u \in A, \quad D_i \in \mathrm{Pic}^0(C).$$

Lemma 11.33 *We have $D_1 \cdot \cdots \cdot D_{g+1} = 0$ in $\mathrm{CH}_0(C^{(g+1)})$.*

These two properties immediately imply that $F^{g+1}\mathrm{CH}_0(A) = 0$, which concludes the proof of lemma 11.31, so also that of theorem 11.29, once we have proved lemmas 11.32 and 11.33. □

Proof of lemma 11.32 Obviously, the product $\cdot$ and the Pontryagin product satisfy

$$\Phi'_*(D_1 \cdot \cdots \cdot D_{g+1}) = (j_* D_1) * \cdots * (j_* D_{g+1})$$

for $D_1, \ldots, D_{g+1} \in \mathrm{Pic}\, C$. Now, by definition, $F^{g+1}\mathrm{CH}_0(A)$ is generated by the $Z_1 * \cdots * Z_{g+1}$ with $Z_i \in \mathrm{CH}_0(A)_{\mathrm{hom}}$. It thus suffices to prove that the cycles $\{u\} * j_* D$, $D \in \mathrm{Pic}^0 C$, $u \in A$ generate $\mathrm{CH}_0(A)_{\mathrm{hom}}$.

But $\mathrm{CH}_0(A)_{\mathrm{hom}}$ is generated by the $\{z\} - \{0\},\ z \in A$. As Φ is surjective, we can write $z = \sum_{1\leq i\leq g} z_i,\ z_i \in C$. But then

$$\begin{aligned}\{z\} - \{0\} &= \{z_1 + \cdots + z_g\} - \{z_2 + \cdots + z_g\} + \{z_2 + \cdots + z_g\} - \{0\}\\ &= \{u\} * (\{z_1\} - \{0\}) + \{z_2 + \cdots + z_g\} - \{0\},\end{aligned}$$

where $u = z_2 + \cdots + z_g$. But $\{z_1\} - \{0\} \in \mathrm{Im}\ j_*(\mathrm{Pic}^0(C))$, since $z_1 \in C,\ 0 \in C$.
More generally, we have

$$\begin{aligned}\{z_k + \cdots + z_g\} - \{0\} &= \{z_k + \cdots + z_g\} - \{z_{k+1} + \cdots + z_g\}\\ &\quad + \{z_{k+1} + \cdots + z_g\} - \{0\}\\ &= \{u_k\} * (\{z_k\} - \{0\}) + \{z_{k+1} + \cdots + z_g\} - \{0\},\end{aligned}$$

with $u_k = z_{k+1} + \cdots + z_g$ and $\{z_k\} - \{0\} \in j_*(\mathrm{Pic}^0(C))$.
By induction on the largest integer l such that

$$z = \sum_{1\leq i\leq l} t_i, \quad t_i \in C,$$

this shows that $\{z\} - \{0\}$ lies in the subgroup generated by the $\{u\} * j_*(D),\ D \in \mathrm{Pic}^0(C),\ u \in A$. $\square$

Proof of lemma 11.33 We have a natural correspondence $Z \subset C^{(g)} \times C^{(g+1)}$ defined by

$$Z = \{(z, z') \in C^{(g)} \times C^{(g+1)} \mid z \subset z'\}.$$

For $z' = \{z_1, \ldots, z_{g+1}\} \in C^{(g+1)}$, we have

$$Z^* z' = \sum_l \{z_1, \ldots \hat{z}_i \ldots, z_{g+1}\} \in \mathcal{Z}_0(C^{(g)}). \tag{11.11}$$

We first show that $Z^* : \mathrm{CH}_0(C^{(g+1)}) \to \mathrm{CH}_0(C^{(g)})$ is an isomorphism. For this, consider the inclusion

$$i_0 : C^{(g)} \to C^{(g+1)}$$

given by $i_0(z) = z + c_0,\ z \in C^{(g)}$. Then, for $z \in \mathrm{CH}_0(C^g)$, we have

$$Z^*(i_{0*}(z)) = z + z',$$

where z' is a cycle of $C^{(g)}$ supported on $C^{(g-1)} \times c_0$. It follows easily that

$$Z^* \circ i_{0*} : \mathrm{CH}_0(C^{(g)}) \to \mathrm{CH}_0(C^{(g)})$$

is an isomorphism. Now, $i_{0*} : \mathrm{CH}_0(C^{(g)}) \to \mathrm{CH}_0(C^{(g+1)})$ is an isomorphism since $\Phi' \circ i_0 = \Phi$ and the two maps

$$\Phi_* : \mathrm{CH}_0\big(C^{(g)}\big) \to \mathrm{CH}_0(A), \quad \Phi'_* : \mathrm{CH}_0\big(C^{(g+1)}\big) \to \mathrm{CH}_0(A)$$

are isomorphisms. Indeed, Φ is birational while Φ' has generic fibre $\mathbb{P}^1$ by Abel's theorem.

Next, using formula (11.11), it is easy to check that $Z^*z' = 0$ in $\mathrm{CH}_0(C^{(g)})$ for a cycle z' of the form $D'_1 \cdot \cdots \cdot D'_{g+1}$ with $D'_k \in \mathrm{Pic}^0(C)$. Since Z^* is an isomorphism, $z' = 0$ for such a cycle. □

11.3.3 Fourier transform

Let A be an abelian variety, and let $\widehat{A} = \mathrm{Pic}^0 A$ be the dual abelian variety. There exists a well-defined line bundle on $A \times \widehat{A}$, called the Poincaré bundle and written $\mathcal{P}$, satisfying the properties:

(i) For $L \in \mathrm{Pic}^0 A$ parametrised by the point $l \in \widehat{A}$, we have $\mathcal{P}_{|A\times l} = L$.
(ii) The restriction of $\mathcal{P}$ to $0 \times \widehat{A}$ is trivial.

The class $c_1(\mathcal{P}) \in H^2(A \times \widehat{A}, \mathbb{Z})$ can be identified with

$$\mathrm{Id} \in H^1(\widehat{A}, \mathbb{Z})^* \otimes H^1(\widehat{A}, \mathbb{Z}) \cong H^1(A, \mathbb{Z}) \otimes H^1(\widehat{A}, \mathbb{Z}) \subset H^2(A \times \widehat{A}, \mathbb{Z}).$$

Let Θ denote the class of $\mathcal{P}$ in $\mathrm{CH}^1(A \times \widehat{A})$, and let us introduce

$$\exp \Theta = \sum_{0 \le i \le 2g} \frac{\Theta^i}{i!} \in \mathrm{CH}(A \times \widehat{A})_{\mathbb{Q}}.$$

Definition 11.34 (see Mukai 1981; Beauville 1983) *The Fourier transform* $\mathcal{F} : \mathrm{CH}(A)_{\mathbb{Q}} \to \mathrm{CH}(\widehat{A})_{\mathbb{Q}}$ *is defined by*

$$\mathcal{F}(Z) = \mathrm{pr}_{2*}(\mathrm{pr}_1^* Z \cdot \exp \Theta).$$

Proposition 11.35 *If*

$$\widehat{\mathcal{F}} : \mathrm{CH}(\widehat{A})_{\mathbb{Q}} \to \mathrm{CH}(A)_{\mathbb{Q}}$$

is the Fourier transform of $\widehat{A}$ *(where we use the biduality* $\widehat{\widehat{A}} = A$*), then*

$$\widehat{\mathcal{F}} \circ \mathcal{F} = (-1)^g (-1)^*,$$

where $-1 : A \to A$ *denotes the map* $z \mapsto -z$.

Proof If $\widehat{\Theta} \in \mathrm{CH}(\widehat{A} \times A)_{\mathbb{Q}}$ is the class of the Poincaré divisor $\widehat{\mathcal{P}}$ of $\widehat{A}$, we need to show that

$$\exp\widehat{\Theta} \circ \exp\Theta = (-1)^g \Gamma_{-1} \in \mathrm{CH}_g(A \times A),$$

where $\circ$ is the composition of correspondences and Γ_{-1} is the graph of the map -1. Consider the triple product $A \times \widehat{A} \times A$, and let p_i denote the projection onto the ith factor, and p_{ij} projection onto the product of the ith and jth factors. We have the equality

$$p_{12}^*\mathcal{P} + p_{23}^*\widehat{\mathcal{P}} = \tilde{\mu}^*\mathcal{P} \in \mathrm{Pic}\, A \times \widehat{A} \times A,$$

where $\tilde{\mu} : A \times \widehat{A} \times A \to A \times \widehat{A}$ is defined by $\tilde{\mu}(z, z', z'') = (z + z'', z')$. This equality comes from the fact that $\widehat{\mathcal{P}}$ is the image of $\mathcal{P}$ under the isomorphism $A \times \widehat{A} \cong \widehat{A} \times A$, and from the fact that for $D \in \mathrm{Pic}^0 A$, we have $\mu^*D = \mathrm{pr}_1^*D + \mathrm{pr}_2^*D \in \mathrm{Pic}^0 A \times A$.

Passing to the exponential, we then obtain

$$p_{12}^*\exp\Theta \cdot p_{23}^*\exp\widehat{\Theta} = \tilde{\mu}^*\exp\Theta \in \mathrm{CH}(A \times \widehat{A} \times A)_{\mathbb{Q}}.$$

Thus, we have

$$\widehat{\Theta} \circ \Theta = p_{13*}(p_{12}^*\exp\Theta \cdot p_{23}^*\exp\widehat{\Theta}) = p_{13*}(\tilde{\mu}^*\exp\Theta).$$

Letting $p : A \times \widehat{A} \to A$ denote the projection onto A, and $\mu : A \times A \to A$ the sum map

$$(z, z') \mapsto z + z',$$

we now observe that the diagram

$$\begin{array}{llcl}
p_{13}: & A \times \widehat{A} \times A & \longrightarrow & A \times A \\
 & \quad\downarrow \tilde{\mu} & & \downarrow \mu \\
p: & A \times \widehat{A} & \longrightarrow & A
\end{array}$$

is Cartesian, and μ is flat, so that by proposition 9.8, we have the equality $p_{13*} \circ \tilde{\mu}^* = \mu^* \circ p_*$. We thus conclude that

$$\widehat{\Theta} \circ \Theta = \mu^* p_* \exp\Theta \in \mathrm{CH}(A \times A)_{\mathbb{Q}}.$$

It remains only to prove that

$$\mu^* p_* \exp\Theta = (-1)^g \Gamma_{-1} \in \mathrm{CH}(A \times A)_{\mathbb{Q}}.$$

But this follows from the equalities

$$p_* \exp\Theta = (-1)^g \{0\} \in \mathrm{CH}(A)_{\mathbb{Q}} \qquad (11.12)$$

and

$$\mu^*\{0\} = \Gamma_{-1} \in \mathrm{CH}(A \times A)_{\mathbb{Q}}. \tag{11.13}$$

The last equality is obvious, since $\mu^{-1}(0) = \{(x, -x) \mid x \in A\}$ and μ is submersive, so that the set-theoretic equality is an equality of cycles.

The equality (11.12) is proved by noting that if $M_k : A \times \widehat{A} \to A \times \widehat{A}$ is the map $(x, y) \mapsto (x, ky)$, then $p \circ M_k = p$, and moreover, $M_k^*\mathcal{P} = k\mathcal{P}$. It follows that $M_k^*\Theta^i = k^i\Theta^i$, and as $M_{k*} \circ M_k^* = k^{2g}$ Id since k^{2g} is the degree of M_k, we find that

$$M_{k*}(\Theta^i) = k^{2g-i}\Theta^i \in \mathrm{CH}^i(A \times \widehat{A})_{\mathbb{Q}}.$$

As $p_* \circ M_{k*} = p_*$, we obtain

$$p_*\Theta^i = 0, \quad i \neq 2g,$$

so that

$$p_* \exp \Theta = p_* \frac{\Theta^{2g}}{2g!} \in \mathrm{CH}(A)_{\mathbb{Q}}.$$

But the same argument applied to the multiplication M'_k defined by $M'_k(z, z') = (kz, z')$ also shows that the 0-cycle $p_* \frac{\Theta^{2g}}{2g!}$ satisfies

$$m_{k*}z = z \in \mathrm{CH}_0(A)_{\mathbb{Q}} \quad \text{for all } k \in \mathbb{Z},$$

where $m_k : A \to A$ denotes multiplication by k. It is then easy to deduce that $z = \alpha\{0\}$, where $\alpha \in \mathbb{Q}$ is given by $\alpha = \deg \frac{\Theta^{2g}}{2g!}$. For this, one uses the Bloch filtration on $\mathrm{CH}_0(A)$. As the successive quotients $\mathrm{Gr}^i_F\mathrm{CH}_0(A)$ are quotients of $A^{\otimes i}$, m_{k*} acts via k^i on $\mathrm{Gr}^i_F\mathrm{CH}_0(A)$, and it follows that a cycle $z \in \mathrm{CH}_0(A)_Q$ such that $m_{k*}z = z$ is necessarily a multiple of $\{0\}$.

To conclude, we easily compute that $\alpha = (-1)^g$. □

The Fourier transform also has the remarkable property of exchanging the Pontryagin product and the intersection product, as follows.

Proposition 11.36 *For $Z, Z' \in CH(A)_{\mathbb{Q}}$, we have the relation*

$$\mathcal{F}(Z * Z') = \mathcal{F}(Z) \cdot \mathcal{F}(Z') \ \ in \ \ \mathrm{CH}(\widehat{A})_{\mathbb{Q}}.$$

Proof By definition, we have

$$Z * Z' = \mu_*(Z \times Z'),$$

where $Z \times Z' = \mathrm{pr}_1^* Z \cdot \mathrm{pr}_2^* Z' \in \mathrm{CH}(A \times A)_{\mathbb{Q}}$. Therefore, writing $p_1,\ p_2$ for the projections of $A \times \widehat{A}$ onto its factors, and $\tilde{\mu} = (\mu, \mathrm{Id}) : A \times A \times \widehat{A} \to A \times \widehat{A}$, we find that

$$p_1^*(Z * Z') = p_1^* \mu_*(Z \times Z') = \tilde{\mu}_*(p_{12}^*(Z \times Z')),$$

where $p_{12} : A \times A \times \widehat{A} \to A \times A$ is the projection onto the two first factors.

This now gives

$$\begin{aligned}\mathcal{F}(Z * Z') :&= p_{2*}(p_1^*(Z * Z') \cdot \exp \Theta) = p_{2*}(\tilde{\mu}_*(p_{12}^*(Z \times Z')) \cdot \exp \Theta) \\ &= p_{2*}\tilde{\mu}_*(p_{12}^*(Z \times Z') \cdot \tilde{\mu}^* \exp \Theta) = q_{3*}(p_{12}^*(Z \times Z') \cdot \tilde{\mu}^* \exp \Theta),\end{aligned}$$

where $q_3 : A \times A \times \widehat{A} \to \widehat{A}$ denotes the projection onto the third factor.

Now we use the relation

$$\tilde{\mu}^* \Theta = p_{13}^* \Theta + p_{23}^* \Theta$$

as above, to conclude that

$$\mathcal{F}(Z * Z') = q_{3*}(q_1^* Z \cdot q_2^* Z' \cdot p_{13}^* \exp \Theta \cdot p_{23}^* \exp \Theta),$$

where the q_i are the projections of $A \times A \times \widehat{A}$ onto its factors.

Applying the projection formula, we then find that the right-hand term is equal to $\mathcal{F}(Z) \cdot \mathcal{F}(Z')$. □

11.3.4 Results of Beauville

Apart from the structure given by the Pontryagin product, the Chow groups $\mathrm{CH}_i(A)_{\mathbb{Q}}$ are equipped with the action of the homotheties

$$m_k^* : \mathrm{CH}(A)_{\mathbb{Q}} \to \mathrm{CH}(A)_{\mathbb{Q}}, \quad k \in \mathbb{Z}^*.$$

The homotheties also act on the cohomology of A, and as $m_k^* = k\,\mathrm{Id}$ on $H^1(A, \mathbb{Z})$ and $H^l(A, \mathbb{Z}) = \bigwedge^l H^1(A, \mathbb{Z})$, m_k^* acts via $k^l \mathrm{Id}$ on $H^l(A, \mathbb{Z})$. Conjecture 11.21 predicts that $\mathrm{CH}^l(A)_{\mathbb{Q}}$ has a filtration $F^i \mathrm{CH}^l(A)_{\mathbb{Q}}$ such that $F^{l+1}\mathrm{CH}^l(A)_{\mathbb{Q}} = 0$, and that a correspondence acting trivially on $H^{2l-i}(A, \mathbb{Z})$ will also act trivially on $\mathrm{Gr}_F^i \mathrm{CH}^l(A)_{\mathbb{Q}}$. The morphisms m_k^* should thus act via $k^{2l-i}\mathrm{Id}$ on $\mathrm{Gr}_F^i \mathrm{CH}^l(A)_{\mathbb{Q}}$, and we conclude that this conjecture actually predicts a decomposition as a direct sum

$$\mathrm{CH}^l(A)_{\mathbb{Q}} = \bigoplus\nolimits_{0 \le i \le l} \mathrm{CH}_i^l(A)_{\mathbb{Q}},$$

where

$$\mathrm{CH}_i^l(A)_{\mathbb{Q}} = \left\{ Z \in \mathrm{CH}^l(A)_{\mathbb{Q}} \mid m_k^* Z = k^{2l-i} Z \text{ for all } k \in \mathbb{Z}^* \right\}.$$

Using the Fourier transform, Beauville proved the following result.

Theorem 11.37 (Beauville 1986a) *Letting* $\mathrm{CH}^l_i(A)_{\mathbb{Q}}$ *be defined as above, we have a decomposition*

$$\mathrm{CH}^l(A)_{\mathbb{Q}} = \bigoplus\nolimits_{l-g\leq i\leq l} \mathrm{CH}^l_i(A)_{\mathbb{Q}}. \tag{11.14}$$

This decomposition is obtained using the Fourier transform $\mathcal{F}$. If $Z \in \mathrm{CH}^l(A)_{\mathbb{Q}}$, write $\mathcal{F}(Z) = \sum_q Y_q$, $Y_q \in \mathrm{CH}^q(\widehat{A})$. Then, using the behaviour of $\mathcal{F}$ with respect to the isogenies $m : A \to A$, $m \in \mathbb{Z}$, Beauville showed that $\widehat{\mathcal{F}}(Y_q) \in \mathrm{CH}^l_s(A)_{\mathbb{Q}}$ with $s = l + q - g$. Thus, we have the decomposition

$$(-1)^g(-1)_*Z = \sum_q Z_q, \quad Z_q = \widehat{\mathcal{F}}(Y_q) \in \mathrm{CH}^l_{l+q-g}(A)_{\mathbb{Q}}.$$

As $0 \leq q \leq g$, we have $l - g \leq l + q - g \leq l$ in this decomposition.

The problem of whether only the positive integers i occur in Beauville's decomposition (11.14), as predicted by conjecture 11.21, remains open.

Exercises

1. *The Bloch–Beilinson filtration and the Nori filtration.* Let X be a smooth projective n-dimensional variety, and let $Y \overset{j}{\hookrightarrow} X$ be an ample hypersurface.
   (a) Using the Lefschetz theorem 1.23, show that if the Hodge conjecture is satisfied, then there exists a cycle $\Gamma \in \mathrm{CH}^{n-1}(X \times Y)_{\mathbb{Q}}$ such that the cycle

   $$\Gamma' := (1, j)_*(\Gamma) \in \mathrm{CH}^n(X \times X)_{\mathbb{Q}}$$

   satisfies the condition that

   $$[\Gamma']^* = [\Gamma]^* \circ j^* : H^l(X, \mathbb{Q}) \to H^l(X, \mathbb{Q})$$

   is equal to the identity for $l \leq n - 1$.
   (b) Let $Y_n \subset \cdots \subset Y_1 \subset Y_0 = X$ be smooth subvarieties such that Y_i is an ample hypersurface of Y_{i-1} for $1 \leq i \leq n$. Let $j_i : Y_i \hookrightarrow X$ denote inclusion. Show similarly that if the Hodge conjecture is satisfied, then for every i there exists a cycle $\Gamma_i \in \mathrm{CH}^{n-i}(X \times Y_i)_{\mathbb{Q}}$ such that the cycle

   $$\Gamma'_i := (1, j_i)_*(\Gamma_i) \in \mathrm{CH}^n(X \times X)_{\mathbb{Q}}$$

satisfies the condition that

$$[\Gamma_i']^* : H^l(X, \mathbb{Q}) \to H^l(X, \mathbb{Q})$$

is equal to the identity for $l \leq n - i$.

(c) Show that if the Bloch–Beilinson conjecture 11.21 is satisfied, then

$$\Gamma_i'^* : \mathrm{CH}^l(X)_{\mathbb{Q}} \to \mathrm{CH}^l(X)_{\mathbb{Q}}$$

acts via the identity on $\mathrm{Gr}_F^s \mathrm{CH}^l(X)_{\mathbb{Q}}$ for $s \geq 2l - n + i$. Deduce that under the same hypothesis, the map

$$\Gamma_i{}^* : F^{2l-n+i}\mathrm{CH}^l(Y_i)_{\mathbb{Q}} \to F^{2l-n+i}\mathrm{CH}^l(X)_{\mathbb{Q}}$$

is surjective.

Assume now that $2l = n$. Then $\dim Y_i = 2l - i$ and

$$\mathrm{CH}^l(Y_i) = \mathrm{CH}_{l-i}(Y_i).$$

Thus, for $2l = n = \dim X$, we have shown the following statement:

If the Hodge conjecture and the Bloch–Beilinson conjecture are satisfied, then the maps

$$\Gamma_i{}^* : F^i\mathrm{CH}_{l-i}(Y_i)_{\mathbb{Q}} \to F^i\mathrm{CH}^l(X)_{\mathbb{Q}}$$

are surjective. In particular, for $i > 0$, we have

$$F^i\mathrm{CH}^l(X)_{\mathbb{Q}} \subset \mathcal{N}_{l-i}\mathrm{CH}^l(X)_{\mathrm{hom},\mathbb{Q}}, \tag{11.15}$$

where $\mathcal{N}_j\mathrm{CH}(X)_{\mathrm{hom}}$ is the Nori filtration (see section 8.3.1). Finally, we have

$$F^l\mathrm{CH}^l(X)_{\mathbb{Q}} \subset \mathcal{N}_0\mathrm{CH}^l(X)_{\mathrm{hom},\mathbb{Q}} = \mathrm{CH}^l(X)_{\mathrm{alg},\mathbb{Q}}. \tag{11.16}$$

(d) Show that the conclusions (11.15) and (11.16) still hold under the same hypotheses, without the condition $2l = n$. (If $2l > n$, then replace X by $X \times X'$, with $\dim X' = 2l - n$. If $2l < n$, then apply (c) with $i = 1$ and induction on dimension.)

2. *Other applications of the Fourier transform in the Chow ring of an abelian variety* (see Beanville 1983, 1986a). Let A be an abelian variety, and let $Z \in \mathrm{CH}^l(A)_{\mathbb{Q}}$.

(a) Show that if the Bloch–Beilinson conjecture 11.21 is satisfied, the components Z_i of the Beauville decomposition

$$Z = \sum_i Z_i, \quad Z_i \in \mathrm{CH}^l_i(A)_{\mathbb{Q}}$$

are zero for $i < 0$.

(b) Deduce that under the same hypothesis, the components Y_q of the decomposition

$$\mathcal{F}(Z) = \sum_q Y_q, \quad Y_q \in \mathrm{CH}^q(\widehat{A})$$

are zero for $q < g - l$ (Beauville's conjecture).

(c) (see Beauville 1983). Show that if the conclusion of (a) (or (b)) is satisfied, then

$$\Gamma_1 * \cdots * \Gamma_k * Z = 0 \text{ in } \mathrm{CH}(A)_{\mathbb{Q}}$$

for $k > l$ and for $\Gamma_i \in \mathrm{CH}(A)_{\mathrm{hom}}$ for all i. (Use propositions 11.35 and 11.36, together with the fact that for $Z_i \in \mathrm{CH}(A)_{\mathrm{hom}}$, we have $\mathcal{F}(Z_i) \in \mathrm{CH}^{>0}(\widehat{A})$.)

(d) Show similarly that if the conclusion of (a) (or (b)) is satisfied, then

$$\Gamma_1 * \cdots * \Gamma_k * Z \in \mathrm{CH}(A)_{\mathrm{alg},\mathbb{Q}} \subset \mathrm{CH}(A)_{\mathbb{Q}}$$

for $k \geq l$ and for $\Gamma_i \in \mathrm{CH}(A)_{\mathrm{hom}}$ for all i.

References

A. Andreotti, T. Frankel (1959). The Lefschetz theorem on hyperplane sections, *Ann. Math.* (2)**69** 713–717.

A. Andreotti, T. Frankel (1969). The second Lefschetz theorem on hyperplane sections, in *Global Analysis, A symposium in honor of K. Kodaira*, Princeton University Press, 1–20.

E. Arbarello, M. Cornalba, P. Griffiths, J. Harris (1985). *Geometry of Algebraic Curves* vol. I, Grundlehren der Math. Wiss. **267**, Springer.

W. Barth (1970). Transplanting cohomology classes in complex projective space, *Amer. J. Math.* **92**, 951–967.

W. Barth, C. Peters, A. Van de Ven (1984). *Compact Complex Surfaces*, Ergebnisse der Math. und ihrer Grenzgebiete 3. Folge, Band 4, Springer.

A. Beauville (1977). Variétés de Prym et jacobiennes intermédiaires, *Ann. Sci. École Norm. Sup.* **10**, 309–391.

A. Beauville (1978). *Surfaces algébriques complexes*, Astérisque **54**, Société mathématique de France.

A. Beauville (1983). Quelques remarques sur la transformée de Fourier dans l'anneau de Chow d'une variété abélienne, in *Algebraic Geometry (Tokyo/Kyoto 1982)*, LNM **1016**, Springer, 238–260.

A. Beauville (1986a). Sur l'anneau de Chow d'une variété abélienne, *Math. Ann.* **273**, 647–651.

A. Beauville (1986b). Le groupe de monodromie des familles universelles d'hypersurfaces et intersections complètes, LNM **1194**, Springer, 8–18.

S. Bloch (1972). Semi-regularity and de Rham cohomology, *Inventiones Math.* **17**, 51–66.

S. Bloch (1973). K_2 and algebraic cycles, *Ann. Math.* **99**, 347–379.

S. Bloch (1976). Some elementary theorems about algebraic cycles on abelian varieties, *Inventiones Math.* **37**, 215–228.

S. Bloch (1979). Torsion algebraic cycles and a theorem of Roitman, *Compositio Math.* **39**, 107–127.

S. Bloch (1980a). On an argument of Mumford in the theory of algebraic cycles, in *Géométrie algébrique, Angers 1979*, Sijthoff-Noordhoff.

S. Bloch (1980b). *Lectures on Algebraic Cycles*, Duke Univ. Math. Series IV.

S. Bloch, A. Ogus (1974). Gersten's conjecture and the homology of schemes, *Ann. Sci. École Norm. Sup.*, 4ème série, **7**, 181–202.

S. Bloch, V. Srinivas (1983). Remarks on correspondences and algebraic cycles, *Amer. J. Math.* **105**, 1235–1253.

S. Bloch, A. Kas, D. Lieberman (1976). 0-cycles on surfaces with $p_g = 0$. *Compositio Math.* **33**, 135–145.

R. Bott (1957). Homogeneous vector bundles, *Ann. Math.* **69**(2), 203–248.

J. Carlson (1987). The geometry of the extension class of a mixed Hodge structure, *Proc. Symp. Pure Math.* **46**, 199–222.

J. Carlson, R. Donagi (1987). Hypersurface variations are maximal, *Inventiones Math.* **89**, 371–374.

J. Carlson, P. Griffiths (1981). Infinitesimal variation of Hodge structure and the global Torelli problem. In *Géométrie algébrique, Angers, 1980* (ed. A. Beauville), Sijthoff-Noordhoff, 51–76.

J. Carlson, D. Toledo (1999). Discriminant complements and kernels of monodromy representations, *Duke Math. J.* **97**(3), 621–648.

J. Carlson, M. Green, P. Griffiths, J. Harris (1983). Infinitesimal variations of Hodge structure (I), *Compositio Math.* **50**, 109–205.

E. Cattani, P. Deligne, A. Kaplan (1995). On the locus of Hodge classes, *J. Amer. Math. Soc.* **8**, 483–506.

H. Clemens (1983a). Homological equivalence, modulo algebraic equivalence is not finitely generated, *Publ. Math. IHES* **58**, 19–38.

H. Clemens (1983b). Double solids, *Adv. Math.* **47**, 107–230.

H. Clemens (1984). Some results about Abel–Jacobi mappings. In Griffiths *et al.* (1984), 289–304.

A. Conte, J. P. Murre (1978). The Hodge conjecture for fourfolds admitting a covering by rational curves, *Math. Ann.* **238**, 461–513.

D. Cox, M. Green (1990). Polynomial structures and generic Torelli for projective hypersurfaces, *Compositio Math.* **73**, 121–124.

P. Deligne (1968). Théorèmes de Lefschetz et critères de dégénérescence de suites spectrales, *Publ. Math. IHES* **35**, 107–126.

P. Deligne (1971). Théorie de Hodge II, *Publ. Math. IHES* **40**, 5–57.

P. Deligne (1972). La conjecture de Weil pour les surfaces $K3$, *Inventiones Math.* **15**, 206–226.

P. Deligne (1975). Théorie de Hodge III, *Publ. Math. IHES* **44**, 5–77.

P. Deligne, P. Griffiths, J. Morgan, D. Sullivan (1975). Real homotopy theory of Kähler manifolds, *Inventiones Math.* **29**, 245–274.

J.-P. Demailly (1996). Théorie de Hodge L^2 et théorèmes d'annulation. In *Introduction à la théorie de Hodge*, Panoramas et Synthèses **3**, Société mathématique de France, 3–111.

R. Donagi (1983). Generic Torelli theorem for projective hypersurfaces, *Compositio Math.* **50**, 325–353.

R. Donagi, M. Green (1984). A new proof of the symmetriser lemma and a stronger weak Torelli theorem, *J. Diff. Geom.* **20**, 459–461.

F. El Zein, S. Zucker (1984). Extendability of normal functions associated to algebraic cycles. In Griffiths *et al.* (1984), 269–288.

H. Esnault, K. Paranjape (1994). Remarks on absolute de Rham and absolute Hodge cycles. *C. R. Acad. Sci. Paris* **319**, série I, 67–72.

H. Esnault, E. Viehweg (1988). Deligne–Beilinson cohomology, in *Beilinson's Conjectures on Special Values of L-Functions* (ed. Rapoport, Schappacher and Schneider), Perspect. Math. **4**, Academic Press, 43–91.

H. Esnault, E. Viehweg (1992). *Lectures on Vanishing Theorems*, DMV Seminar Band 20, Birkhäuser.

H. Esnault, E. Viehweg, M. Levine (1997). Chow groups of projective varieties of very small degree, *Duke Math. J.* **87**, 29–58.

N. Fakhrudin (1996). Algebraic cycles on generic abelian varieties, *Compositio Math.* **100**, 101–119.

W. Fulton (1984). *Intersection Theory*, Ergebnisse der Math. und ihrer Grenzgebiete 3. Folge, Band 2, Springer.

R. Godement (1958). *Topologie algébrique et théorie des faisceaux*, Hermann.

M. Green (1984a). Koszul cohomology and the geometry of projective varieties, *J. Diff. Geom.* **19**, 125–171.

M. Green (1984b). Koszul cohomology and the geometry of projective varieties II, *J. Diff. Geom.* **20**, 279–289.

M. Green (1984c). The period map for hypersurface sections of high degree of an arbitrary variety, *Compositio Math.* **55**, 135–156.

M. Green (1989a). Components of maximal dimension in the Noether–Lefschetz locus, *J. Diff. Geom.* **29**, 295–302.

M. Green (1989b). Griffiths' infinitesimal invariant and the Abel–Jacobi map, J. Diff. *Geom.* **29**, 545–555.

M. Green (1998). Higher Abel–Jacobi maps, *Documenta Mathematica*, extra volume, ICM II, 267–276.

M. Green, S. Müller-Stach (1996). Algebraic cycles on a general complete intersection of high multi-degree of a smooth projective variety, *Compositio Math.* **100**, 305–309.

M. Green, J. Murre, C. Voisin (1993). *Algebraic Cycles and Hodge Theory*, CIME course, LNM **1594**, Springer.

P. Griffiths (1968). Periods of integrals on algebraic manifolds, I, II, *Amer. J. Math.* **90**, 568–626, 805–865.

P. Griffiths (1969). On the periods of certain rational integrals I, II, *Ann. Math.* **90**, 460–541.

P. Griffiths (1979). A theorem concerning the differential equations satisfied by normal functions associated to algebraic cycles, *Amer. J. Math.* **101**, 94–131.

P. Griffiths (1983). Infinitesimal variations of Hodge structures (III): determinantal varieties and the infinitesimal invariant of normal functions, *Compositio Math.* **50**, 267–324.

P. Griffiths (1984). Curvature properties of the Hodge bundles. In Griffiths *et al.* (1984), 29–49.

P. Griffiths, J. Harris (1978). *Principles of Algebraic Geometry*, Wiley-Interscience.

P. Griffiths, J. Morgan (1981). *Rational Homotopy Theory and Differential Forms*, Progress in Math. **16**, Birkhäuser.

P. Griffiths *et al.* (1984). *Topics in Transcendental Algebraic Geometry* (ed. P. Griffiths), Annals of Math. Studies **106**, Princeton University Press.

R. Hain (1987a). The geometry of the mixed Hodge structure on the fundamental group, *Proc. Symp. Pure Math.* **46**(2), 247–281.

R. Hain (1987b). The de Rham homotopy theory of complex algebraic varieties I, *K-Theory* **1**, 271–324.

R. Hain (1987c). *Higher Albanese Manifolds*, LNM **1246**, Springer.

J. Harris (1979). Galois groups of enumerative problems, *Duke Math. J.* **46**, 685–724.

R. Hartshorne (1977). *Algebraic Geometry*, Graduate Texts in Math. **52**, Springer.

H. Hironaka (1964). Resolution of singularities of an algebraic variety over a field of characteristic zero I, II, *Ann. Math.* **79**, 109–326.

H. Inoze, M. Mizukami (1979). Rational equivalence of 0-cycles on some surfaces of general type with $p_g = 0$, *Math. Ann.* **244**, 205–217.

U. Jannsen (1989). *Mixed Motives and Algebraic K-Theory*, LNM **1400**, Springer.

U. Jannsen (1994). Motivic sheaves and filtrations on Chow groups, in *Motives, Proc. Symp. Pure Math.* **55**(1) (ed. by U. Jannsen, S. Kleiman, J.-P. Serre), 245–302.

N. Katz (1973). Étude cohomologique des pinceaux de Lefschetz, *SGA* 7, exposé 18, LNM **340**, Springer.

S. Kleiman (1968). Algebraic cycles and the Weil conjectures, in *Dix exposés sur la théorie des schémas*, North-Holland, 359–386.

S. Kobayashi (1970). *Hyperbolic Manifolds*, Dekker.

K. Kodaira (1963a). On compact analytic surfaces III, *Ann. Math.* **78**, 1–40.

K. Kodaira (1963b). On stability of compact submanifolds of complex manifolds, *Amer. J. Math.* **85**, 79–94.

J. Kollár (1990). Lemma p. 134 in *Classification of Irregular Varieties* (ed. E. Ballico, F. Catanese, C. Ciliberto), LNM **1515**, Springer.

J. Kollár (1996). *Rational Curves on Algebraic Varieties*, Ergebnisse der Math. und ihrer Grenzgebiete, Springer.

R. Laterveer (1996). Algebraic varieties with small Chow groups, *J. Math. Kyoto Univ.* **38**, 673–694.

J. Lewis (1995). A generalization of Mumford's theorem, *Illinois J. Math.* **39**, 288–304.

Yu. Manin (1968). Correspondences, motives and monoidal transforms, *Math. USSR Sbornik* **6**, 439–470.

J. Milnor (1963). *Morse Theory*, Annals of Mathematical Studies **51**, Princeton University Press.

J. Morgan (1978). The algebraic topology of smooth algebraic varieties, *Publ. Math. IHES* **48**, 137–204.

S. Mukai (1981). Duality between $D(X)$ and $D(\hat{X})$ with its application to Picard sheaves, *Nagoya Math. J.* **81**, 153–175.

D. Mumford (1968). Rational equivalence of 0-cycles on surfaces, *J. Math. Kyoto Univ.* **9**, 195–204.

D. Mumford (1976). *Algebraic Geometry I: complex projective varieties*, Grundlehren der Math. Wisse., Springer.

D. Mumford (1988). *The Red book of Varieties and Schemes*, re-issue LNM **1358**, Springer.

J. P. Murre (1985). Applications of algebraic K-theory to the theory of algebraic cycles. In *Proc. Conf. Algebraic Geometry, Sitjes 1983*, LNM **1124**, Springer, 216–261.

J. P. Murre (1990). On the motive of an algebraic surface, *J. Reine Angew. Math.* **409**, 190–204.

J. P. Murre ( ). On a conjectural filtration on the Chow groups of an algebraic variety, *Indag. Math.* N.S. **4**(2), 177–188.

M. Nori (1993). Algebraic cycles and Hodge theoretic connectivity, *Inventiones Math.* **111**, 349–373.

K. Paranjape (1994). Cohomological and cycle theoretic connectivity, *Ann. Math.* **140**, 641–660.

C. Peters (1977). On certain examples of surfaces with $p_g = 0$ due to Burniat, *Nagoya Math. J.* **66**, 109–119.

I. Piatetski-Shapiro, I. Shafarevich (1971). A Torelli theorem for algebraic surfaces of type K3, *Math. USSR Izvestiya* **35**, 530–572.

Z. Ran (1993). Hodge theory and the Hilbert scheme, *J. Diff. Geom.* **37**, 191–198.

A. Roitman (1972). Rational equivalence of 0-cycles, *Math. USSR Sbornik* **18**, 571–588.

A. Roitman (1980). The torsion of the group of 0-cycles modulo rational equivalence, *Ann. Math.* **111**, 553–569.

W. Rudin (1966). *Real and Complex Analysis*, McGraw-Hill.

M. Saito (1999). Arithmetic mixed Hodge modules, preprint 1999.

S. Saito (1996). Motives and filtrations on Chow groups, *Inventions Math.* **125**, 149–196.

C. Schoen (1993). On Hodge structures and non-representability of Chow groups, *Compositio Math.* **88**, 285–316.

B. Shiffman, A. J. Sommese (1985). *Vanishing Theorems on Complex Manifolds*, Progress in Math. **56**, Birkhäuser.

E. Spanier (1966). *Algebraic Topology*, Tata McGraw-Hill.

D. Sullivan (1978). Infinitesimal calculations in topology, *Publ. Math. IHES* **47**, 269–331.

E. Viehweg (1995). *Quasi-projective Moduli for Polarized Manifolds*, Ergebnisse der Math. und ihrer Grenzgebiete, Band 30, Springer.

C. Voisin (1986). Théoréme de Torelli pour les cubiques de $\mathbb{P}^5$, *Inventiones Math.* **86**, 577–601.

C. Voisin (1988a). Une précision concernant le théorème de Noether, *Math. Ann.* **280**, 605–611.

C. Voisin (1988b). Une remarque sur l'invariant infinitesimal des fonctions normales, *C. R. Acad. Sci. Paris*, **307**, série 1, 157–163.

C. Voisin (1989). Composantes de petite codimension du lieu de Noether–Lefschetz, *Comment. Math. Helvetici* **64**, 515–526.

C. Voisin (1992). Sur les zéro-cycles de certaines hypersurfaces munies d'un automorphisme, *Ann. Scuola Norm. Sup. Pisa*, serie 4, **19**, 473–492.

C. Voisin (1994). Variations de structure de Hodge et zéro-cycles sur les surfaces générales, *Math. Ann.* **299**, 77–103.

C. Voisin (1999a). A generic Torelli theorem for the quintic threefold. In *New Trends in Algebraic Geometry* (ed. Hulek, Catanese, Peters and Reid), Lond. Mathematical Society Lecture Note Series **264**, Cambridge University Press, 425–463.

C. Voisin (1999b). Some results on Green's higher Abel–Jacobi map, *Ann. Math.* **149**, 451–473.

C. Voisin (2002b). Nori's connectivity theorem and higher Chow groups, *J. Inst. Math. Jussieu* **1**(2), 307–309.

C. Voisin (2002a). Hodge Theory and Complex Algebraic Geometry I. Cambridge University Press.

Index

Printed in the United States
By Bookmasters